# Benchmark Papers In Genetics Series

## Editor: David L. Jameson — University of Houston

GENETICS AND SOCIAL STRUCTURE: Mathematical Structuralism in Population Genetics and Social Theory / *Paul Ballonoff*
GENES AND PROTEINS / *Robert P. Wagner*
DEMOGRAPHIC GENETICS / *Kenneth M. Weiss and Paul A. Ballonoff*
MUTAGENESIS / *John W. Drake and Robert E. Koch*
EUGENICS: Then and Now / *Carl Jay Bajema*
CYTOGENETICS / *Ronald L. Phillips and Charles R. Burnham*
STOCHASTIC MODELS IN POPULATION GENETICS / *Wen-Hsiung Li*
EVOLUTIONARY GENETICS / *D. L. Jameson*
GENETICS OF SPECIATION / *D. L. Jameson*
HUMAN GENETICS: A Selection of Insights / *William J. Schull and Ranajit Chakraborty*
HYBRIDIZATION: An Evolutionary Perspective / *Donald A. Levin*
POLYPLOIDY / *R. C. Jackson and Donald P. Hauber*
EXPERIMENTAL POPULATION GENETICS / *Roger Milkman*
SOMATIC CELL GENETICS / *Richard L. Davidson*
QUANTITATIVE GENETICS, PART I: Explanation and Analysis of Continuous Variation / *W. G. Hill*
QUANTITATIVE GENETICS, PART II: Selection / *W. G. Hill*

Related Titles in Benchmark Papers in Systematic and Evolutionary Biology
ARTIFICIAL SELECTION AND THE DEVELOPMENT OF EVOLUTIONARY THEORY / *Carl Jay Bajema*
NATURAL SELECTION THEORY: From the Speculations of the Greeks to the Quantitative Measurements of the Biometricians / *Carl Jay Bajema*
SEXUAL SELECTION THEORY PRIOR TO 1900 / *Carl Jay Bajema*

# QUANTITATIVE GENETICS
## PART II
## Selection

Edited by
**W. G. HILL**
Institute of Animal Genetics
University of Edinburgh

A Hutchinson Ross Benchmark® Book

**Van Nostrand Reinhold Company**
**Scientific and Academic Editions**
New York   Cincinnati   Stroudsburg
Toronto   London   Melbourne

Copyright © 1984 by **Van Nostrand Reinhold Company Inc.**
Benchmark Papers in Genetics, Volume 15
Library of Congress Catalog Card Number: 83-12534
ISBN: 0-442-23218-7

All rights reserved. No part of this book covered by the copyrights hereon may be reproduced or used in any form or by any means—graphic, electronic, or mechanical, including photocopying, recording, taping, or information storage and retrieval systems—without permission of the publisher.

Manufactured in the United States of America.

Published by Van Nostrand Reinhold Company Inc.
135 West 50th Street
New York, New York 10020

Van Nostrand Reinhold Company Limited
Molly Millars Lane
Wokingham, Berkshire RG11 2PY, England

Van Nostrand Reinhold
480 Latrobe Street
Melbourne, Victoria 3000, Australia

Macmillan of Canada
Division of Gage Publishing Limited
164 Commander Boulevard
Agincourt, Ontario MIS 3C7, Canada

15  14  13  12  11  10  9  8  7  6  5  4  3  2  1

**Library of Congress Cataloging in Publication Data**
Main entry under title:
Quantitative genetics.

  (Benchmark papers in genetics; v. 15)
  Includes indexes.
  Contents: pt. 1. Explanation and analysis of
continuous variation—pt. 2. Selection.
  1. Quantitative genetics—Addresses, essays,
lectures.  I. Hill, W. G.  II. Series. [DNLM:
1. Genetics—Collected works. 2. Biometry—Collected
works. W1 BE516 v.15/QH 452.7 Q16]
QH452.7.Q365  1984  575.1  83-12534
ISBN 0-442-23219-5 (v. 1)
ISBN 0-442-23218-7 (v. 2)
ISBN 0-442-23217-9 (set)

# CONTENTS

| | |
|---|---|
| Series Editor's Foreword | ix |
| Preface | xi |
| Contents by Author | xiii |
| Introduction | 1 |

## NATURE OF SELECTION RESPONSE

Editor's Comments on Papers 1 Through 4 — 8

1. **CASTLE, W. E.:** The Mutation Theory of Organic Evolution, from the Standpoint of Animal Breeding
 *Science* **21**:521-525 (1905) — 14

2. **JENNINGS, H. S.:** Experimental Evidence on the Effectiveness of Selection
 *Am. Nat.* **44**:136-145 (1910) — 19

3. **STURTEVANT, A. H.:** *An Analysis of the Effects of Selection*
 Carnegie Institution of Washington Publication No. 264, 1918, pp. 36-54 — 29

4. **CASTLE, W. E.:** Piebald Rats and Selection, a Correction
 *Am. Nat.* **53**:370-376 (1919) — 49

## STATISTICAL PREDICTIONS OF SELECTION RESPONSE

Editor's Comments on Papers 5 Through 9 — 58

5. **LUSH, J. L.:** Progeny Test and Individual Performance as Indicators of an Animal's Breeding Value
 *J. Dairy Sci.* **18**:1-8, 16-19 (1935) — 64

6. **HAZEL, L. N.:** The Genetic Basis for Constructing Selection Indexes
 *Genetics* **28**:476-490 (1943) — 76

7. **FALCONER, D. S.:** The Problem of Environment and Selection
 *Am. Nat.* **86**:293-298 (1952) — 91

8. **DICKERSON, G. E., and L. N. HAZEL:** Effectiveness of Selection on Progeny Performance as a Supplement to Earlier Culling in Livestock
 *J. Agric. Res.* **69**:459-476 (1944) — 97

9. **HENDERSON, C. R.:** General Flexibility of Linear Model Techniques for Sire Evaluation
 *J. Dairy Sci.* **57**:963-972 (1974) — 115

*Contents*

## GENETICAL PREDICTIONS OF SELECTION RESPONSE

**Editor's Comments on Papers 10 Through 13**     128

**10**    FISHER, R. A.: The Fundamental Theorem of Natural Selection     135
*The Genetical Theory of Natural Selection,* Clarendon Press, Oxford, 1930, pp. 30–37.

**11**    HALDANE, J. B. S.: A Mathematical Theory of Natural and Artificial Selection. Part VII. Selection Intensity as a Function of Mortality Rate     143
*Cambridge Philos. Soc. Proc.* **27:**131-136 (1930)

**12**    COMSTOCK, R. E., H. F. ROBINSON, and P. H. HARVEY: A Breeding Procedure Designed to Make Maximum Use of Both General and Specific Combining Ability     149
*Agron. J.* **41:**360-367 (1949)

**13**    ROBERTSON, A.: A Theory of Limits in Artificial Selection     157
*R. Soc. (London) Proc.* **B153:**234-249 (1960)

## RESULTS FROM SELECTION EXPERIMENTS

**Editor's Comments on Papers 14 Through 20**     174

**14**    DUDLEY, J. W.: 76 Generations of Selection for Oil and Protein Percentage in Maize     186
*Proceedings of the International Conference on Quantitative Genetics,* E. Pollak, O. Kempthorne, and T. B. Bailey, Jr., eds., Iowa State University Press, Ames, 1977, pp. 459–473

**15**    MATHER, K.: Variation and Selection of Polygenic Characters     201
*J. Genet.* **41:**159-175, 183-193 (1941)

**16**    LERNER, I. M., and E. R. DEMPSTER: Attenuation of Genetic Progress Under Continued Selection in Poultry     229
*Heredity* **5:**75-91, 94 (1951)

**17**    ROBERTSON, F. W.: Selection Response and the Properties of Genetic Variation     246
*Cold Spring Harbor Symp. Quant. Biol.* **20:**166-177 (1955)

**18**    FALCONER, D. S.: The Genetics of Litter Size in Mice     258
*J. Cell. Comp. Physiol.* **56** (suppl. 1)**:**153-164, 167 (1960)

**19**    CLAYTON, G. A., J. A. MORRIS, and A. ROBERTSON: An Experimental Check on Quantitative Genetical Theory. I. Short-Term Responses to Selection.     270
*J. Genet.* **5:**131-151 (1957)

**20**    BELL, A. E., C. H. MOORE, and D. C. WARREN: The Evaluation of New Methods for the Improvement of Quantitative Characteristics     291
*Cold Spring Harbor Symp. Quant. Biol.* **20:**197-211 (1955)

## SELECTION AND MAINTENANCE OF GENETIC VARIATION

**Editor's Comments on Papers 21 Through 24**     308

**21**    WRIGHT, S.: The Roles of Mutation, Inbreeding, Crossbreeding and Selection in Evolution     313
*Int. Cong. Genet., 6th, Proc.* **1:**356-366 (1932)

| | | | |
|---|---|---|---|
| 22 | **ROBERTSON, A.:** | Selection in Animals: Synthesis<br>*Cold Spring Harbor Symp. Quant. Biol.* **20:**225-229 (1955) | 324 |
| 23 | **BULMER, M. G.:** | The Effect of Selection on Genetic Variability<br>*Am. Nat.* **105:**201-209, 211 (1971) | 329 |
| 24 | **LANDE, R.:** | The Maintenance of Genetic Variability by Mutation in a Polygenic Character with Linked Loci<br>*Genet. Res. (Cambridge)* **26:**221-235 (1976) | 338 |

## NATURE OF QUANTITATIVE GENETIC VARIATION

**Editor's Comments on Papers 25 Through 28**     354

| | | | |
|---|---|---|---|
| 25 | **CLAYTON, G., and A. ROBERTSON:** | Mutation and Quantitative Variation<br>*Am. Nat.* **89:**151-158 (1955) | 360 |
| 26 | **LINNEY, R., B. W. BARNES, and M. J. KEARSEY:** | Variation for Metrical Characters in *Drosophila* Populations. III. The Nature of Selection<br>*Heredity* **27:**163-174 (1971) | 368 |
| 27 | **"STUDENT":** | A Calculation of the Minimum Number of Genes in Winter's Selection Experiment<br>*Ann. Eugen.* **6:**77-82 (1934) | 380 |
| 28 | **THODAY, J. M.:** | Location of Polygenes<br>*Nature* **191:**368-370 (1961) | 386 |

**Author Citation Index**     389
**Subject Index**     393
**About the Editor**     397

# SERIES EDITOR'S FOREWORD

The study of any discipline assumes the mastery of the literature of the subject. In many branches of science, even one as new as genetics, the expansion of knowledge has been so rapid that there is little hope of learning of the development of all phases of the subject. The student has difficulty mastering the textbook, the young scholar must tend to the literature near his own research, the young instructor barely finds time to expand his horizons to meet his class-preparation requirements, the monographer copes with wider literature but usually from a specialized viewpoint, and the textbook author is forced to cover much the same material as previous and competing texts to respond to the user's needs and abilities.

Few publishers have the dedication to scholarship to serve primarily the limited market of advanced studies. The opportunity to assist professionals at all stages of their careers has been recognized by Hutchinson Ross and by a distinguished group of editors knowledgeable in specific portions of papers that demonstrate both the development of knowledge and the atmosphere in which that knowledge was developed. There is no substitute for reading great papers. Here you can learn how questions are asked, how they are approached, and how difficult and essential it is to obtain definitive answers and clear writing.

Dr. Hill has approached one of the most difficult areas of genetics. Significant literature covers well over a century of writing, and the interaction of this field with the development of statistical theory, plant and animal breeding, and biochemical and molecular biological approaches results in a complexity difficult for observers from other fields to appreciate. The field is important to modern food production, to health care, and to the social sciences; moreover, it provides a basis for much of the theoretical understanding of evolutionary processes. In the first volume Dr. Hill gives us an understanding of the basis of quantitative variation, and in the second he chooses from the extensive literature on selection experiments. I find it exciting to reread these papers with his commentary at hand. His scholarship will be appreciated by his colleagues.

DAVID L. JAMESON

# PREFACE

This volume is Part II of a collection of Benchmark Papers in Quantitative Genetics. In Part I, "The Explanation and Analysis of Continuous Variation," I pointed out that the original intention was to produce a single volume, but that I found it impossible to span the field adequately in the limited space. The papers in Part I, which essentially deal with the static description of populations in quantitative genetic terms, are intended to serve as groundwork for the papers on selection that are included here. The papers chosen for this volume are substantially more recent, and I have paid less attention to the historical development. No doubt, for some tastes, the reprints are still too old (this is not molecular biology); perhaps for others, too young. I hope the reader finds them all interesting.

My intention has been to cover both the theoretical basis and some of the experimental results and to feature as many as possible of the major workers. Perhaps my own interests and prejudices show through in the preponderance of papers on theory and on animals, but studies are included that show that the predictions actually work in practice. Fortunately, I have been able to include a greater number of papers than in the first volume, and I hope I have achieved some breadth of coverage. Even so, only lengthy excerpts could be taken from a few, and there are many more papers I would have liked to include—indeed, whole books. The choosing was always an interesting problem.

I again wish to acknowledge many people for their help: David Jameson for inviting me to prepare these collected papers and for his advice; the authors and publishers for permission to reprint their material and, in the case of many authors, for helpful comments; Douglas Falconer and Joe Felsenstein for criticism and advice on the manuscript; and Jackie Bogie for her excellent typing of it. I am especially grateful to my many colleagues, who provide such a stimulating environment in Edinburgh and who have acted as constructive sounding boards for my suggestions of papers.

WILLIAM G. HILL

# CONTENTS BY AUTHOR

Barnes, B. W., 368
Bell, A. E., 291
Bulmer, M. G., 329
Castle, W. E., 14, 49
Clayton, G. A., 270, 360
Comstock, R. E., 149
Dempster, E. R., 229
Dickerson, G. E., 97
Dudley, J. W., 186
Falconer, D. S., 91, 258
Fisher, R. A., 135
Haldane, J. B. S., 143
Harvey, P. H., 149
Hazel, L. N., 76, 97
Henderson, C. R., 115
Jennings, H. S., 19

Kearsey, M. J., 368
Lande, R., 338
Lerner, I. M., 229
Linney, R., 368
Lush, J. L., 64
Mather, K., 201
Moore, C. H., 291
Morris, J. A., 270
Robertson, A., 157, 270, 324, 360
Robertson, F. W., 246
Robinson, H. F., 149
'Student', 380
Sturtevant, A. H., 29
Thoday, J. M., 386
Warren, D. C., 291
Wright, S., 313

# QUANTITATIVE GENETICS, PART II

# INTRODUCTION

The study of selection of quantitative traits encompasses both evolution by natural selection and improvement of domestic plants and animals by artificial selection. Although evolution may have proceeded by periods of rapid change followed by near stasis, the major traits have presumably always shown continuous or nearly continuous distributions, and there remain substantial overlaps of distribution among species. The power of selection to change a population, if not to develop new species, has been convincingly demonstrated by experimentalists and breeders; a few generations can suffice to change a population to a new mean level well outside its original range.

Part I of these volumes of Benchmark Papers in Quantitative Genetics, "The Explanation and Analysis of Continuous Variation," dealt with the evidence, the historical development, and the acceptance of the multifactorial Mendelian model, and in particular with prediction of the correlation among relatives. Those results are an essential background for the discussions of selection. As a consequence, although papers in Part I were taken only to the mid-1960s, more recent papers are included here, with less concentration on the early historical development. Another, more fortunate difference between the volumes is that the papers that represent important contributions to selection were, on average, shorter than those in other areas, so rather more papers are included. Even so, they cannot be expected to include *all* the notable developments.

Evidence for the effectiveness of selection came from two quite different sources. The first was implicit: the apparent ability of natural selection to produce new modifications and to increase adaptation, leading to Darwin's and Wallace's theories of evolution. The second was more explicit: the experiments of Castle and colleagues (Paper 1), the Illinois corn oil study, and many others showed that substantial changes in a population could occur, even though Johannsen and others such as Jennings (Paper 2) had been unable to get responses in pure lines. There were alternative explanations: changes to new variants

*Introduction*

occurred by mutation, or responses were due simply to increases in frequency of favorable genes. It is obvious to us now that both causes are major contributors, the relative importance depending on the time span of the selection, but the topic generated substantial debate. The general resolution of the differences, in favor of changes in frequency, occurred by about 1920, and is discussed in the papers of Sturtevant (Paper 3) and Castle (Paper 4). These papers are included in the first section, "Nature of Selection Response," and they put selection into the multifactor Mendelian framework that by then was essentially accepted as an explanation of continuous distribution of quantitative traits, correlation among relatives, inbreeding depression, and heterosis, all topics considered in Part I.

There was, nevertheless, neither a theory for predicting responses to artificial selection practised in a population, nor any mathematical demonstration that natural selection could lead to substantial evolution of a population, nor any prediction of the rate at which it might occur. The major advances in the evolutionary context were made by Haldane, Fisher, and Wright, culminating in their publications of the early 1930s (Fisher, 1930; Haldane, 1932; Wright, 1931). It is not, of course, possible to reproduce these works in full in the present volume, but an attempt is made to give a flavor of their studies by including particular aspects most closely related to the inheritance of quantitative traits. Most of the major developments in the context of applications to artificial selection were made by workers concerned with animal breeding.

Prediction equations in quantitative genetics can be at two levels: those based on changes of gene frequency at individual loci, perhaps then summed over loci to show changes in the trait, and those based essentially on regressions of progeny on parent performance assuming, for example, each is normally distributed. The necessary statistics for the latter, such as heritability, do not require any knowledge of the numbers, effects, and frequencies of genes influencing the trait. Predictions can be based on quantities observed directly in the population, but are of value only so long as heritability, for example, remains constant. Since selection, by changing gene frequencies, must change heritability, this latter assumption would seem invalid, but in practice it usually holds well for a number of generations. In large animals with slow generation turnover this time is plenty long enough for the formulae to be useful.

Papers in which selection theory is developed at the statistical level, in terms of variances, covariances among relatives, heritabilities, and regressions, are therefore included in Section II, "Statistical Predictions of Selection Theory," while those dealing primarily with changes at the individual locus level are deferred to Section III, "Genetical Predictions of Selection Response." The ordering is somewhat arbitrary, but the statistical papers inherently deal with the more short-term problems. The topics covered span most of the major topics in the application of quantitative genetics to animal and plant breeding; however, because the developments have mostly been on the animal model, none that take their motivation directly from plant breeding problems are included. An important consideration is the relative efficiency of alternative methods of selection, classically of selection on individual performance versus selection on progeny performance. This problem is discussed by Lush (Paper 5), and Paper 8 by Dickerson and Hazel extends the analysis in an important way to consider the optimum rate to turn over generations. Obviously there is a trade-off between reducing the proportion selected and increasing the generation interval. In any improvement program there are multiple objectives, and the method for dealing with these as correlated traits in a selection index is described by Hazel in Paper 6. Falconer shows in Paper 7 how the ideas of genetic correlations can be extended to performance in different environments. Finally, Henderson, in Paper 9, reviews and develops what has become the integrating procedure for assessing the genetic merit of individuals, Best Linear Unbiased Prediction (BLUP).

Section III includes an abstract (Paper 10) from Fisher's (1930) book, *The Genetical Theory of Natural Selection,* on the fundamental theorem of natural selection. This is the most widely known formula in the application of quantitative genetics to evolution and has generated much research into its applicability. Regrettably, only that abstract and the short paper by Haldane (Paper 11) could be included from Fisher's and Haldane's pioneering work on evolution by natural selection, and Haldane's paper is of more important application in showing how artificial selection changes gene frequencies. The remaining papers in this section deal with two particular but important problems in artificial selection: Comstock, Robinson, and Harvey (Paper 12) show how overdominance can be utilized by reciprocal recurrent selection (RRS), and A. Robertson (Paper 13) discusses problems of selection limits due to fixation of genes in finite populations. These two papers broke new ground in the application of quantitative genetics to animal and plant improvement and stimulated much experimental work.

There is an extensive and still growing literature on selection experiments in quantitative genetics. In a steadily accumulating body of information, it is not obvious which contributions are the most important, for different experiments are never complete replicates of

## Introduction

each other. The majority of those chosen for inclusion in Section IV, "Results from Selection Experiments," were conducted in the 1950s, when laboratory animals were being used to test quantitative genetic selection theory and as models for farm livestock in experiments on the inheritance of growth and reproduction. Indeed, all the experiments described in this section were carried out in institutions concerned with agricultural research.

Perhaps the best known single selection experiment is the Illinois corn oil project, which started before the turn of the century and still continues. It is represented here by a recent report by Dudley, Paper 14. The *Drosophila* experiments conducted by Mather (Paper 15) were important not just for their descriptive value, but because they led him to hypotheses about the nature of quantitative (biometrical) genetic variation and the relations among genes on the chromosome.

Poultry have been longest subjected to intense artificial selection in breeding programs based on quantitative genetic principles. Lerner and Dempster (Paper 16) were early exponents of the principles and conducted important experiments on poultry. They were among the first to identify problems of selective plateaux in farm animal populations, and the results are outlined in their paper. Problems of fitness-associated limits were also a notable feature of F. W. Robertson's and Reeve's thorough studies (reported by Robertson in Paper 17) on growth in *Drosophila*. A major aim of selection experiments is to investigate the nature of inheritance of traits of importance, of which the preceding are examples. Falconer's paper (Paper 18) on the genetics of litter size is a further example, in which results were obtained that were not predicted *a priori* by the theory because they depended on the biology of the specific trait under investigation.

Another major aim of selection experiments is to test the validity of quantitative genetic theory. Clayton, Morris, and A. Robertson (Paper 19) undertook a direct test of various aspects of the theory using *Drosophila*, and, fortunately, they found good agreement between expectation and observation. Bell, Moore, and Warren (Paper 20) used laboratory animals in model selection experiments to test the efficiency of alternative breeding programs, a nice example of an intermediate step in the path from theory to experiment to practice.

In Section V, "Selection and Maintenance of Genetic Variation," we return to theory—not to directional selection as in previous sections, but to interactions among loci and traits, and to the role of selection in the maintenance of variation. Paper 21 is a summary by Wright of his basic evolutionary views, written in 1932, discussing work he had just published in more detail and was subsequently to expand at much greater length. A. Robertson (Paper 22) reviews the relationships between metric characters and fitness and considers alternative models for maintaining variation for different kinds of traits. The papers of Bulmer (Paper 23) and of Lande (Paper 24) are concerned with the effects of selection on the amount of genetic variance a trait exhibits; in both studies many of the results are couched in terms of linkage disequilibrium. Bulmer's analysis applies to directional and to stabilizing selection and Lande's only to stabilizing selection, but Lande considers the balance between gain of variation by mutation and loss by selection. These papers have been important in stimulating interest in analysis of quantitative genetic variation by workers in evolutionary and ecological genetics.

The final section covers experimental results and techniques for predicting how much variation is produced by mutation, how variation is maintained by selection, and the numbers, effects, and locations in the chromosome of the genes affecting quantitative traits. Some of these problems have been touched on in earlier papers (e.g., those by Mather and F. W. Robertson), but the papers in Section VI, "Nature of Quantitative Variation," go into more detail. An estimate of the new variation arising from mutation is obtained by Clayton and A. Robertson (Paper 25). Direct evidence of the effect of stabilizing selection on reducing the fitness of extreme individuals is produced by Linney, Barnes, and Kearsey in Paper 26 (data on this topic are much less numerous than models). The Illinois corn oil experiment was mentioned previously, and as early as 1934, "Student" (Paper 27) used the results to estimate the numbers of genes affecting the trait. Although his results are probably just as unreliable as others, his presentation and approach to the problem are of interest. In the final paper, Thoday (Paper 28) outlines a method for identifying the effects and positions of genes influencing quantitative traits. Indeed, it is an attempt to describe the formal genetics of a quantitative trait.

The overall coverage is bound to be patchy, and there may be whole areas omitted. Early experimental papers tended to be very detailed and lacked any theoretical basis. There were also many theoretical developments that could have been covered, particularly on linkage and on finite populations, as well as experiments testing predictions in relation to population size. The major topic of changing rates of evolution, on which much heat is expended, has also been ignored. Here the intention has been to include not just the important topics, but also papers from as many as possible of the important workers on selection for quantitative traits.

For further reading and references, the textbooks by Falconer (1981) and Mather and Jinks (1982) and the symposium volumes edited by Pollak et al. (1977) and Robertson (1980) are recommended,

*Introduction*

and there are many relevant papers in the Proceedings of the Second World Congress on Genetics Applied to Livestock Production (1982). A more complete review of texts in the general area of quantitative genetics is given in the introduction to Part I.

A criticism that can be leveled against quantitative genetics as a discipline, as opposed to other branches of genetics, is that it is simply a "black-box" approach. An individual or a population is described not in terms of its components, the genes and their biological actions, but in terms of, for example, input-output parameters such as the regression of progeny on parent performance or realized heritability. Some attempts can be made to describe the formal genetics—for example, in the approach made by Thoday—but if there are many loci, a complete description becomes a practical impossibility. Is the science of quantitative genetics and its application to problems of selection then of any value? It is, because predictions can be made from the "black-box" theory, they can be and have been tested by experiments, and animal and plant breeders have put them to practical use. It is hoped that the papers included in this volume will help to convince any skeptical readers.

## REFERENCES

Falconer, D. S., 1981, *Introduction to Quantitative Genetics,* 2nd ed., Longmans, London.
Fisher, R. A., 1930, *The Genetical Theory of Natural Selection,* Clarendon Press, Oxford.
Haldane, J. B. S., 1932, *The Causes of Evolution,* Longmans, Green, London.
Mather, K., and J. L. Jinks, 1982, *Biometrical Genetics,* 3rd ed., Chapman and Hall, London.
Pollak, E., O. Kempthorne, and T. B. Bailey, Jr., eds., 1977, *Proceedings of the International Conference on Quantitative Genetics,* Iowa State University Press, Ames.
Robertson, A., ed., 1980, *Selection Experiments in Laboratory and Domestic Animals,* Commonwealth Agricultural Bureaux, Slough.
Wright, S., 1931, Evolution in Mendelian Populations, *Genetics* **16**:97-159.

# NATURE OF SELECTION RESPONSE

# Editor's Comments
# on Papers 1 Through 4

1   **CASTLE**
    *The Mutation Theory of Organic Evolution, from the Standpoint of Animal Breeding*

2   **JENNINGS**
    *Experimental Evidence on the Effectiveness of Selection*

3   **STURTEVANT**
    Excerpt from *An Analysis of the Effects of Selection*

4   **CASTLE**
    *Piebald Rats and Selection, a Correction*

 

It was many years after the rediscovery of Mendel's laws before the model of many discrete Mendelian genes of individually small effect associated with environmental influences was accepted as the basis of the inheritance of continuously distributed quantitative traits. The multifactorial model was first proposed, but later dismissed, by Pearson (1904) because he obtained predictions of correlations among relatives at variance with those obtained previously by Galton. The generalization by Yule (1906) to include other than completely dominant loci and environmental deviations solved this problem, but acceptance of the model as an explanation of correlations among relatives did not come until later, following the classical analyses of Weinberg (e.g., 1910), Fisher (1918), and Wright (1921 series). Experimental evidence for multifactorial inheritance was provided in the analyses of crosses in terms of individual loci by Nilsson-Ehle (1909) and East (1910). Finally, the models developed by Jones (1917) could explain heterosis and predict inbreeding effects when coupled with Wright's (1921) analysis, which pre-Mendelian models could not.

The foregoing reports (except Nilsson-Ehle's) are reprinted in Part I of these Benchmark Papers, and the history is reviewed in more detail. With minor exceptions however, these papers were concerned solely with the static description of a population and did not deal with problems of evolution by natural selection or of artificial selection. The effects of selection and the means by which they could be explained by the multifactorial model are covered, at least partly, in this section. Again, as in the first volume, the discussion relies heavily on Provine's (1971) excellent text, The *Origins of Theoretical Population Genetics.*

Darwin had argued that natural selection of continuous variations would produce evolutionary change. Direct proof of the effectiveness of selection in changing a population was lacking, however, although the results of animal and plant breeders gave indirect evidence. The first clear demonstration was provided by W. E. Castle of Harvard University, who selected for an increased incidence of polydactyly in guinea pigs. The results of these experiments were first reported briefly (Paper 1) and subsequently in more detail (Castle, 1906). Paper 1 also sets out Castle's views of the evolutionary mechanism.

Castle first noted a single polydactylous male. Further breeding of the parents resulted in no more polydactylous progeny from the dam, but five more from the sire, all but one in matings to descendents of the sire; in all, the dam had 30 offspring, the sire, 147 (incorrectly stated as 139 in Paper 1 but corrected subsequently [Castle, 1906]). After four generations of selection, the incidence had markedly increased: for the six males mated from this generation, each having at least 32 offspring, the percentages of polydactylous offspring were 88 percent, 34 percent, 81 percent, 83 percent, 99 percent, and 84 percent (Castle, 1906). Castle also noted that the proportion of polydactylous individuals that had a well-developed and functional fourth toe, as opposed, for example, to a small, turned up, or hanging toe, increased over generations as the total incidence increased. The trait was seen not to behave as would be expected if it had been determined by a single Mendelian gene. Although polydactyly is not a classic continuously distributed trait, the response to selection in incidence and expression observed by Castle certainly illustrated the power of selection to change a population by means other than single substitution at a single locus, to a level well outside the initial range. Castle and co-workers subsequently undertook a number of other major selection experiments, notably on color pattern and extent of color in rats. Reports of some of these experiments are given in Castle's text (1911), and further references are given by Sturtevant (Paper 3) and Castle (Paper 4).

The genetic explanation of the responses to selection in the laboratory or in nature was to become a controversial subject. Castle (Paper 1) argued that all changes arose as a consequence of mutation—that is, of the production of new variants in the germ line not previously present in the population. This view was later to be challenged. Let us first turn to evidence from other experiments.

The most important of these, preceding those of Castle, were

*Editor's Comments on Papers 1 Through 4*

undertaken by W. Johannsen (1903), working in Denmark with beans. A translation of his monograph is available (Gall and Putschar, 1955).

Johannsen first showed that selection of large and small beans from a mixed population led to offspring deviating in the same direction, but by a somewhat lesser amount, than predicted by Galton's law of regression. When, in the following year, he grew several beans from each plant, or pure line, he concluded there was no relationship between the size of the parental bean and those on resulting plant. Therefore, he wrote, "selection has had no reliably demonstrable influence on the types of the pure line—the regression is complete, quite up to the type of the line"; further, "the usual well-known result of selection—successive progenies in the direction of selection in the course of a few generations—depends . . . on the progressive purification with each generation of the deviating line concerned. And it will now be easily understood that the action of selection cannot be carried out beyond fixed limits . . . . " (Johannsen, 1903; cited by Provine, 1971, p. 94). In fact, Pearson and Weldon (1903) showed that the offspring-parent correlation within lines in Johannsen's data was not negligible, as he claimed, but was actually 0.35. As we now know, had Johannsen's lines really been pure, he would indeed have achieved a zero correlation unless there were any environmental covariances. However, his results on the constancy of pure lines posed problems for Pearson and the proponents of Galton's law of regression.

Further evidence on the lack of effectiveness of selection within pure lines was provided by H. S. Jennings in the United States, working with *Paramecium*. His main results were reported in detail (Jennings, 1908); a subsequent shorter and more popular account is included as Paper 2. Unfortunately, he quotes none of his actual data for, as Provine (1971) points out, they were a lot less clearcut than Jennings reported them. The environmental variability observed in the *Paramecium* in fact made any certain conclusions more a statement of faith. Nevertheless, Jennings's paper is included because it sets out the arguments of the pure-line protagonists in a clear way, the main premises being summarized on pp.139–140. He argues for the undoubted analytical power of the use of pure lines, but is less convincing in his conclusion that natural selection can do no more than find the best type originally present in the population (save for rare mutations). Another notable protagonist of the pure-line theory was R. Pearl, who was unable to obtain responses to selection for egg production in poultry (Pearl and Surface, 1909). As subsequent experimental work has shown (see Paper 16), this trait shows extensive environmental variability, and good experimental design is required to detect responses.

Castle's experiments, notably those on selection for the extent of pigmentation in populations of rats (Castle and Phillips, 1914), were, however, producing new stable phenotypes in populations with a mean expression well outside the range in the base population, and it was becoming clear that the pure-line hypothesis was untenable (except of course in what we would now know as completely inbred populations, or in clones). There remained, however, the need to explain the genetic mechanism responsible for the responses to selection. One alternative was that the changes were due solely to new mutations arising in the course of selection. This was the view put forward by Castle in Paper 1. The other alternative was that the trait was determined by a very large number of segregating genes, a hypothesis becoming accepted after 1910 as the explanation of quantitative variability, such that all individuals in the founder population had a negligible chance of having the best possible genotype. Selection coupled with reassortment in reproduction would, however, enable the frequency of improved types to increase, while further segregation would result in new types outside the original range, and so on.

The arguments for the latter hypothesis were put forward most clearly by A. H. Sturtevant (1918), a member of Morgan's group, in the discussion of a substantial selection experiment he had conducted in *Drosophila*, which, together with the summary, is reproduced as Paper 3. Morgan had already pointed out that the effects of some Mendelian genes could be small and that these could explain the slow continuous changes in evolution predicted by Darwin (Morgan, 1916). The discussion to Sturtevant's paper is particularly useful not only in summarizing the arguments, but also in reviewing the great bulk of experimental work on selection conducted to that time, including Sturtevant's own study.

The problems Sturtevant outlined were: "1. Does selection use germinal differences already present, or differences that arise during the experiment, or both? 2. In case it uses new differences, does it cause them to occur more frequently, and does it influence their direction? 3. Are differences, already present or arising *de novo*, more likely to occur in the locus of the gene under observation, or in other loci?" (p.36). Sturtevant points out that new mutations are rare, argues that they are no more likely to occur in the genes under selection, and concludes that selection must therefore act on factors already present. He goes on to argue that the increased variability observed in the $F_2$ over the $F_1$ of a cross can be due to segregation at many loci at which different genes are present in the lines. Following a review of the literature, he dismisses the possibility that the genes "change" by "contamination" in the $F_1$; he uses the multifactorial hypothesis to explain the published data purporting to show such contamination.

*Editor's Comments on Papers 1 Through 4*

Castle responded (in Paper 4) by completely accepting Sturtevant's arguments that the individual gene is not changed by the process of selection. This paper is notable because of both the change of Castle's stance and the experiment, conducted at Wright's suggestion, which Castle found conclusive. In this experiment he took two selected lines of rats divergent for the extent of the hooded pattern, one nearly all black, the other nearly all white, and backcrossed these repeatedly to a third line, in each case selecting out hooded individuals. The backcrossing would remove modifiers, and if the hooded gene *per se* had not changed, he would expect to find the same distribution of the extent of the hooded pattern from the low and high lines. This indeed was what he obtained (his Tables I and II).

There was now a fairly well established body of experimental evidence that showed that selection could gradually produce substantial changes in a population to an extent far outside the original range, and that these changes were due primarily to increase of frequency of favorable genes at a large number of loci. There was no conflict between the Mendelian and Darwinian views.

A mathematical basis for these predictions was to follow at the hands of Fisher, Haldane, and Wright, and only then was the Mendelian explanation of the evolution of quantitative traits to become accepted. We cannot review all their work in this volume, but some parts of it are presented in subsequent sections.

## REFERENCES

Castle, W. E., 1906, *The Origin of a Polydactylous Strain of Guinea Pigs*, Carnegie Institution of Washington Publication No. 49, pp. 17-29.

Castle, W. E., 1911, *Heredity in Relation to Evolution and Animal Breeding*, Appleton, New York.

Castle, W. E., and J. C. Phillips, 1914, *Piebald Rats and Selection*, Carnegie Institution of Washington Publication No. 195.

East, E. M., 1910, A Mendelian Interpretation of Variation That Is Apparently Continuous, *Am. Nat.* **44:**65-82.

Fisher, R. A., 1918, The Correlation Between Relatives on the Supposition of Mendelian Inheritance, *R. Soc. (Edinburgh) Proc.* **52:**399-433.

Gall, H., and E. Putschar, 1955, Concerning Heredity in Populations and in Pure Lines, in *Selected Readings in Biology for Natural Sciences 3*, University of Chicago Press, Chicago, pp. 172-215.

Jennings, H. S., 1908, Heredity, Variation and Evolution in Protozoa. 2. Heredity and Variation of Size and Form in Paramecium, with Studies of Growth, Environmental Action and Selection, *Am. Philos. Soc. Proc.* **47:**393-546.

Johannsen, W., 1903, *Uber Erblichkeit in Populationen und in Reinen Linien*, Gustav Fischer, Jena.

Jones, D. F., 1917, Dominance of Linked Factors as a Means of Accounting for Heterosis, *Genetics* **2:**466-479.

Morgan, T. H., 1916, *A Critique of the Theory of Evolution*, Princeton University Press, Princeton.

Nilsson-Ehle, H., 1909, Investigations on Crosses of Oats and Wheat, *Lunds University Årsskrift*, n. s., ser. 2, vol. 5, no. 2.

Pearl, R., and F. Surface, 1909, Is There a Cumulative Effect of Selection? *Z. Indukt. Abstamm. Verebungsl.* **2:**257-275.

Pearson, K., 1904, Mathematical Contributions to the Theory of Evolution. XII. On a Generalized Theory of Alternative Inheritance, With Special Reference to Mendel's Laws, *R. Soc. (London) Philos. Trans.* **A203:**53-86.

Pearson, K., and W. F. R. Weldon, 1903, Inheritance in *Phaseolus vulgaris*, *Biometrika* **2:**499-503.

Provine, W. B., 1971, *The Origins of Theoretical Population Genetics*, University of Chicago Press, Chicago.

Weinberg, W., 1910, Further Contributions to the Theory of Inheritance, *Arch. Rassen Ges.-Biol.* **7:**35-49, 169-173.

Wright, S., 1921, Systems of Mating, *Genetics* **6:**111-178.

Yule, G. U., 1906, On the Theory of Inheritance of Quantitative Compound Characters on the Basis of Mendel's Laws—A Preliminary Note, *Report 3rd Int. Conf. Genet. 1906*, Royal Horticultural Society, pp.140-142.

# THE MUTATION THEORY OF ORGANIC EVOLUTION, FROM THE STANDPOINT OF ANIMAL BREEDING

### W. E. Castle
*Assistant Professor of Zoology, Harvard University*

The mutation theory, as I understand it, is not designed to replace Darwin's theory of natural selection, nor is it capable of replacing that theory. Natural selection must still be invoked to choose between different organic forms, preserving the more efficient, destroying the less efficient. The question raised by this new theory is, What sort of forms are subjected to the action of natural selection? Is there a complete gradation of forms between two extreme conditions and is natural selection called upon to choose from this whole series the one which is organically most efficient, or is the task simpler and is the choice made merely between two widely separated conditions of the ideal series? Thus, we find within a species two varieties, one larger than the other. Have they diverged by gradual cumulation of minute differences in size, or by a single step? These alternative views are known, respectively, as the selection theory and the mutation theory. Both views were recognized by Darwin as possibilities, though he seems to have attached more importance to the process of gradual modification. Most of his followers have given attention exclusively to this process, but a few, like Bateson and de Vries, have regarded modification by steps as the more important process, if not the only one efficient in the formation of new species. Bateson ('94) has called modifications of this sort discontinuous variations, but de Vries ( :01– :03) calls them mutations, and the latter designation seems likely to be generally adopted.

Darwin rightly attached great importance to the variations of domesticated animals and plants as throwing light on the origin of species. He recognized that there is no essential difference between breeds and species, and that if we can ascertain how breeds originate we can infer much as to the origin of species. He made an extensive study of breeds of animals as well as of plants, but no one has followed this up or even recognized its great importance until within very recent years. What we need to know is how, precisely, are new breeds formed. We know that they are forming under our very eyes all the time and that this has been going on since the earliest historic times and no doubt a great deal longer, yet the method eludes us.

The successful practical breeder, the man who originates breeds, is a keen observer, a man of unusual intelligence and skill and of infinite patience. Yet if we ask him how, in general, he does his work, or how a particular result was obtained, we rarely get a satisfactory answer. This is sometimes because, for commercial reasons, it is well to leave a cloud of obscurity surrounding the origin of a successful breed, lest its production be duplicated. More often, however, it is because the breeder himself does not know how the result was attained. He may be able to tell us that such and such animals were mated, such and such of their offspring selected, and after a certain length of time the breed was established and put on the market. But this, after all, gives us little information as to the real nature of the material used and the processes involved in the formation of the new breed. The aims of the biologist are so different from those of the practical breeder that to solve the theoretical problems involved in the formation of breeds the biologist must himself turn breeder, and see new organic forms arise out of material with which he is thoroughly familiar, and under conditions which he can control. So little work of this kind has yet been done that its fruits are scarcely ready to be gathered. Generalizations can as yet be made only tentatively, based on cases dangerously few, or on the rather uncertain and often contradictory testimony of practical breeders and the half-truths told by stock registers.

So far, however, as these various sorts of evidence go, they indicate that the material used by breeders for the formation of new breeds consists almost exclusively of mutations. The breeder does not set to work with some purely imaginary form in mind, toward which he seeks by selection gradually to mold his material. He commonly either *discovers* the new breed already created and represented by one or more exceptional individuals among his flock, or else he seeks by cross-breeding to combine in a single race characters which he finds already existing separately in different races. In both cases he deals with mutations, *i. e.*, with characters unconnected by a series of transition stages with the normal form. An illustration from my own experience may help to make this clear. A little more than four years ago I obtained a number of ordinary smooth-coated guinea-pigs and began breeding them with a particular experiment in mind. Among nine young produced by a certain pair, there was one which had a supernumerary fourth digit on one of its hind feet. Neither of the parents had such a digit, nor had I ever heard of the existence of such a character before, either in any of the wild Caviidæ, or among domesticated cavies or guinea-pigs. Further, I have been able to find no reference to such a thing in the literature of the group, though I have several times since found this same mutation in other herds of guinea-pigs. The mother of my four-toed pig never produced another similar individual, though she was the mother in all of thirty young. The father, however, who sired in all 139 young, had five other young with extra toes, but these were all by females descended from himself, so that it seems certain that the mutation had its origin in this particular male. By breeding together the four-toed young and selecting only the best of their offspring I was able within three generations to establish a race with a well developed fourth toe on either hind foot. This race was not *created* by selection, though it was *improved* by that means. Like the poet, in the proverb, it was *born,* not made. Any amount of selection practised on other families of my guinea-pigs would probably never produce a four-toed race, for though carefully watched through as many as seven generations no four-toed pig has appeared among them.

In a second family of my guinea-pigs, which, like the other, was for the purpose of a particular experiment inbred, a different mutation made its appearance. A few individuals were found to have hair about twice as long as that of their parents and grandparents. Intermediate conditions did not occur. Long-haired individuals mated together were found to produce only long-haired young, so that a new breed was already fully established without the exercise of any selection. It was found, in short, that the long-haired character is a Mendelian recessive in relation to the normal short coat, so that matings between long-haired and short-haired animals produce only short-haired young. But these young bred *inter se* produce a definite proportion (about one fourth) of long-haired young, and if no selection is practised among their offspring, but all are allowed to breed freely, the race will continue to contain this proportion of long-haired individuals.

If such a mutation as this occurred in a state of nature, and such a possibility we can scarcely question, a dimorphic species would be the immediate result, containing two varieties alike in every respect except length of hair, in which they would be sharply separated. The two varieties would coexist in the same habitat and might continue to interbreed freely without the destruction or necessary modification of either. Natural selection would now come into operation to choose between the two that one which was more advantageous and the other condition would be gradually eliminated from the race. Or if the two conditions were each the better in a different habitat, then by the gradual destruction of the other in that habitat the two varieties would become geographically separated, though they might continue to coexist in an intermediate zone.

The second method which I mentioned for the artificial production of new breeds is to combine in one race characters already found in different races. This is accomplished through cross-breeding and is made possible by the facts (1) that mutations are alternative in heredity to the normal condition, and (2) that one mutation is entirely independent of another in heredity. If, for example, we cross long-haired with short-haired guinea-pigs, we get, among the second-generation offspring, a mixture of long-haired and of short-haired animals, but, as a rule, no intermediates. Further, if in the original cross one parent was four-toed and the other three-toed, then in the second generation offspring we get all possible combinations of the characters involved in the cross, viz., long-haired four-toed animals, long-haired three-toed, short-haired four-toed and short-haired three-toed. From this array of forms the breeder may now select the particular combination of characters which suits his purpose.

Can we doubt that in nature a similar choice is offered between every mutation and its opposite, combined or uncombined with every other mutation then present in the race?

It is true that cross-breeding may affect to a greater or less extent the nature of the characters involved in a cross, but this sometimes facilitates the creation of desirable breeds, for it serves to induce new mutation, which in some cases is progressive, in others regressive. For example, in guinea-pigs, a cross between a coal-black animal and an albino may restore in the young the ancestral, or 'agouti,' coat consisting of black hairs ticked with reddish yellow, or in other cases may result in the production of a black-white spotted animal. By selection either of these conditions may be perpetuated in a distinct breed. The one is a regressive or reversionary change, the other progressive in that it leads to the production of a new type of pigmented coat.

On the whole, it appears that the formation of new breeds begins with the discovery of an exceptional individual, or with the production of such an individual by means of cross-breeding. Such exceptional individuals are mutations.

An examination of stock registers points in the same direction. The beginnings of new breeds are small. Pedigrees lead back to a few remarkable individuals or to a single one, as in the Ancon sheep. But given the exceptional individual, and a new breed is as good as formed. The few generations which the breeder usually employs in 'fixing' or establishing the breed and during which he practises close breeding serve principally to free the stock from undesirable alternative characters, not to modify the characters retained.

Modification of characters by selection, when sharply alternative conditions (*i. e.*, mutations) are *not* present in the stock, is an exceedingly difficult and slow process, and its results of questionable permanency. Even in so-called 'improved' breeds, which are supposed to have been produced by this process, it is more probable that the result obtained represents the summation of a series of mutations rather than of a series of ordinary fluctuating variations. For mutations are permanent; variations transitory. A moment's reflection will indicate the probable reason. Variations which are distributed symmetrically about a modal condition, so as to produce when graphically expressed a frequency of error curve, represent the result of a number of causes acting independently of each other. These causes are principally external, consisting in varying conditions of food-supply, temperature, density, moisture, light, etc. These conditions alter from generation to generation, and so do effects dependent upon them. Mutations, on the other hand, have an internal origin, in the hereditary substance itself. They are relatively independent of the environment, being affected only by such causes as affect the nature of the hereditary substance itself, one of which apparently is cross-breeding.

There are, however, frequently found in breeds of domesticated animals conditions which are *not* sharply alternative in heredity to the corresponding characters of other breeds. It is an open question whether such conditions could be maintained if cross-breeding were freely allowed with animals of a different character. If not, they could scarcely become racial characters, under the action of natural selection. The race would then become, not sharply dimorphic or polymorphic, as is the case where inheritance is sharply alternative, but subject to extremely great fluctuating variations. It is open to question whether blending characters of this sort found in many breeds may not have been created by selection from masses of fluctuating variations. It will be important to know further whether or not these extreme fluctuating series have had their origin in mutations. Not improbably, as de Vries has in part suggested, one-sided variation curves indicate the occurrence of mutations of this sort.

# 2

*Copyright ©1910 by The University of Chicago*
Reprinted from *Am. Nat.* **44**:136-145 (1910)

## EXPERIMENTAL EVIDENCE ON THE EFFECTIVENESS OF SELECTION[1]

PROFESSOR H. S. JENNINGS

JOHNS HOPKINS UNIVERSITY

In studying the problems of evolution in the common infusorian *Paramecium,* I found that by methodical and progressive selection striking results can be reached.

From a wild culture it is possible by progressively selecting in two opposite directions to obtain finally two lots, one of which is many times as large as the other, and the differences between the two are permanent and hereditary. By properly regulated selection a great variety of permanently differentiated lots are obtainable.

Throughout this work Galton's law of regression was found to hold; that is, the progeny of extreme parents inherited the peculiarities of their parents, but in a less marked degree. Furthermore, the results were such as to lend themselves readily to interpretation as exemplifying Galton's law of ancestral heredity.

Thus the effectiveness of selection was clearly demonstrated. But just what sort of effectiveness does the theory that selection is the dynamic factor in evolution demand? It demands that selection shall so act that it might finally produce progress from *Amœba* up to man. It must produce, from a given condition, something that did not before exist in that given condition.

Has selection so acted in this case? To answer this question, we must evidently know precisely what exists in the condition with which we start. We therefore next work with the progeny of a single individual—forming a "pure line," the characteristics of which we thoroughly know.

Now we try the effects of selection on this pure line.

[1] A paper read before the American Society of Naturalists, December 29, 1909.

Not the faintest trace of effect is produced, even by long-continued methodical selection for hundreds of generations. The race or line is absolutely permanent, so far as the appearance of any hereditary differences are concerned. The individuals of the line do indeed differ greatly among themselves, but these differences are not inherited; they furnish absolutely no foothold for selection.

Examination showed that *Paramecium* consists of many such races, differing among themselves slightly, but each race as unyielding as iron. And the extreme races found in the wild culture are precisely the extremes obtainable by long-continued selection.

The effects of selection have then consisted simply and solely in isolating races that already existed. It had produced nothing new; there had been no progress that would form a step, however slight, in the journey from *Amœba* to man.

When I had reached this point I looked about and found that others had been having similar experiences. The investigator who discovers these things for himself finds perhaps that

> Es ist eine alte Geschichte
> Doch bleibt sie immer neu.

And the second line is as true as the first, for to one who has put months and years on such attempts to accomplish results by methodical selection, its utter powerlessness comes with new and surprising force. Johannsen in working with beans and barley, Hanel with *Hydra*, had found many pure lines existing in nature, but as in my own case, each pure line was absolutely unyielding.

But we know that others *had* found selection effective; a whole series of cases comes at once to our lips. Galton in studying peas and men; Fritz Müller with maize; de Vries with maize and with buttercups, MacCurdy and Castle with guinea pigs and rats—all these had reported definite progress as a result of methodical selection.

Why this difference? Is there one law for the Jews, another for the Gentiles?

Looking into the matter with care, we find that the results with our own material are, after all, like those of the investigators mentioned if we treat it in the same way. None of these workers first isolated their pure races. If we begin with a mixture we can, in beans, in barley, in *Paramecium,* in *Hydra,* by a methodical process of slow selection make gradual progress in a certain direction. But our selection is only a process of purification, and when we finally get a pure race, selection is utterly powerless to go farther. We should have been completely in the dark as to the real effect of selection if we had not carried through rigidly the "pure line" idea.

Is it possible then that we have in this pure line idea an instrument of the greatest importance for analysis? Is it perhaps the key which every one must have in order to understand the results of selection? May it be indeed one of those fundamental ideas which, like the idea of mutation, is fitted to clear and crystallize a confused and turbid mixture? Is it possibly of sufficient importance to deserve agitating a little before the American Society of Naturalists?

Let us put these questions to the practical test; let us apply the idea as an instrument for the dissection of the classic cases which seem to demonstrate the efficacy of selection in producing change of type.

Johannsen in his recent book has used the pure line concept as an instrument for analysis of the entire field of variation, heredity and evolution, and to him is due the credit of first perceiving the importance of this concept, when sharply defined, as such an instrument for research and presentation. The work of Johannsen, I believe, will remain one of the landmarks of progress in this field. But my own analysis has been independent of Johannsen's and diverse from it, developing inevitably from what I have myself seen, so that I may venture still to present some of its results.

But how can we apply the pure line idea to organisms whose lines are *not* pure; organisms that interbreed freely; organisms in which the characters of a given line split off, separate, and become exchanged for those of other lines, in the way characteristic of Mendelian inheritance?

The pure line idea here becomes a little elusive, a little abstract. But possibly it is still helpful as an instrument of analysis; let us try it. In order not to emphasize purity where impurity is the rule, let us substitute Johannsen's term *genotype* for "pure line"—defining the genotype as a set of individuals which, so long as they are interbred, produce progeny that are characteristically uniform in their hereditary features, not systematically splitting into diverse groups.

Now, how can we determine whether the genotype concept, with its consequences for the effects of selection, applies to organisms with biparental inheritance? Reflection shows that if it does, certain general propositions are true: if these propositions are found to hold, the genotypic explanation of the effects of selection is confirmed.

1. The first proposition is this: Organisms in which selection has shown itself effective are composed of many genotypes; of many races that are diverse in their hereditary characters. This we know to be true.

2. Second, from such a mixture of genotypes it is possible to isolate by selection any of the things that are present—perhaps in a great number of different combinations.

3. But from such a mixture it is *not* possible to get by methodical selection anything not present (save when rare mutations have occurred).

4. Therefore it is not possible to get by methodical selection anything lying outside the extremes of the genotypic characters already existing.

This is perhaps practically our most important proposition. For in order that selection shall produce progression from *Amœba* to man, it is evidently necessary that it should give us characters lying beyond the extremes of what already exists.

5. Our fifth proposition is that in the case of genotypes that cross-breed readily, we may get an indefinite number of combinations of all that lies between the extremes of the existing genotypes—the variety of combinations realized depending on the rules of inheritance.

Now, if we test by these propositions the classic cases of effective selection, what is the result?

Galton's work with peas and with men yields at once to the analysis, giving precisely the results which the genotypic idea requires. A by-product of the analysis is the practical evaporation of the laws of regression and of ancestral inheritance so far as their supposed physiological significance is concerned;[2] they are found to be the product mainly of a lack of distinction between two absolutely diverse things—between non-heritable fluctuations on the one hand, and permanent genotypic differentiations on the other.

The experiments of Müller and de Vries on maize yield with equal readiness. In these cases the male parents are unknown; the freest sort of crossing was occurring, and what selection did was pick out the progeny of extreme male genotypes, till the result approached the limit of the most extreme existing genotype under the cultural conditions.

MacCurdy and Castle's experiments in changing by selection the color-patterns of rats and guinea-pigs dealt

[2] This significance was supposed to lie in showing that the characteristics of the progeny depend upon the characteristics of the ancestors for many generations back, in ways that are definable. "Mr. Galton's view of the effect of regression follows inevitably from the general theory of chance, if we regard the character of an individual as a phenomenon due to a series of complex groups of causes, *among which are the characters of each ancestor*." (W. F. R. Weldon, *Biometrika*, I, p. 370.)

The law of ancestral inheritance does not hold in pure lines, even in a statistical sense, as has repeatedly been shown. The progress of a long series of extreme ancestors does not differ from those of a series of average ancestors.

with races of complicated descent; they plunge us at once into all the difficulties due to interweaving, blending and transfer of characters from one genotype to another. But if we stick closely to the general propositions already stated, we shall have a guide. MacCurdy and Castle got by selection all sorts of conditions lying between the extremes with which they started. But did they get anything lying outside these extremes, as would be required in order to show that we can by selection make evolutionary progress? As I read their results, they did *not*. Their experiments are most important for many problems of variation and inheritance, but they do not give us evidence that methodical selection can produce anything beyond combinations of what already exists; hence they do not help us in getting from *Amœba* to man.

The work of the German breeders who have for years practised methodical selection for improvement of agricultural races clears up at once under the genotype idea, as the analyses of Fruwirth, v. Rümker and others show us. Continued methodical selection is often necessary, but what it does is to purify a contaminated race—a process which, owing to the laws of inheritance, may require several generations.

I have spoken only of those experiments which seem at first view to show the efficacy of selection; brevity requires me to pass without mention over investigations which, while not carried on with pure lines, support and reinforce the conclusions drawn from such work. Such for example are the fundamental experiments of Tower, the recent work of Pearl, of Shull, and many of the experiments of de Vries.

Thus far the dissecting knife of the pure line idea succeeds admirably in letting the light into the obscure workings of selection. Any one who uses it with precision will find what an important advance its exact formulation by Johannsen marks over even such an analysis as that given by de Vries. In the work of de Vries, as in that of many recent authors, the selection idea is appraised at essentially its true value, but much in its action is left obscure. The reader is surprised at the accounts of experiments in which selection does accomplish marked results, though according to the general theory, it should not; one is left puzzled in judgment as to what we may expect from it. With the sharply formulated pure line concept as a guide, most of this obscurity disappears.

And then, to keep us from resting on our oars; to give us humility and spur us to further work—we come to the one case in which the pure line idea fails to bring clearness. This is de Vries's experiment with buttercups. Here, after selection the extreme was moved far beyond that before selection. Before selection the extreme number of petals was eleven; after selection it was thirty-one. Before selection no single individual had an average number of petals above six; after selection the average of all was above nine, and some had an average of thirteen! It is true that there are "mitigating circumstances" here; the work was not done with pure lines, and the variations dealt with are not of the ordinary fluctuating sort (as de Vries points out); change in cultural conditions doubtless played also a large part. Possibly repetition with thorough analytical experimentation will show that something besides selection has brought about the great changes. But at present the case stands sharply against the generalizations from the pure line work. It is the only such case that I have found.

To sum up, one finds not only that his own results and those of many other modern workers give the pure line interpretation, but also that all other cases that had seemed to point the other way yield readily to the genotypic analysis—save one. If we ride rough shod over this case, as not yet sufficiently studied, then we may draw a tentative conclusion as follows:

The pure line or genotype idea is the one to see clearly and grasp firmly in experimental investigations on selection. Many even of the modern experiments remain obscure in their significance simply because the workers have not grasped this concept, have not shown the relation of their results to it. Further, in presenting one's own work, or in interpreting the accounts of others, the genotype concept is the instrument of precision to take in hand. The results of the analysis made by its aid indicate that most or all of the experiments in methodical selection have consisted in shifting about, isolating and recombining preexisting, permanent hereditary differentiations, giving results that were interpreted as revealing the law of actually progressive evolution, though in reality they had no relation to such a law.

To our conclusion as to the analytical value of the pure line idea we may expect strenuous opposition on the part of that last small remnant (if there be such a remnant) of the biometrical school that still submits to the dictation of Pearson[3]—for by one of those sardonic paradoxes through which nature revenges herself, the men who from outside have lectured biology on the necessity of becoming exact are the strongest opponents of exact experimental and biological analysis—seeming to feel that mathematical treatment renders other kinds of exactness undesirable.[4] Those who find the genotype idea useful may then prepare themselves for one of those justly famous bludgeonings from the dictator of the whilom orthodox biometrical school; this is the last honorable mark of distinction which stamps the investigator as a thorough and exact analyst of things biological.[5]

[3] Note how quickly the biometricians that devote themselves to careful biological investigations fall away from the Pearsonian faith. Darbishire, Davenport, Tower, Shull, Johannsen, Pearl; are there any biologists of achievement that still hold with Pearson?

[4] Pearson in 1901 informs us that evolution is a field "where no tabulation of individual instances can possibly lead to definite conclusions" (*Biometrika*, I, p. 344). This was the year of the appearance of de Vries's *Mutationstheorie*, and of the revival of Mendelism. Compare in definiteness and value the conclusions drawn from the work inaugurated in these two lines, based as it was precisely on the "tabulation of individual instances," with those from the biometrical work, with its careful avoidance of "individual instances."

[5] To name the men who have been subjected to Pearson's most savage

A word more on certain general questions. Can we conclude that if selection has no dynamic effect in changing existing genotypes, that therefore it need not be reckoned with in evolution? Or must we conclude that if it is to be reckoned with at all, selection has opportunity to act only on large leaps in evolution; that evolution takes place by such leaps, and not by imperceptibly small changes?

Such evidence as the pure line work gives implies neither of these things. The differences between the diverse pure lines *have* arisen in some way, if evolution occurs, and once these differences have arisen, they are open to the operation of selection as are any other differences. What the pure line work shows (agreeing in this with other lines of evidence) is that the changes on which selection may act are few and far between, instead of abundant; that they are found not oftener than in one individual in ten thousand, instead of being exhibited on comparing any two specimens; that a large share of the differences between individuals are not of significance for selection or evolution—these being precisely the differences measured as a rule by the biometrician's "coefficient of variation." Thus the work of natural selection is made infinitely more difficult and slow; but logically it is still possible.

Nor does the pure line work assist natural selection, as some have hoped from the mutation work, by making the steps in evolution greater in amount. On the contrary, the work with genotypes brings out as never before the minuteness of the hereditary differences that separate the various lines. These differences are the smallest that can possibly be detected by refined measurements taken in connection with statistical treatment. Johannsen found his genotypes of beans differing constantly merely by weights of two or three hundredths of a gram in the average weight of the seed. Genotypes

assaults is to name the men that have done most to advance our knowledge of heredity. The cases of Castle and of Bateson will occur to every zoologist.

of *Paramecium* I found to show constant hereditary differences of one two-hundredth of a millimeter in length. Hanel found the genotypes of *Hydra* to differ in the average number of tentacles merely by the fraction of a tentacle. That even smaller hereditary differences are not described is certainly due only to the impossibility of more accurate measurements; the observed differences go straight down to the limits set by the probable error of our measures. Genotypes so differing have not risen from one another by large mutations. The genotypic work lends no support to the idea that evolution occurs by large steps, for it reveals a continuous series of the minutest differences between great numbers of existing races.

All together, I think we may say that the pure line or genotype concept presents an instrument of analysis which is worthy, on the basis of what it has thus far done, of a thorough tryout for future work, and no one interested in these questions can afford to neglect it. This conclusion is quite independent of the concrete results reached; the efficacy of selection in modifying genotypes may be demonstrated to-morrow, but the demonstrators will need to show precisely the relation of their results to the pure line concept.

**3**

Reprinted from pages 36-54 of *Carnegie Institution of Washington Publication No. 264*, 1918, 68p.

# AN ANALYSIS OF THE EFFECTS OF SELECTION

## A. H. Sturtevant

## GENERAL DISCUSSION.

### THE SELECTION PROBLEM: QUESTIONS AT ISSUE.

It appears to the writer that the three questions below are the chief ones at issue in the discussion of the selection problem:

1. Does selection use germinal differences already present, or differences that arise during the experiment, or both?
2. In case it uses new differences, does it cause them to occur more frequently, and does it influence their direction?
3. Are differences, already present or arising *de novo*, more likely to occur in the locus of the gene under observation, or in other loci?

It is not, I think, questioned by any one that selection may effect either gradual or sudden change in the mean character of mixed races, or that it may even, occasionally, produce such an effect in pure races if a mutation in the desired direction happens to occur.

*1. Does selection use germinal differences that are already present, or differences that arise during the experiment?*

Everyone who has bred animals or plants is familiar with the fact that different strains, even when rather closely related, differ in all sorts of minor points—size, proportions of organs, shade of color, resistance to disease, fertility, temperament, rate and habit of growth—in fact, in almost any respect that one investigates. This can only mean that such strains differ genetically; and since the kinds of differences are usually so numerous, they probably usually have many genetic differences—*i. e.*, they differ in respect to many factors. In any race not normally self-fertilizing or closely inbred, crosses between individuals of different constitution must then be frequent. And such crosses must, on the assumption that the original differences were Mendelian, lead to the production of a population more or less heterozygous for factors that produce minor effects on all sorts of characters. The assumption that the differences are Mendelian rests on the observed facts, (1) that demonstrably Mendelian factors may produce effects on practically any kind of character studied, and effects of practically any observable degree; and (2) that non-Mendelian inheritance has never been demonstrated, except for a few cases of plastic characters in plants and cases of infectious diseases.[1] Other kinds of inheritance may exist; but the available data indicate that they must be extremely rare. Therefore the chances are that any observed difference between two strains is Mendelian.

If these conclusions be accepted, it follows that any strain not very closely inbred is likely to be heterozygous for factors influencing many characters. Selection for these characters will then be effective in isolating favorable combinations of such "modifying factors."

---

[1] One may refuse to call these cases of inheritance if he chooses to define that term so as to exclude them.

Mendelian differences are still arising by mutation and may arise in a selection experiment as well as anywhere else; and those that arise in such an experiment are as likely to affect the character under observation as are any Mendelian differences taken at random. It is therefore probable that selection sometimes makes use of variations that arise during the course of the experiment, or, rather, that variations which may be available do arise.

The question is, what is the relative frequency of the two kinds of available factor differences—those already present and those that arise *de novo*? The answer is found by investigation of the data on selection in inbred lines and in crossbred lines. In closely inbred strains there are not likely to be many factor differences present when selection is begun, while in crossbred lines these differences are likely to be numerous.

That selection is usually effective in crossbred lines is a well-known fact, demonstrated many times with many different organisms. Not many experiments have been carried out on closely inbred material, but those of Johannsen (1903), MacDowell (1917), and the present paper (p. 11) show that selection may be without effect in such lines. In two of these cases selection was effective until the lines became highly inbred. But mutations influencing the characters under observation have been obtained in the selection experiments of Castle and Phillips (1914), Morgan (Morgan, Sturtevant, Muller, and Bridges, 1915, p. 205), Lutz (1911), and those reported in this paper (p. 31).[1]

Apparently, then, selection produces its effects chiefly through isolation of factors already present, but occasionally available mutations do arise during the course of the experiment.

*2. Does selection cause mutations, or influence their direction?*

The usual selection experiment consists in breeding from individuals that are extreme in some respect. This extreme character may be environmental in origin, or it may be caused by germinal differences. In the first case, no geneticist is likely seriously to maintain that selection will have any effect whatever. In case the extreme character is germinal in origin, selection will of course be effective in eliminating certain genetic types. Moreover, given a combination of genes that produce the character in a certain degree, we are evidently in a better position to reach a further stage than if we have the character less well developed. For how long a tail will be when it gains an inch evidently depends on how long it was before it gained that inch. But it seems incomprehensible that selection of individuals of a constitution favor-

---

[1] Evidence derived from forms that reproduce asexually is also available in studying this question, for such reproduction commonly prevents recombination, and therefore gives results comparable with those obtained from homozygous strains. Some of the evidence obtained from studies on asexually produced Protozoa (*e. g.*, Calkins and Gregory, 1913; Jennings, 1916; Middleton, 1915) has shown that selection may be very successful in changing such forms. But it is very doubtful if these animals are comparable with the Metazoa in the method of distribution of their chromatin. It seems not improbable that in some cases recombination may here be possible in asexual reproduction.

able to the development of a given character can make more likely the occurrence of factorial variations affecting that character, or variations affecting it in a given direction. As a matter of fact, there is no evidence for such a conclusion. The occurrence of mutations is ordinarily such an extremely rare phenomenon that it would be very difficult to obtain statistically significant data in the matter. Moreover, when one is selecting for a character, one is examining his animals or plants for that character with unusual care, so that any mutations in that character are very likely to be observed and tested, provided they are in the direction in which selection is being carried out. It follows from these considerations that extremely careful controls are required before any data on these questions can have any significance.

*3. Are variations more likely to occur in the locus of the gene under observation, or in other loci?*

In *Drosophila* over 25 different and independent mutant factors affect the color of the eye. In mice there are 7 or more independent factors affecting coat-color. According to Little (1915) there are 2 and probably 3 independently segregating factors that affect spotting in these animals. There are at least 14 and probably more definite genes (in different loci) that affect bristle number in *Drosophila*, not counting the "modifying factors" studied by MacDowell and the writer.

In view of these and many similar facts, it is certain that changes in a given character may be brought about by changes in many different parts of the germ-plasm. If selection of a given mutant race, say hooded rats or Dichæt *Drosophila*, is likely to cause or to isolate mutations in the gene that differentiates that race from the normal type (*i. e.*, the hooded factor or the Dichæt factor) rather than in any other factors, it follows that mutant allelomorphs must be more variable than "normal" ones. For, by analogy with mice, hooded rats are homozygous for the normal allelomorphs of several possible factors affecting spotting; and Dichæt flies are certainly homozygous for the normal allelomorphs of at least 13 mutant factors that affect bristle number. It may be true that mutant factors are on the average more variable than their normal allelomorphs; but no evidence to that effect is at hand; and owing to the great difficulty of statistical treatment of the frequency of mutations alluded to above, such evidence will be very difficult to obtain.[1]

In the absence of such evidence, it is more probable that variations will appear in other factors, since there are many of them to vary, but commonly only one that is responsible for the difference under observation. That changes of the one factor itself may occur in selection experiments, however, has been shown by Castle (Castle and Wright, 1916) and the writer (p. 31). It does not follow that selection has caused these variations or that they are more likely to occur than are variations in other factors.

---

[1] Evidence has been obtained by Emerson (1917), who used unusually favorable material, that shows clearly that different allelomorphs may at times differ greatly in their mutability.

## CONTAMINATION OF ALLELOMORPHS.

When two races that differ in quantitative characters are crossed, it is frequently observed that $F_1$ is fairly uniform, and that $F_2$ shows an increase in variability together with the production of forms intermediate between the parent races and often different from the $F_1$. There are two current methods of accounting for these cases:

(1) The two races are assumed to have differed in a number of Mendelian factors affecting the character in question. The observed result is then explained as due to the recombinations of these factors.

(2) The two races are assumed to have differed in only one factor affecting the character in question, and the new types observed in $F_1$ are supposed to be due to "contamination" in the $F_1$ hybrid, that is, allelomorphs present in the heterozygote are supposed to have influenced each other, so that they do not come out unchanged.

The fundamental principle of the first explanation—that more than one factor may influence the same character—is admitted by all Mendelians. But many of the adherents of that explanation are unwilling to admit that "contamination of allelomorphs" has ever been experimentally demonstrated. Let us then examine the evidence that is brought forward in support of that assumption.

The following quotations are the chief ones bearing on the question that I have been able to find in recent literature:

"The currently accepted explanation (of size inheritance), which its supporters choose to call 'Mendelian,' rests upon the idea of gametic purity in Mendelian crosses. It assumes that Mendelian unit-characters are unchangeable and unvarying, and that when they seem to vary this is due to a modifying action of other unit-characters (or factors) . . . . The idea of unit-character constancy is a pure assumption. In numerous cases unit-character inconstancy has been clearly shown, as in the plumage and toe characters of poultry according to the observations of Bateson and Davenport, and the coat-characters and toe-characters of guinea-pigs in my own observations. Unit-character inconstancy is the *rule* rather than the exception." (Castle, 1916*b*, p. 209.)

". . . . I have shown in numerous specific cases that when unlike gametes are brought together in a zygote they mutually influence each other; they partially blend, so that after separation they are less different than they were before. The fact remains to be accounted for that partial blending does occur (1) when polydactyl guinea-pigs are crossed with normals (Castle, 1906); (2) when long-haired guinea-pigs are crossed with short-haired ones (Castle and Forbes, 1906); and (3) when spotted guinea-pigs or rats are crossed with those not spotted (MacCurdy and Castle, 1907). Davenport has furnished numerous instances of the same thing in poultry; indeed, he has shown that "imperfection of dominance" and of segregation are the rule rather than the exception in Mendelian crosses in poultry." (Castle, 1916*d*, p. 253.)

". . . . The English unit-character had changed quantitatively in transmission from father to son. This seems to us conclusive evidence against the idea of unit-character constancy, or 'gametic purity.'" (Castle and Hadley, 1915.)

". . . . We are often puzzled by the failure of a parental type to reappear in its completeness after a cross—the merino sheep or the fantail pigeon, for example. These exceptions may still be plausibly ascribed to the interference of a multitude of factors, a suggestion not easy to disprove; though it seems to me equally likely that segregation has been in reality imperfect." (Bateson, 1914.)

Fractionation is referred to by Bateson in this same paper as probably due to imperfect segregation. Illustrations are Dutch rabbit and Picotee and other sweet peas. (See p. 298.)

"Accordingly we seem limited to the conclusion that a slowly blending gene is involved in the cross between early flowering and late flowering peas, that the blending after one generation of heterozygosis may be small in amount, but after three generations it is in the majority of cases practically complete, so that the commonest 'constant' class in the entire hybrid population is one strictly intermediate between the modes of the parental varieties. This interpretation is entirely in harmony with the observed modification through crossing of many Mendelizing characters, as observed by Davenport, Bateson, and many others in poultry, guinea-pigs, swine, and other animals, as well as in plants." (Castle, 1916b, p. 215.)

Hayes (1917) states on the basis of his experiments with variegated maize:

". . . . One might conclude that certain heterozygous combinations produce germinal instability which exhibits itself either as imperfect segregation, gametic contamination, or sporophytic variation."

In these quotations the following cases have been cited as evidence in favor of contamination, and therefore calling for investigation:[1]

1. Polydactyl guinea-pigs (Castle, 1906).
2. Long-haired guinea-pigs (Castle and Forbes, 1906).
3. Spotted guinea-pigs and rats (MacCurdy and Castle, 1907).
4. English rabbits (Castle and Hadley, 1915).
5. Poultry, plumage and toe characters (Bateson and Davenport).
6. Merino sheep.
7. Fantail pigeons.
8. Dutch rabbits.
9. Picotee and other types of sweet peas.
10. Flowering time in peas (Hoshino, 1915).
11. Unspecified case in swine.
12. Variegated pericarp in maize (Hayes, 1917).

Before we can discuss some of these cases intelligently it is necessary that we make sure what Castle means by the terms "gametic purity" and "unit-character." Unless these terms are understood in such a way as to eliminate from consideration the idea of recombination of independent factors there is, of course, nothing to discuss. If by gametic impurity or inconstancy of unit-characters is meant that recombination of modifying factors occurs, the existence of such phenomena must be granted at once—this is, in fact, the main contention of the school of "pure line" advocates or "mutationists." I think the two following quotations from Castle are sufficient to show that there need be no disagreement on the question of defining these terms:

"What we want to get at, if possible, is the objective difference between one germ-cell and another, as evidenced by its effect upon the zygote, and it is

---

[1] The rough-coated guinea-pig was formerly cited (e. g., Castle and Phillips, 1914), but is now never used. This is because Wright (Castle and Wright, 1916) has shown the results to be due to multiple factors.

the constancy or inconstancy of these objective differences that I am discussing. If these are quantitatively changeable from generation to generation, then change in the variability of the zygote composing a generation might arise *without factorial recombinations.*"[1] (Castle, 1914a.)

"The head, the hand, the stomach, stomach-digestion, these are not unit-characters so far as any one knows. But if a race without hands were to arise and this should Mendelize in crosses with normal races, then we should speak of a unit-character or unit-factor for 'hands,' loss of which or variation in which had produced the abnormal race. But in so doing we should refer not to the hand as an anatomical part of the body nor to the thousand and one factors concerned in its production, but merely to *one hypothetical factor* to which we assign the failure of the hand to develop in a particular case. It is immaterial whether we call this a *unit-character* or *unit-factor* or use both terms interchangeably . . . . ." (Castle, 1916b, p. 100.)

### 1. POLYDACTYL GUINEA-PIGS.

The most extensive data on this case are apparently in the paper (Castle, 1906) cited in the quotation already given. The extra-toe character was at first irregular in appearance, but was improved by selection. In five generations, without very close inbreeding, a practically uniform race was obtained. When crosses to normal were made, the $F_1$ results varied from nearly all normal to nearly all polydactylous. $F_2$ contained both normal and extra-toed individuals. It is pointed out by Castle in this paper that the results are very similar to those obtained by Bateson from polydactylous fowls. Bateson's comment on that case is given below.

In the absence of any definite data regarding $F_2$ counts, the case as reported is entirely explicable on the multiple-factor view. Castle himself said of it, five years after the publication of the above paper:

"An alternative explanation is possible, viz., that the development of the fourth toe depends upon the inheritance of several independent factors, and that the more of these there are present, the better will the structure be developed. The correctness of such an interpretation must be tested by further investigation." (Castle, 1911, p. 101, footnote.)

So far as I have discovered, such further investigations have not yet been reported, although five years later this case is listed as No. 1 among those that demonstrate contamination of allelomorphs.

### 2. LONG-HAIRED GUINEA-PIGS.

The reference given for this case (Castle and Forbes, 1906) seems to contain the most recent and complete data regarding it.

Angora guinea-pigs appeared in a short-haired stock, apparently as segregated recessives. On crossing to short and extracting, there were produced some animals of intermediate hair-length, and some unusual ratios. But similar intermediates appeared in another strain of shorts, apparently uncrossed with angoras, thus making it highly probable that we are dealing here with a factor already present in the

---

[1] Italics mine.

race, and not produced by the cross of angora×short. The unusual ratios are based on quite small numbers, and the authors admit that there are difficulties in separation of the three classes, apparently due to overlapping. Moreover, we are given the results only in total, not from each mating separately.

Castle himself has said of this case: " . . . a single unit-character is concerned. Crosses in such cases involve no necessary change in the race, but only the continuance within it of two sharply alternative conditions." (Castle, 1911, p. 39.)

### 3. Spotted Guinea-Pigs and Rats.

The reference given for these cases is MacCurdy and Castle (1907). I am unable to find in that paper any evidence regarding guinea-pigs that bears on the question of contamination. Nothing but selection experiments are reported. There is, so far as I am aware, no evidence of significance in this connection in the more recent literature on spotting in guinea-pigs.

The evidence referred to from rats is apparently that obtained from crosses between hooded and Irish races. Hooded rats extracted from such crosses had more extensive colored areas than the uncrossed hooded rats. The data given by Castle and Phillips (1914) and analyzed by MacDowell (1916) show that this is true only when the hooded race is a "minus" one. The "plus" hooded race becomes *less* pigmented when crossed to Irish (or to self). MacDowell has shown that these results conform very closely to the expectations based on the multiple-factor view.

The later evidence on the case of the hooded rat is discussed elsewhere in this paper.

### 4. English Rabbits.

The data for this case are contained in two papers (Castle and Hadley, 1915a, 1915b), in each of which the full presentation is made. The spotting of the English rabbit is a dominant character and is somewhat variable. A single heterozygous male, of the grade designated 2, was mated to a number of Belgian hares. 187 English young were produced, of mean grade 2.43, and of these $F_1$ English, a buck of grade 3.75 (only one $F_1$ English was of higher grade), was then mated to the same Belgian hare females. 189 English young, of mean grade 2.92, were produced.

This case presents no difficulties for the multiple-factor view, since no evidence is given that indicates the original English buck to have been homozygous for all modifying factors, or that prevents us from supposing the Belgian mother of the $F_1$ buck to have transmitted more plus modifiers to him than were present in his father. Under the circumstances, it would have been very surprising if the two lots of young had been of the same mean grade.

## 5. Plumage and Toe Characters in Poultry.

We are referred to the observations of Bateson and Davenport for these cases. In one instance it is stated that Davenport has shown that "imperfection of dominance" and of segregation are the rule in poultry. The question of imperfection of dominance is not apropos in this connection. As Castle has said, regarding another case:

". . . . if black is crossed with brown, the crossbreds are apt to develop in their coats more brown pigment granules than do homozygous or pure blacks. Nevertheless, we have no reason to question the entire purity of the gametes, both dominant and recessive, formed by such cross-bred black animals. It is the dominance, not the segregation, which is imperfect." (Castle, 1911, p. 91.)

That $F_1$ results do not bear on the question has been shown by Bateson (1909), who says with regard to polydactylous fowls:

"It might be pointed out that when, as in these examples, the abnormal result is clearly perceptible in $F_1$, no question arises as to the occurrence of an imperfect segregation. The peculiarity is evidently zygotic, and is caused either by some feature of zygotic organization, or by the influence of external circumstances." (Bateson, 1909, p. 251.)

Moreover, in any case involving irregularities in dominance, imperfect segregation in crosses between different breeds would be very difficult to demonstrate.

## 6. Merino Sheep.

No reference to the data in this case are given. I have been unable to discover anything more definite than a few general statements by practical breeders regarding the effects of crossing Merinos.

Bateson admits, in the passage quoted above, that this and the next case "may be ascribed to the interference of a multitude of factors."

## 7. Fantail Pigeons.

This case has been studied by Morgan (Morgan, Sturtevant, Muller, and Bridges, 1915, p. 186). The fantail type did not reappear in the comparatively small $F_2$ generation, but individuals not far from the fantail were obtained; and when the $F_1$ hybrids were mated to fantails, several of the offspring fell within the range of the fantail race. Bateson's "failure of a parental type to reappear in its completeness after a cross" is, then, scarcely applicable to this case.

## 8 and 9. Dutch Rabbits and Cases in Sweet Peas. Fractionation.

These are the specific cases cited as illustrations of Bateson's theory of "fractionation" or "subtraction stages," of which he states that 'it is to be inferred that these fractional degradations are the consequences of irregularities in segregation." In the case of the sweet pea, Bateson has pointed out that white flowers and the extreme dark flowers of the deep purple Black Prince were among the earliest variations to appear, while the intermediate forms have arisen later, as he suggests by fractionation. It would seem to follow that they have arisen in heterozygous forms, for otherwise the fact that the larger variants appeared first would be of no significance. There is, I think, no evidence to show that the later variations did actually arise in heterozygous forms, either in sweet peas or in rabbits. These factors are all inherited separately, and this fact would seem to rule them out of consideration if one adopts the chromosome theory of inheritance or if one appeals to multiple allelomorphs as evidence in favor of the variability of genes. In short, we have no evidence regarding the origin of these forms, and their present behavior seems to indicate that they are not due to fractionation. The only evidence in favor of such a hypothesis is the somatic appearance of the characters.

### 10. Flowering Time in Peas.

Castle (1916a, p. 324) has summarized this case as follows:

"Hoshino (1) recognizes that gametic contamination results from crossing early and late flowering varieties; (2) recognizes also that variation may occur among the cross-bred families, as well as in different pure lines of the uncrossed races, as regards the 'quality,' value, or potency of the same gene; (3) although Hoshino does not refer to the fact, his observations show clearly that genetic variation of a gradual or fluctuating sort occurs in at least one of the varieties which he crossed.

". . . . What I want to suggest is that in these several agencies we have a sufficient explanation of the variation observed in Hoshino's $F_2$, $F_3$, and $F_4$ generations, without invoking a two-factor hypothesis (as Hoshino has done), one factor being enough."

Castle's argument is that a difference in one pair of genes is sufficient to account for the result, if contamination be assumed; and that one difference is a simpler assumption than two. I have argued here that such an assumption is *not* simpler, unless we can find positive evidence that contamination ever occurs. In the present case, then, we must turn to the evidence that led Hoshino to suppose contamination to have occurred.

Hoshino crossed an early-flowering pea and a late-flowering one. The $F_1$ was nearly as late as the late parent; $F_2$, obtained by self-fertilizing $F_1$, approximated fairly closely to 3 late : 1 early, but the two classes were somewhat more variable than the corresponding parent varieties, and apparently overlapped slightly. Hoshino self-fertilized 236 of these $F_2$ plants and obtained 46 families that he classified as constant, *i. e.*, supposedly homozygous. This is a fair approximation to the 1 in 4 expected if two pairs of genes are responsible for the result. Hoshino shows that two pairs of genes will, in fact, account for most of the results obtained. There are certain facts not thus accounted for, but Hoshino shows (p. 265) that "secondary" modifiers (*i. e.*, modifiers producing only small effects) will account for all these facts, with a single exception. Three families were obtained from $F_2$ plants that must, on the two-factor view, have been of the same constitution. These plants were heterozygous for one pair of genes only. They produced, in $F_4$, the same type of later constant (homozygous) families, but differed slightly in the flowering times of the earlier constant families produced. According to Hoshino's view, if the earlier types differed the later ones should have differed in the same direction, because they must have received the same "secondary modifiers." This objection is not valid, for specific modifiers that act only in the presence of certain other genes are well known (see especially Bridges, 1916), and are sufficient to account for the differences observed. This argument is the only one that Hoshino gives to support his conclusion that contamination must have occurred. We must then conclude that the case does not furnish positive evidence for contamination, since it is explicable without recourse to that hypothesis.[1]

### 11. Unspecified Case in Swine.

This case is cited by Castle (1916b, p. 215), but no references or authorities are given. It appears, however, from the legend under figure 93 (opposite p. 139) that the belted character is the one referred to. The only data bearing on this case that I have found are presented by Spillman (1907), and consist of information supplied largely by practical swine-breeders. Spillman himself interpreted the case as one in which two factor-pairs are involved. The data also suggest the possibility that we are dealing with a case of "imperfect dominance" similar to those in poultry. At best, the data are meager and indefinite.

### 12. Variegated Pericarp in Maize.

The paper of Hayes (1917) referred to above should be studied in connection with those of Emerson, particularly his full paper (Emerson, 1917), dealing with the same character. These two workers have shown that there is a remarkable series of multiple allelomorphs in this case, and Emerson has shown very clearly that some of these allelomorphs mutate quite frequently—the only established instance of the sort.

---

[1] We are not here directly concerned with Castle's contention that Hoshino's results prove the effectiveness of selection within a pure line. I can not, however, refrain from a few comments on that contention. Castle states (1916a, p. 324), in connection with the differences in flowering-time between the offspring of early and late flowering sister-plants: "From long experience in studies of rats with such small differences as are here indicated I have no hesitation in concluding that fluctuating variation of genetic significance is here in evidence." One wonders how experience in dealing with differences in pigmentation in rats can give an observer special ability in determining by inspection the significance of three-tenths of a day difference in the flowering time of peas. With respect to Castle's calculations from Hoshino's data, it may be pointed out that the greatest favorable difference recorded, 1.27 days, is incorrect, and should read 0.26 day. In view of the fact that there is no guarantee that the material used was homozygous, I have thought it scarcely worth while to recalculate all the differences, or to determine their probable errors; but it is certain that the probable error of each difference is of the same order of magnitude as the average difference itself, *i. e.*, about 0.3 day.

Hayes has, by selection from a mixed population, established four different grades of variegation (including self-colored and colorless) that breed true and that represent four allelomorphs. The two intermediate types, "mosaic" and "pattern," are the ones of special interest in the present connection. When these two types were crossed, the mosaic type was dominant, but there was an increase in variability in $F_1$ and some individuals with more pigment than either parent were obtained. The parent races had been selfed and selected for about six generations before the cross was made. In view of the great amount of heterozygosis that seems to be normally present in maize, and the large number of chromosome pairs (20?), this seems to be hardly sufficient to make certain that both races were pure for their modifiers. The increased variability of $F_1$ is therefore not surprising; and that phenomenon would of course be expected to be followed by a still greater increase in variability in $F_2$. Such an increase was, in fact, observed, and is the chief basis for Hayes's conclusion that contamination may occur. The data are not sufficient to demonstrate that new allelomorphs arise more often in heterozygotes than in homozygotes; and even if it be shown that they do so, it does not follow that there has been contamination of allelomorphs. There are too many unknown factors involved in the production of these new allelomorphs for such a conclusion to be valid without very careful controls.

It appears from the foregoing review that the cases cited as illustrations of contamination of allelomorphs or imperfect segregation are all explicable on the multiple-factor view, or rest on extremely indefinite data.

One series of data bearing on the question has been presented in this paper (p. 32), and has been interpreted as giving evidence against contamination. Three other cases have been worked out by Muller (1916) and Marshall and Muller (1917). Muller kept three mutant characters of *Drosophila* in heterozygous condition for about 75 generations. The factors were kept constantly in flies heterozygous for their normal allelomorphs, so that the characters remained unseen for a long time.

Muller extracted one of these characters (dachs) from this stock, and measured the tarsi, using the length of thorax as a standard of comparison. Dachs flies are characterized by shortened tarsi; and the flies from the heterozygous stock were found to have tarsi actually a trifle shorter than those found in a stock that had been kept pure for dachs. This result was not very conclusive, chiefly because it was based on a very few flies.

Marshall and Muller made much more extensive studies with the wing characters, curved and balloon, derived from the same heterozygous stock. They obtained a similar result; the wings were no nearer the normal than were those of curved and of balloon flies that had been kept in pure stocks. These results, taken in connection with the data presented above for bristle number in flies from lines heterozygous for Dichæt, furnish definite evidence against contamination of allelomorphs in heterozygous forms.

### Castle's Experiments with Hooded Rats.

Perhaps the best known selection experiment is that carried out by Castle and various collaborators (Castle and Phillips, 1914, Castle and Wright, 1916, etc.) with hooded rats. The theoretical conclusions reached by Castle are not in agreement with those arrived at by various other investigators, including the author, although for the most part the data obtained are very similar. Castle's results have been discussed by Muller (1914a) and MacDowell (1916), who have shown in detail that all the data known to them were explainable on the multiple-factor view, without recourse to such hypotheses as contamination of factors or production of factorial variations by selection. One point has, I think, not been sufficiently emphasized by them, namely, that the rat experiments are hard to evaluate properly until we are in possession of more accurate data regarding the pedigrees. Since these two criticisms were written, Castle (Castle and Wright, 1916) has given some additional data, which he has used, in a reply (Castle, 1917) to MacDowell's paper, as arguments against the latter's conclusions.

With regard to the question of pedigrees, to take up these questions in order, the main point on which information is desired is: How closely inbred were the rats, both before and after the beginning of the selection experiment? The following quotations contain most of the available evidence on this matter:

"Since the entire stock is descended from a very few individuals (less than a dozen), and we have at no time hesitated to mate together brother and sister, provided they varied in the same direction, but have always used the most extreme individuals (plus or minus) which were available, to mate with each other, it follows that very close inbreeding must have occurred throughout the experiment." (Castle, 1914b.)

"It is impossible for a colony of 33,000 rats to be produced from an original stock of less than a dozen animals, with constant breeding together of these which are alike in appearance and pedigree, and with continuous selection of extremes in two opposite directions, without the production of pedigrees which in the course of each selection experiment interlock generation after generation and finally become in large part identical with each other. This has been repeatedly verified in individual cases, but is incapable of a more generalized statement or of demonstration in generalized form. At least I am unable to devise such demonstration." (Castle, 1916d.)

Elsewhere (Castle and Phillips, 1914, p. 20) it is stated that *part* of the original stock consisted in a mixed lot of trapped rats that "had probably arisen by the crossing of an escaped albino rat with wild ones." We do not know where the rest of the stock came from, and we do not know how the animals used to start the selection experiments were derived from these sources. We do not know how many individuals were used to start the selection experiment; and we do not know anything as to the relationship between the rats in the two series (plus and minus). And, finally, we have only very indefinite data as to what system of breeding was followed during the experiment. All this information is very much needed, if we are to know how to interpret the results. It is conceivable that each series was split up into a number of separate lines, and that these have been crossed from time to time. Such a system would result in bringing together modifying factors more slowly than would a system of very close inbreeding. It is, of course, very improbable that any such system has been followed; and such an assumption is by no means necessary for a multiple-factor interpretation of the results. But definite information is very desirable, as is indicated by an analogous case.

In connection with certain work that the writer has been carrying on with Mr. J. W. Gowen, pedigrees of the two famous thoroughbred race-horses, Sysonby and Artful, have been tabulated. These pedigrees are both practically complete for 10 ancestral generations. They constitute a fair random sample of pedigrees in the breed, for Sysonby was of pure English blood, while Artful had many American-bred ancestors. The two pedigrees show no name in common until we reach the fifth ancestral generation. In that generation there are three names that appear in both pedigrees. But by the time we reach the tenth ancestral generation, approximately 90 per cent of the 1,024 names in Artful's pedigree appear also in the first ten generations of Sysonby's pedigree. And the result would certainly be even more striking if the pedigrees were studied for a few more generations, or if two English-bred horses were compared. Here, then, we have a clear case of "interlocking" pedigrees. Yet in spite of the long inbreeding (12 to 20 or more generations, with scarcely any out-crosses) which the breed has undergone, there are still a large number of bay or brown and of chestnut race-horses, besides a few grays and blacks. Of the four Mendelian factor pairs (see Sturtevant, 1912) for which the race was originally heterozygous, it has become homogeneous only in that the roan factor has been eliminated.[1] Clearly, selection for any one of the colors now present would still be effective in eliminating the others. The breed, which we may suppose to be inbred to something like the same degree as Castle's hooded rats, is still very far from a "pure line."

The new data presented by Castle and not taken up by MacDowell consist of two points: The crosses of extracted hoodeds (from plus

---

[1] Even in the early days roan race-horses were not at all common. Both roan and gray have been selected against.

## AN ANALYSIS OF THE EFFECT OF SELECTION.

race×wild) to wild, and the relations of the "mutant" series to the selected series.

When the plus race was crossed to wild, and $F_2$ hoodeds were extracted, it was found that in these extracted animals the mean grade was lighter (less "plus") than that of their selected grandparents. This, as MacDowell pointed out, is the expectation on the multiple-factor view. But Castle now states that when these extracted hoodeds are again crossed to wild, and hooded is extracted once more, the twice-extracted hoodeds are about midway in mean grade between their extracted grandparents and the uncrossed plus race. As he says, the wild race might have been expected to bring these animals still farther away from the plus race if modifying factors were involved. Evidently it is very important that we know as much as possible about the wild rats used in these experiments, in order that we may know what they were likely to carry in the way of modifying factors. These rats, we are told, all came from the same stock, which was trapped at the Bussey Institution in large numbers and was reared for two generations in the laboratory. "In making the second set of crosses, the extracted individual has, wherever possible, been crossed with its own wild grandparent." An examination of the table given shows that not more than 102 of the 256 twice-extracted hoodeds *can* have been produced in this way, unless individuals of the same sex were mated together. Just how many of the 102, and which ones, does "wherever possible" include? How many wild rats were used in the original crosses? These questions are important, because it is evident from a study of the data that the result emphasized by Castle is due almost entirely to the descendants of one original plus-line female; 41 of the 73 once-extracted hoodeds were $F_2$'s from this female; and their mean grade was 3.05, as against 3.3 for the remaining $F_2$'s, and 3.17 for the generation as a whole. The twice-extracted hoodeds tracing to this female were of mean grade 3.47, while those from the other original hoodeds were again of approximately grade 3.3. Further data regarding the pedigree and other descendants of the mates of this female and of her grandchildren are very much needed. Information regarding the ancestry of the female herself would also be interesting.

It should also be pointed out that this case, accepted at its face value, is difficult to explain on the view that the hooded-rat results are produced solely by variations in the hooded factor itself. On that view the changes brought about by crossing are usually referred to contamination of the factors in the heterozygote. But that interpretation leaves entirely unexplained the results of the *first* cross to wild. If the hooded factor is contaminated by its allelomorph, the once-extracted hoodeds should be *darker* than their grandparents, whereas in reality they are *lighter*, as would be expected on the multiple-factor view. Castle has met this objection in the following manner (Castle and Wright, 1916):

"This suggests the idea that that loss (of 'plus' character) may have been due to physiological causes non-genetic in character, such as produce increased size in racial crosses; for among guinea-pigs (as among certain plants) it has been found that $F_1$ has an increased size due to vigor produced by crossing and not due to heredity at all. This increased size persists *partially* in $F_2$, but for the most part is not in evidence beyond $F_1$. I would not suggest that the present case is parallel with this, but it seems quite possible that similar non-genetic agencies are concerned in the striking regression of the first $F_2$ and the subsequent reversed regression in the second $F_2$."

This comparison seems to me to be rather far-fetched, and I am quite unable to understand the hypothesis of "non-genetic physiological causes." That they are "physiological" is, of course, obvious; but they depend for their appearance on the pedigree of the animal, and they are persistent to $F_2$, so why "non-genetic"? The results from size crosses are entirely explicable on the basis of Mendelian modifying factors, so why need one appeal to vague "non-genetic," yet transmissible, factors? And is not such an appeal, in principle, an appeal to modifying factors? It certainly involves the assumption that the grade depends on transmissible material other than the hooded factor itself.

In the tenth generation of Castle's plus selection series there appeared two rats of considerably higher grade than any individuals of that series previously recorded. These individuals were shown (Castle and Phillips, 1914, pp. 26–31) to differ from the plus race by a single dominant factor. This has been taken by MacDowell to indicate that a new modifying factor arose by mutation. But Castle has now presented evidence indicating that the mutation occurred in the hooded locus itself. When homozygous "mutants" were crossed to wild rats, $F_2$ consisted in self-colored rats and rats of the same grade as the mutant series—no hooded individuals. (Castle and Wright, 1916.) Castle (1916) concludes from this evidence: "This serves to confirm the general conclusion that throughout the entire series of experiments with the hooded pattern of rats we are dealing with quantitative variations in one and the same genetic factor." Now, the "mutant" variation differs from the other results obtained by Castle in two respects: It appeared suddenly, as a definite and very slightly variable character, and it fails, when crossed to self, to give normal hooded in $F_2$. Because of the first point, it is probable that it arose during the experiment as a new variation; because of the second, it is probable that it is a variation in the hooded factor itself. Since these conclusions as to its nature are based entirely on the points in which it differs from the remainder of the results, it is difficult to see how Castle's case for these results is in any way improved. On the contrary, if this is the behavior to be expected of a new variation arising in the hooded factor, then the "mutant" variation is evidently the only case of that sort that Castle has reported.

### GENERAL CONCLUSIONS.

That many characters may be influenced by more than one pair of genes has long been recognized, and this is the essence of the multiple-factor view. That genes exist which require the action of other genes before they produce visible effects has also been long known. Furthermore, that there are genes which produce very slight visible effects is now another commonplace. Given these three facts, and the hypothesis (which is supported by much specific evidence) that most races are heterozygous for a number of such genes is all that is required to complete the conception that is held by most adherents of the view that multiple factors or modifying genes are responsible for the results of selection.

In specific cases, the existence of definite modifying genes has been demonstrated by Dexter, Bridges, Muller and Altenburg, and the author. All other data in question fit in with the view that selection ordinarily acts only by isolating modifiers.

Modification of factors by selection, crossing, fractionation, or similar means is undemonstrated in any given case, and has been shown not to occur in other cases that are typical of the results usually obtained. Factors do change, and more than two forms are possible for certain loci; but there is no known method of inducing such changes; and they are ordinarily quite rare and definite.

### SUMMARY.

(1) Dichæt is a dominant character, the gene being lethal when homozygous (yellow-mouse case). The gene is in the third chromosome.

(2) Dichæt flies are more variable in bristle number than are not-Dichæts. This variability is partly environmental, partly genetic.

(3) Selection was effective in isolating both plus and minus Dichæt lines.

(4) A cross between two separate inbred plus lines gave no increase in variability and no increase in parent-offspring correlation. Therefore the two lines were presumably of very similar constitution, though independent in origin.

(5) A cross between an inbred plus line and an inbred minus line gave the results characteristic of such crosses—increased variability in $F_2$ and increased parent-offspring correlation.

(6) Linkage tests demonstrated that modifying genes exist in the selected lines. Several lines were shown to differ in one or more second-chromosome modifiers, and at least one of these modifiers was shown to cross over from the speck gene.

(7) In one case at least one third-chromosome modifier was shown to exist and to cross over from Dichæt, which must lie to the left of it.

(8) Two third-chromosome lethals were obtained. These were shown to be new mutations, not due to fractionation of the Dichæt gene.

(9) A new allelomorph of Dichæt, called Extended, appeared in a plus selected line. It is argued that this mutation was not due to fractionation of the Dichæt gene, and was not influenced by the selection that was carried on.

(10) Another character, somatically indistinguishable from Extended, was shown to be due to a recessive second-chromosome gene.

(11) A study of unselected Dichæts, and of the not-Dichæts produced by long-continued mating together of Dichæts, is shown to furnish evidence against the view that allelomorphs are contaminated in heterozygotes.

(12) A general discussion of the selection problem is divided into three parts: (a) an attempt is made to clear up certain current misunderstandings and disagreements as to what questions are really at issue; (b) cases cited as evidence for contamination of allelomorphs are discussed in detail, and the conclusion is drawn that contamination is unproved and is an unnecessary hypothesis, with some direct evidence against it; (c) certain specific objections are raised to arguments made by Castle on the basis of his experiments with hooded rats.

## BIBLIOGRAPHY.

BATESON, W.
    1909. Mendel's principles of heredity. 2d impression, Cambridge.
    1914. Address of the president of the British Association. Science, n. s., 40.

BRIDGES, C. B.
    1915. A linkage variation in *Drosophila*. Jour. Exper. Zool., 19.
    1916. Non-disjunction as proof of the chromosome theory of heredity. Genetics, 1.

CALKINS, G. N., and L. H. GREGORY.
    1913. Variations in the progeny of a single ex-conjugant of *Paramecium caudatum*. Jour. Exper. Zool., 15.

CASTLE, W. E.
    1906. The origin of a polydactylous race of guinea-pigs. Carnegie Inst. Wash. Pub. 49.
    1911. Heredity in relation to evolution and animal breeding. New York.
    1914a. Multiple factors in heredity. Science, 39.
    1914b. Variation and selection; a reply. Zeitschr. Abst. Vererb., 12.
    1916a. New light on blending and Mendelian inheritance. Amer. Nat., 50.
    1916b. Genetics and eugenics. Cambridge, Mass.
    1916c. Report in Carnegie Inst. Wash. Year Book No. 15.
    1916d. Can selection cause genetic change? Amer. Nat., 50.
    1917. Piebald rats and multiple factors. Amer. Nat., 51.

—— and A. FORBES.
    1906. Heredity of hair-length in guinea-pigs and its bearing on the theory of pure gametes. Carnegie Inst. Wash. Pub. 49.

—— and P. B. HADLEY.
    1915a. The English rabbit and the question of Mendelian unit-character constancy. Amer. Nat., 49.
    1915b. Same. Proc. Nat. Acad. Sci., 1.

—— and J. C. PHILLIPS.
    1914. Piebald rats and selection. Carnegie Inst. Wash. Pub. 195.

—— and S. WRIGHT.
    1916. Studies of inheritance in guinea-pigs and rats. Carnegie Inst. Wash. Pub. 241.

DEXTER, J. S.
    1914. The analysis of a case of continuous variation in *Drosophila* by a study of its linkage relations. Amer. Nat., 48.

EMERSON, R. A.
    1917. Genetical studies of variegated pericarp in maize. Genetics, 2.

HAYES, H. K.
    1917. Inheritance of a mosaic pericarp pattern color of maize. Genetics, 2.

HOSHINO, Y.
    1915. On the inheritance of the flowering time in peas and rice. Journ. Coll. Agr. Tohoku Imper. Univ., Sapporo, Japan, 6.

JENNINGS, H. S.
    1916. Heredity, variation, and the results of selection in uniparental reproduction in *Difflugia corona*. Genetics, 1.

JOHANNSEN, W.
    1903. Ueber Erblichkeit in Populationen und in reinen Linien. Jena.

LITTLE, C. C.
    1915. The inheritance of black-eyed white spotting in mice. Amer. Nat., 49.

LUTZ, F. E.
    1911. Experiments with *Drosophila ampelophila* concerning evolution. Carnegie Inst. Wash. Pub. 143.

MACCURDY, H., and W. E. CASTLE.
    1907. Selection and cross-breeding in relation to the inheritance of coat-pigments and coat-patterns in rats and guinea-pigs. Carnegie Inst. Wash. Pub. 70.

MacDowell, E. C.
 1915. Bristle inheritance in *Drosophila*. I. Extra bristles. Jour. Exper. Zool., 19.
 1916. Piebald rats and multiple factors. Amer. Nat., 50.
 1917. Bristle inheritance in *Drosophila*. II. Selection. Jour. Exper. Zool., 23.

Marshall, W. W., and H. J. Muller.
 1917. The effect of long-continued heterozygosis on a variable character in *Drosophila*. Jour. Exper. Zool., 22.

Middleton, A. R.
 1915. Heritable variations and the results of selection in the fission rate of *Stylonychia pustulata*. Jour. Exper. Zool., 19.

Morgan, T. H., A. H. Sturtevant, H. J. Muller, and C. B. Bridges.
 1915. The mechanism of Mendelian heredity. New York.

Muller, H. J.
 1914a. The bearing of the selection experiments of Castle and Phillips on the variability of genes. Amer. Nat., 48.
 1914b. A gene for the fourth chromosome of *Drosophila*. Jour. Exper. Zool., 17.
 1916. The mechanism of crossing over. Amer. Nat., 50.
 1917. An Oenothera-like case in *Drosophila*. Proc. Nat. Acad. Sci., 3.

Pearson, K.
 1911. On the probability that two independent distributions of frequency are really samples from the same population. Biometrika, 8.

Spillman, W. J.
 1907. Inheritance of the belt in Hampshire swine. Science, n. s., 26.

Sturtevant, A. H.
 1912. A critical examination of recent studies on color inheritance in horses. Journ. Genet., 2.

# PIEBALD RATS AND SELECTION, A CORRECTION

## W. E. Castle

In a recent important publication Dr. Sturtevant makes "an analysis of the effects of selection" in which he ably maintains the current view that the single gene is not changed by processes of systematic selection. His argument rests on a careful experimental study of the behavior of the character "dichaet" in *Drosophila*, followed by a general discussion of other work, my own in particular. I am represented as completely opposed to his view, and so I have been at times, but such is not the case at present. I agree so fully with his general conclusion that I want to obviate needless discussion based on the misapprehension.

I thought two years ago that I had evidence that a single gene had changed in the course of a selection experiment, this gene being concerned in producing the hooded pattern of rats. I now find this view rendered untenable by further experiments, the results of which are in course of publication. These results show that the supposed changes in a single gene are more probably due to changed residual heredity, which very likely may consist wholly of other "modifying" genes.

The crucial experiment was one suggested by Dr. Sewall Wright. The divergent hooded races, "plus" and "minus," resulting from selection, were to be crossed repeatedly with a third race, the hooded character being recovered as a recessive in $F_2$ following each cross and its variability compared with that of the uncrossed race. It was believed that if multiple modifying genes were involved, repeated crossing with a pure third race would tend to remove these, in which case the extracted hooded character being deprived of its plus modifiers would be substantially identical with the hooded character deprived of its minus modifiers, as seen respectively in hooded recessives derived from the plus and from the minus crosses. Well, they *are* substantially identical, but it has taken some time and a good deal of trouble to establish the fact. First we had to secure a satisfactory third race to use in the crosses, one free from contamination of any sort by crosses. This we sought in a wild race. But ordinary wild rats will not breed under laboratory conditions. So we resorted to trapping immature wild rats from a single locality and using these as a foundation stock. Crosses with the plus race were then started successfully, but the corresponding experiment with the minus race was hard to get going and so has lagged behind the plus crosses. A report on the result of the plus crosses was made in 1916 (Castle and Wright). The crosses with the minus race were not then sufficiently advanced to show what their outcome would be and this was still true when reply was made to the criticism of MacDowell, as it had been previously when reply was made to Muller and to Pearl, and subsequently when I addressed the Washington Academy of Science on the rôle of selection in evolution (1917). But since then the minus crosses have given what seems to be conclusive evidence that the single gene had not been altered by selection, although the inherited complex responsible for the hooded character had steadily been altered in opposite directions and these alterations were permanent in the sense that they represented racial modes, stable so long as the race was not outcrossed.

I still have on hand a few representatives of the plus and of the minus races which because of their low fecundity it has been impossible to select further for several generations. The two races are very different in appearance. The plus race shows no white except on the under side and sometimes along the flank. The minus race shows no black except a short hood lying anterior to the shoulders, and in an occasional individual a small black spot or two in the middle of the back or on the tail. Yet the variability of each race is still considerable; as measured by our "grades" it has not appreciably diminished in recent generations. The somatic differences entailed by the selection experiments with the hooded character of rats are seemingly greater than those secured by Sturtevant or by MacDowell in regard to bristle number in *Drosophila*, yet I doubt not they may be explained on similar grounds.

Crossing with a wild race affects very differently the plus and the minus selected races. See Tables I and II. The plus race was much less affected than the minus race. Its mean grade was lowered, by three successive crosses with the wild race, not over three quarters of a grade. The standard deviation was about doubled by the first cross. That is the variability of the hooded character, when extracted in $F_2$ from the first wild cross, was about twice as great as the variability of the hooded character in the uncrossed plus selected race. In the second and third crosses the variability declined somewhat, but was still considerably greater than that of the uncrossed race. It was indeed very similar to that of the plus race in the first seven generations of the plus selection experiment. (See Castle and Wright, p. 186.)

TABLE I

RESULTS OF REPEATEDLY CROSSING THE PLUS SELECTED RACE WITH A WILD RACE

|  | Mean Grade | Standard Deviation | Number of Hooded Young |
|---|---|---|---|
| Control, uncrossed plus race, generation 10 | +3.73 | .36 | 776 |
| *Once* extracted hooded $F_2$ young | +3.17 | .73 | 73 |
| *Twice* extracted hooded $F_2$ young | +3.34 | .50 | 256 |
| *Thrice* extracted hooded $F_2$ young | +3.04 | .64 | 19 |

TABLE II

RESULTS OF REPEATEDLY CROSSING THE MINUS SELECTED RACE WITH A WILD RACE

|  | Mean Grade | Standard Deviation | Number of Hooded Young |
|---|---|---|---|
| Control, uncrossed minus race, generation 16 | −2.63 | .27 | 1,980 |
| *Once* extracted hooded $F_2$ young | −.38 | 1.25 | 121 |
| *Twice* extracted hooded $F_2$ young | +1.01 | .92 | 49 |
| *Thrice* extracted hooded $F_2$ young | +2.55 | .66 | 104 |

The crosses of the minus race were started six generations later in the course of the selection experiments, with animals of generation 16, minus selection series. They show effects much more striking than those of the plus crosses. See Table II. The minus selected race had now attained a mean of −2.63. A single cross, with the same wild race used in the crosses of the plus series, lower the grade to −.38, extinguishing all the changes in mean grade made by sixteen generations of selection, and leaving the extracted hooded character in a highly variable state (standard deviation 1.25, nearly five times what it had been before). A second cross with the same wild race converted the extracted hooded individuals for the most part into a plus group, mean +1.01, but with variability somewhat decreased, standard deviation .92. A third cross with the wild race has given extracted hooded individuals *exclusively plus in character*, range from + 1.00 to + 3.50, mean + 2.55. The variability has simultaneously fallen to .66, which is only about one third greater than that of the minus race in the first five generations of the selection experiment. (See Castle and Wright.) One family containing fourteen thrice extracted hooded individuals has a mean grade for the hooded individuals of + 3.05, which is practically identical with the grade of the thrice extracted hooded individuals resulting from the plus crosses (Table I).

It thus appears that three or at most four crosses with a wild race suffice to obliterate all the racial differences which had been induced by ten generations of selection in the case of the plus race and sixteen generations in the case of the minus race. The plus race was changed almost immediately by a single cross, but the change was small (a fact which misled me until the results of the minus crosses were secured). The changes with the minus race were so great that they could not be fully secured by less than three or possibly four successive crosses (eight generations of offspring). The wild race, which we used in our crosses, evidently had a residual heredity much more like that of our plus-selected than like that of our minus-selected race. When the hooded gene from either race was introduced by repeated crosses into this residual heredity, the result was to produce hooded races of very similar grade, a little lower in grade than the plus selected race, but very much higher in grade than the minus selected race.

It thus becomes clear that the changes which had occurred in the hooded character as a result of selection were *detachable* changes and are probably in nature independently inherited modifying factors. This is a view which Phillips and I gave as one of two possible interpretations of the results which we published in 1914. Morgan, Muller, MacDowell and others have insisted that this was the only reasonable interpretation which could be given, but I have not been satisfied with this conclusion in advance of a really crucial experiment, such as I believe has now been performed. Meanwhile the probability that the theory of multiple modifying factors is correct as a general explanation of similar cases has been greatly strengthened by the work of Muller, Bridges, Sturtevant and others, showing that genetic factors, having a definite demonstrable position in linkage systems, influence in a particular way the somatic manifestation of characters varying quantitatively or qualitatively. I accept their interpretations as correct in the light of our present knowledge.

I should feel like apologizing for my own obtuseness in not reaching a similar conclusion sooner, did I not recall with satisfaction how much clearer the rôle of selection now stands revealed than it did when these experiments were begun, and to the clearing up of the situation I shall at least hope that this rat work has contributed something, if only by provoking investigation.

The "Mutation Theory" of DeVries gave us a picture of selection as an agency temporarily effective in producing racial changes, but with those changes gradually vanishing as soon as the selection ceased. Johannsen denied within "pure lines" even temporary effectiveness of selection. A strictly logical use of Johannsen's conclusions would have limited their application to such organisms as he studied, self-fertilizing ones completely homozygous for all genetic factors and subject apparently to no new changes in such factors. But the doctrine was straightway extended in the views of most geneticists to selection of every sort and he was treated as a traitor to Mendelism who saw any utility in selection or advocated its use as a means of improving the inherited characters of animals or plants.

The situation is wholly different to-day. Through the investigations of Jennings and his pupils on protozoa, of Stout on *Coleus,* and of Shammel on citrous fruits, the fact is clear that even within clones genetic changes may and do frequently occur and that systematic selection will serve to isolate these and thus lead to racial improvement. Those who have tried systematic selection in the case of cross fertilizing organisms have in some cases noted the occurrence of "mutations" with such frequency as to make progressive change under selection easily obtainable. Emerson and Hayes, in the case of certain pericarp color patterns of maize, find "mutations" so common that a wide range of variability results and selection is able to isolate, from such material, types "relatively stable," but very diverse in appearance. Modifying factors are not involved in Emerson's explanation of his results, but rather such instability of a single gene as leads to frequent mutation. Selection experiments with the variegated coat-patterns of mammals seem to involve less abrupt but otherwise similar changes, but modifying factors rather than repeated mutation seems to be the explanation required in view of the results of crosses reported in this paper.

That selection by one means or another is an effective agency in producing racial changes is not questioned to-day, as it was ten years ago.[1] The only question now at issue is whether the single gene is changeable. I am inclined to think, with Sturtevant, that while single genes do occasionally change producing multiple allelomorphs, a much more common occurrence is change in visible characters through modifying factors. Whether the direction of genetic variation is controllable, other than by the manipulation of modifying genes or the discovery of multiple allelomorphs remains to be determined. The evidence at present is largely negative. It is undeniable that liability to genetic variation is much greater in some organisms than others, much greater as regards some kinds of character than as regards others, but whether we can produce variability of a genetic character is quite a different question. We certainly at present have to follow nature's lead rather than to lead nature, as regards the course of evolutionary change.

W. E. CASTLE

BUSSEY INSTITUTION,
HARVARD UNIVERSITY.

[1] Sturtevant's presentation of my views is a bit unfair in that it seems to imply that whenever I have spoken of "variation in a unit character," I have consistently meant variation in a single gene, whereas in discussing the case of the English rabbit, I have expressly reserved judgment on this point. In a large part of my experimental work, the question under investigation has been—do the visible characters which conform with Mendel's law in transmission suffer modification in crosses or as a result of selection? The present generation of geneticists has apparently forgotten that this was ever a debatable question. We all admit now that contamination occurs in crosses and that modification may be effected by selection, and we seek only to explain *how* the contamination is brought about (as by modifying factors) or *how* the modification is produced in the course of systematic selection (as by the isolation of modifiers in homozygous state). But in the days when the doctrine of gametic purity was under discussion, such "contamination" or "modification" was not admitted.

When Sturtevant denies the occurrence of "contamination," he uses the term in a very restricted sense, not as I have used it in the foregoing sentence, nor as it was formerly used in Mendelian discussions. What he means is not change in the visible character, as the hooded character of rats, but change in a single gene which is known absolutely to limit the manifestation of the hooded character in any form. I agree with his view that there is no conclusive evidence that this single gene had changed in the course of selection experiments, except in the case of our "mutant" race.

## BIBLIOGRAPHY

Castle, W. E.
- 1916. Can Selection Cause Genetic Change? AMER. NAT., 50.
- 1917. Piebald Rats and Multiple Factors. AMER. NAT., 51.
- 1917a. The Rôle of Selection in Evolution. *Jour. Wash. Acad. Sci.*, 7, p. 369.

Castle, W. E., and Hadley, P. B.
- 1915. The English Rabbit and the Question of Mendelian Unit-character Constancy. *Proc. Nat. Ac. Sci.*, 1.

Castle, W. E., and Phillips, J. C.
- 1914. Piebald Rats and Selection. Carnegie Inst. Wash., Publ. No. 195.

Castle, W. E., and Wright, Sewall.
- 1916. Studies of Inheritance in Guinea-pigs and Rats. Carnegie Inst. Wash., Publ. No. 241.

Emerson, R. A.
- 1917. Genetical Studies of Variegated Pericarp in Maize. *Genetics*, 2.

Hayes, H. K.
- 1917. Inheritance of a Mosaic Pericarp Pattern Color of Maize. *Genetics*, 2.

Jennings, H. S.
- 1916. Heredity, Variation, and the Results of Selection in Uniparental Reproduction in *Difflugia corona*. *Genetics*, 1.

MacDowell, E. C.
- 1916. Piebald Rats and Multiple Factors. AMER. NAT., 50.
- 1917. Bristle Inheritance in *Drosophila*. *Jour. Exp. Zool.*, 23.

Stout, A. B.
- 1915. The Establishment of Varieties in *Coleus* by the Selection of Somatic Variations. Carnegie Inst. Wash., Publ. No. 218.

Sturtevant, A. H.
- 1918. An Analysis of the Effects of Selection. Carnegie Inst. Wash., Publ. No. 264.

# STATISTICAL PREDICTIONS OF
# SELECTION RESPONSE

# Editor's Comments
# on Papers 5 Through 9

**5  LUSH**
Excerpts from *Progeny Test and Individual Performance as Indicators of an Animal's Breeding Value*

**6  HAZEL**
*The Genetic Basis for Constructing Selection Indexes*

**7  FALCONER**
*The Problem of Environment and Selection*

**8  DICKERSON and HAZEL**
*Effectiveness of Selection on Progeny Performance as a Supplement to Earlier Culling in Livestock*

**9  HENDERSON**
*General Flexibility of Linear Model Techniques for Sire Evaluation*

On the assumption of a continuous distribution of genotypic values and their components, such as breeding value, and of environmental deviations, selection responses ($R$) can be predicted by straightforward regression arguments—for example, by the classic formula $R = h^2 S = i h^2 \sigma_p$, where $h^2$ is the heritability, $S$ the selection differential, $i$ the selection intensity ($= S/\sigma_p$), and $\sigma_p$ the phenotypic standard deviation. The critical point is that the gene effects and frequencies do not enter into the calculations. Strictly, predictions hold for only single generations, but experimental results, such as those discussed in a later section, show that linear responses tend to continue longer—that is, that changes in variance are small for several generations (see also Falconer, 1981, Ch. 12). Thus the predictions of response and related quantities made solely from knowledge of variances and covariances are referred to here as statistical predictions. It is on these predictions that many animal- and, to a lesser extent, plant-breeding programs are based.

The most influential developments in this area were made by J. L. Lush and his school at Iowa State College from around 1935, and three important papers from this phase are included. Some of that work is reviewed in the various editions of *Animal Breeding Plans* (Lush, 1937, 1945). A further discussion and example of Lush's contributions is included in Part I.

In Paper 5, Lush considers the relative accuracy of progeny test and individual performance as predictors of breeding value. The foregoing equation for predicting response can be generalized to $R = irh\sigma_p$. Here $r$ is the accuracy of selection or correlation of the selection criterion with genotype, strictly breeding value, and is thus a critical parameter in comparing alternative selection criteria. Lush compares values of $r$ for individual selection $(r = h)$ and for progeny test. The paper is important both because it compares the alternative methods in the way still generally accepted as proper and because it shows that use of a few progeny records may not be as good a predictor of the animal's breeding merit as records on itself, a lesson still not universally accepted. Some parts of the paper were omitted because of lack of space, but a brief summary is inserted in the text.

Lush refers to the problem of combining information from two or more sources—that is, the progeny test and the dam's own record—but does not quantify the argument (p.17). Following Hazel's work on multitrait selection indices (Paper 6), Lush considered in another well-known paper (Lush, 1947) the problem of combining individual and family information together more generally and formally.

The problem of simultaneous improvement of a population for several traits, or even of description of the joint inheritance of several traits, had received little formal treatment in early studies of quantitative inheritance. The first paper in which a selection index was proposed to optimize multitrait improvement was written by H. F. Smith (1936), described as "A Discriminant Function for Plant Selection," but Smith acknowledges his indebtedness to R. A. Fisher and writes in his paper that the theory "is little more than a transcription of his suggestions." Subsequently but independently, L. N. Hazel, a student and later a colleague of Lush at Iowa State University, developed essentially the same index for use in farm animals (Paper 6). This paper has become more widely known, both because greater use of indices has been made in animals and because the concepts of genetic and phenotypic correlations used by Hazel have become widely adopted. Hazel's paper is also important because in it he outlines procedures for estimating genetic correlations.

The methods developed by Hazel are still used, the only major change in presentation being in the use of matrix notation and in the generalization to include information on relatives. There is now a very extensive literature (for reviews, see, for example, Henderson [1963],

Turner and Young [1969], and, without use of matrix notation, Falconer [1981]). Within the framework of the index, all linear combinations of information on the individual or his relatives for one or more traits can be put together to improve genetically any single trait or any combination of traits expressed as economic merit. At about the same time that Hazel published the index methodology, Hazel and Lush (1942) compared the efficiency of the index with alternative means of selection, showing, albeit for the most simple case of independent traits, that the index was the most efficient method.

Hazel's analysis of genetic correlations is concerned solely with different traits—for example, live weight and market score—of the pig. D. S. Falconer showed in Paper 7 that the same formulation could be applied to performance in different environments; his ideas developed from an experiment concerned with growth of mice on two different diets (Falconer and Latyszewski, 1952). In such a case it is not possible to define a phenotypic correlation, but the genetic correlation, estimated from selection experiments or analysis of family data, estimates the amount of genotype $\times$ environment interaction, or lack thereof. Falconer's paper could have been included in Part I because it deals with correlation among relatives, but as the paper is written in terms of direct and correlated responses to selection and is concerned with a well-known selection experiment, it seemed more appropriate to include it here. There are further papers by Falconer in Part I and in this volume (Paper 18).

In summary, Falconer's approach has two important features. First, it enables the magnitude of any genotype $\times$ environment interaction to be quantified. Second, it enables the standard formulae for multiple traits in quantitative genetics, including equations for predicting response, to be employed in the case of observations on two or more environments—for example, to compute the relative efficiencies of selection in one or several environments.

Papers 5 through 7 deal essentially with the accuracy of selection, whether using relatives' information for a single trait or data on two or more traits or two or more environments. Rates of genetic progress per unit of time (for example, year) depend also on the rate of turnover of generations, and prediction formulae have to take account of the overlap of generations, necessary in species where the reproductive rate of females is low. The first major discussion and analysis of the problems of predicting and optimizing animal rates of progress was by Dickerson and Hazel. They first present a general formula [their Eq. (1)], showing that the rate of progress is, with overlapping generations, dependent on the ratio of mean genetic selection differential of parents to mean generation interval, and then consider a number of examples, pointing out how selection intensity, accuracy, and generation interval have to be balanced. In particular, their analysis of the problem of two-stage selection—for example, on individual performance followed by progeny performance—remains as complete as any to this day.

Dickerson has made many other major contributions to quantitative inheritance and animal breeding. Notable among these are his discussions of "genetic slippage," lack of progress in poultry populations (Dickerson, 1955) and his proposals for the design and interpretation of experiments to select breeds of livestock, to incorporate them into crosses, and to optimize improvements within them (Dickerson, 1969).

When family information is incorporated into a selection index, the standard methods of Hazel (Paper 6) and Lush (1947) strictly apply when animals are contemporaneous and there are equal numbers in each family in each environmental group. In this special case, estimates of fixed effects (such as farm, pen, or parity) do not affect comparisons among individuals. More generally, however, it is formally necessary to estimate simultaneously the fixed effects and the breeding values of the animals. With the latter assumed to be randomly sampled from a genotypic distribution, the general problem is seen to be one of analysis of a mixed model, in which the breeder's objective is to predict the random effects or breeding values.

The proper formulation and development of the statistical procedures is almost entirely due to C. R. Henderson of Cornell University. Their origins date back to analyses in his doctoral thesis, completed at Iowa State University (Henderson, 1948). The motivation for the methods has largely been in dairy sire evaluation by progeny test, where daughters are distributed over many herds and where the sires of contemporaries cannot be taken to be a random sample of the population. Henderson did not initially realize that the mixed-model methodology was appropriate for dairy sire evaluation, and subsequently there were further delays in adopting the methods because considerable computing power is needed. The procedure is now usually termed Best Linear Unbiased Prediction (BLUP); it is being increasingly adopted worldwide for assessing dairy sires and is finding other applications.

Other of Henderson's papers might have been included in this volume; in Henderson, Kempthorne, Searle, and Von Krosigk (1959), he presented some of the basic formulae, but the application was to estimating genetic trends, rather than sire effects. A further discussion appeared in his review (1963) of selection indices, but he did not discuss the specific application to sire evaluation. Perhaps the most influential and complete description of the BLUP procedure and its application to sire evaluation was given by him in a long paper (1973). More appropriate for inclusion is a shorter review paper of the following year (Paper 9): it outlines the history, the basic methodology, including the mixed model equations, and the applications to sire evaluation. Subsequently, Henderson (1975) considered how selection effects could be taken account of in BLUP, but some of his arguments are rather obscure. Perhaps the most important later development in BLUP procedures was Henderson's (1976) discovery that the inverse $(A^{-1})$ of the relationship among sires could be easily found from the pedigree structure without any matrix inversion [for a general review, see, for example, Thompson (1979)].

The papers included in this section cover a substantial part of the use of statistical predictions—that is, where the population is described solely in terms of variances and covariances. While the predictions are necessarily short term, they are the foundation of most applications of quantitative genetics to animal breeding and, to a lesser extent, to plant breeding. Some of the limitations of the theory—namely, in not taking full account of changes of variance as a consequence of selection—will be discussed later (Paper 23).

A number of other topics and specific papers would have been included if space had been available. Among these are reviews of selection theory in the books of Lush (1937, 1945) or Lerner (1950) and, more specifically, the discussion by Dempster and Lerner (1947) on the use of part-record selection for egg-laying poultry. Classic papers on the role of selection intensity, generation interval, and selection of dairy sires on pedigree-versus-progeny test were written by Rendel and Robertson (1950) and Robertson and Rendel (1950), and a very thorough review of the mathematical principles underlying selection progress and indices was given by Cochran (1951). Because the arguments are couched in terms of effects at individual loci, Fisher's fundamental theorem is included in the next section (Paper 10), but the derivation can be made in direct statistical terms by computation of selection intensity, heritability, and variance. The deficiency of papers on plants in this section is regrettable, but reflects the much smaller attention paid to quantitative genetic prediction formulae and methods in plant breeding.

## REFERENCES

Cochran, W. G., 1951, Improvement by Means of Selection, in *Proceedings of the Second Berkeley Symposium on Mathematical Statistics and Probability*, J. Neyman, ed., University of California Press, Berkeley, pp. 449–470.

Dempster, E. R., and I. M. Lerner, 1947, The Optimum Structure of Breeding Flocks. I. Rate of Genetic Improvement Under Different Breeding Plans, *Genetics* **32:**555-566.

Dickerson, G. E., 1955, Genetic Slippage in Response to Selection for Multiple Objectives, *Cold Spring Harbor Symp. Quant. Biol.* **20:**213-224.

Dickerson, G. E., 1969, Experimental Approaches in Utilizing Breed Resources, *Anim. Breed. Abstr.* **37:**191-202.

Falconer, D. S., 1981, *Introduction to Quantitative Genetics,* 2nd ed., Longmans, London.

Falconer, D. S., and M. Latyszewski, 1952, The Environment in Relation to Selection for Size in Mice, *J. Genet.* **51:**67-80.

Hazel, L. N., and J. L. Lush, 1942, The Efficiency of Three Methods of Selection, *J. Hered.* **33:**393-399.

Henderson, C. R., 1948, *Estimation of General, Specific and Maternal Combining Abilities in Crosses Among Inbred Lines of Swine,* Ph. D. Thesis, Iowa State University, Ames.

Henderson, C. R., 1963, Selection Index and Expected Genetic Advance, in *Statistical Genetics and Plant Breeding,* W. D. Hanson and H. F. Robinson, eds., Nat. Acad. Sci., Nat. Res. Counc. Publ. 982, pp. 141-163.

Henderson, C. R., 1973, Sire Evaluation and Genetic Trends, in *Proceedings of the Animal Breeding and Genetic Symposium in Honor of Dr. Jay L. Lush,* American Society of Animal Science, and American Dairy Science Association, Champaign, Illinois, pp.10-41.

Henderson, C. R., 1975, Best Linear Unbiased Estimation and Prediction in a Selection Model, *Biometrics* **31:**423-499.

Henderson, C. R., 1976, A Simple Method for Computing the Inverse of a Numerator Relationship Matrix Used in Prediction of Breeding Values, *Biometrics* **32:**69-83.

Henderson, C. R., O. Kempthorne, S. R. Searle, and C. M. von Krosigk, 1959, The Estimation of Environmental and Genetic Trends from Records Subject to Culling, *Biometrics* **15:**192-218.

Lerner, I. M., 1950, *Population Genetics and Animal Improvement,* Cambridge University Press, Cambridge.

Lush, J. L., 1937, *Animal Breeding Plans,* Iowa State College Press, Ames.

Lush, J. L., 1945, *Animal Breeding Plans,* 3rd ed., Iowa State College Press, Ames.

Lush, J. L., 1947, Family Merit and Individual Merit as Bases for Selection, *Am. Nat.* **81:**241-261, 362-379.

Rendel, J. M., and A. Robertson, 1950, Estimation of Genetic Gain in Milk Yield by Selection in a Closed Herd of Dairy Cattle, *J. Genet.* **50:**1-8.

Robertson, A., and J. M. Rendel, 1950, Use of Progeny Testing with Artificial Selection in Dairy Cattle, *J. Genet.* **50:**21-31.

Smith, H. F., 1936, A Discriminant Function for Plant Selection, *Ann. Eugen.* **7:**240-250.

Thompson, R., 1979, Sire Evaluation, *Biometrics* **35:**339-353.

Turner, H. N., and S. S. Y. Young, 1969, *Quantitative Genetics and Sheep Breeding,* Macmillan of Australia, Melbourne.

# PROGENY TEST AND INDIVIDUAL PERFORMANCE AS INDICATORS OF AN ANIMAL'S BREEDING VALUE

JAY L. LUSH
*Iowa State College, Ames, Iowa*

It has long been known by breeders that sometimes a few offspring indicate an animal's real breeding value more accurately than the animal's own appearance or performance does. The rediscovery of Mendelism and the discovery a little later (by Bateson and others) of complementary genes and of other epistatic relations among non-allelomorphic genes, provided a simple explanation for this. Occasionally there are cases where a single offspring furnishes more convincing proof of its parent's genotype than could ever be obtained by studying the parent itself. So far as color is concerned, this would be the case with a purebred red calf born in one of the black breeds of cattle.

These are special cases, deserving attention to be sure, but perhaps they are more significant as *exceptions* to a general rule than as *examples* of one? Some writers on animal breeding topics have considered them to be examples. For instance we are sometimes told that a progeny test is "genotypic" and therefore dependable, whereas a cow's own record is "phenotypic" and therefore not dependable. But no fundamental distinction is really involved, since the progeny test itself consists of some kind of an average of "phenotypic" records of the offspring. To suppose that the process of averaging would purge these phenotypic records of all their error and would bring out only the pure gold of genotypic truth seems to require too much faith in the potency of arithmetic!

The present paper presents the results of an inquiry into the biometrical relations which determine the relative accuracy of progeny test and parent's own performance as indicators of the parent's real breeding value. The complete answer contains many variables but most of these have little influence. An answer accurate enough for practical purposes can be presented in usably simple form. For convenience the problem is phrased in terms of dairy cattle breeding thus: Is a cow's own record or the average record

Received for publication May 31, 1934.

Journal Paper, No. J173, from the Iowa Agricultural Experiment Station. Project No. 31.

of n of her daughters the more dependable indicator of the cow's own breeding value?

## METHODS USED AND DEFINITIONS

These biometrical relations were explored by Wright's method of path coefficients (10). Figure 1 shows those relations. The small letters indicate path coefficients which measure (provided the cause and effect relationships are portrayed correctly in the diagram) the importance of the causes of variation in the effect. The path coefficient (h) from heredity (H) to observed characteristic or yield (C) is the ratio of the standard deviation of C due to H to the total standard deviation of C. The path coefficients h and e may be used to express the relative importance of heredity and environment on production records in a particular population of data. Because variation is measured by squares of deviations h and e will not add up to unity. Where there is no correlation between heredity and environment, the squared path coefficients, $h^2$ and $e^2$, will come near to expressing "the relative importance of heredity and environment" in the form in which that idea is ordinarily encountered in practical problems. A definite numerical value for $h^2$ is a description of the population from which it was derived and may not, without other knowledge, be safely applied to populations where the variations in environment or in heredity may be quite different. For prediction purposes, path coefficients are regression coefficients expressed on a scale in which each standard deviation is one unit.

Most of the terms in Figure 1 are self-explanatory. Phenotype is used here to mean what the outwardly observed record would be if it were possible to keep all environmental circumstances rigidly controlled. It differs from the actually observed record in that the latter has been modified by the variations in environment to which the various phenotypes were exposed. Non-additive effects of environment on the phenotype (such as concern Hogben so much in his recent "Nature and Nurture" (5)) may be minimized by expressing the records on scales which show in the most nearly additive way the effects of variations in environment and in heredity. For example there is some evidence to indicate (but perhaps not enough to prove conclusively?) that dairy production records should be expressed in logarithms of the yield instead of in pounds if the most efficient use is to be made of the information in those records. However, for the present question, non-additive combination effects of phenotype and environment may be considered as included in the environment, since they tend to decrease the correlation between records and genotypes.

"Genotype" is obtained by assigning each gene such a value that the simple sum of all the genes in each individual will give the "genotype" for that individual subject to the "least squares rule," that (1) the average genotypic value of the population shall coincide with its average phenotypic value and (2) the sum of the squares of the individual deviations of "genotype" from "phenotype" shall be a minimum.* This "genotype" seems to be the same as Fisher's "expected value" of which he says (4, page 32), using human stature for illustration: "This expected value will not necessarily represent the real stature, though it may be a good approximation of it, but its statistical properties will be more intimately involved in the inheritance of real stature than the properties of that variate itself."

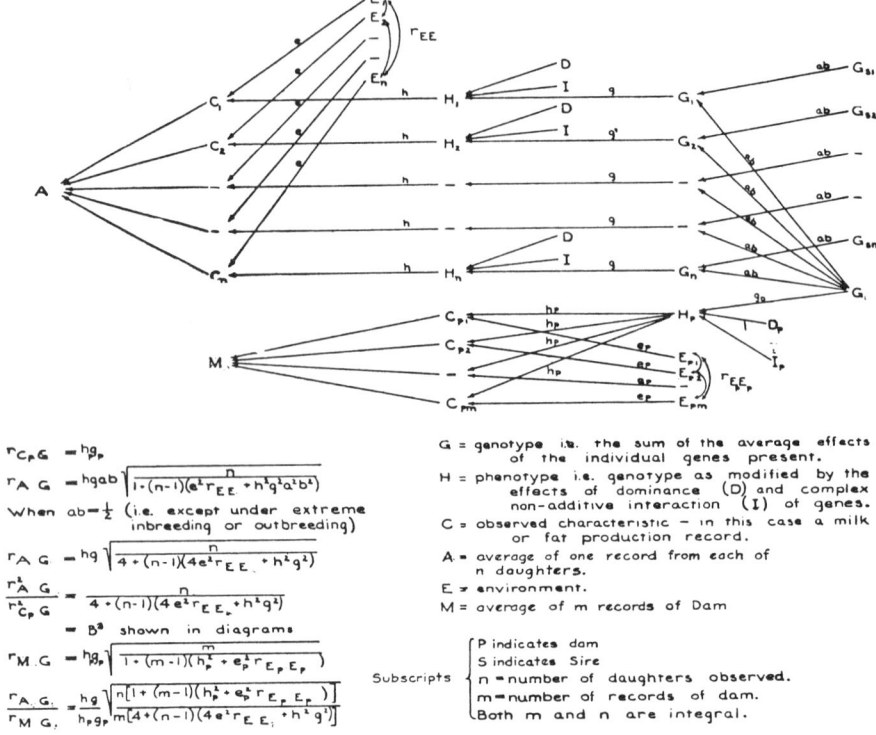

FIG. 1. PATH COEFFICIENT DIAGRAM SHOWING BIOMETRIC RELATIONS BETWEEN PARENT'S RECORDS AND PARENT'S BREEDING VALUE (G) AND THE BIOMETRIC RELATIONS BETWEEN THE PROGENY AVERAGE AND THE PARENT'S BREEDING VALUE.

These genotypic values will be perfectly correlated with Turner's definition (9, page 45) of "transmitting abilities," in so far as dominance is the only cause of discrepancy between genotype and phenotype, but Turner's definition makes no provision for non-additive combination effects of genes which are not allelomorphic to each other. For convenience in this article such non-additive combination effects of non-allelomorphic genes will be

* I am indebted to Dr. Sewall Wright for this method of expressing as much as possible of hereditary differences in terms which contribute to the correlations between relatives.

called epistasis or epistatic effects, although this is a slightly broader meaning than that for which "epistatic" was first used in genetics. In the usage of practical animal breeders, "nicking" most nearly has the same meaning but many breeders also apply the term, "nicking," to phenomena which are nothing but unusual Mendelian recombinations of genes each of which may have nearly additive effects.

"Genotype" as used here is perfectly correlated with breeding value if the latter term is understood to express the average merit of the offspring produced by mating this animal to mates which are a truly representative sample of their breed. If the mates are not a truly representative sample and if there is much epistatic interaction or if there is much dominance and mating is far from random, it must sometimes happen that an individual will produce offspring of higher (or lower) merit in one kind of herd than it would in other herds where the most frequent gene-combinations are quite different. In such cases an animal's breeding value as measured by actual records of progeny, will deviate from what its breeding value would be if its mates were truly representative of their whole breed. However such offspring will not transmit such dominance deviations or epistatic deviations, since those deviations are functions of the *gene combination* rather than of the individual genes considered separately. The *combination* is not transmitted but segregates at the reduction division and the genes reassemble in other combinations in the next generation. Hence "genotype" as used here to express the sum of the average effects of the constituent genes *is* "breeding value" if one considers mating the animal to representative animals of its breed, though it may deviate from "breeding value" if the latter term is used to mean the contribution which this animal could make to descendants produced by breeding it to some isolated and non-typical sub-group of the breed.

"Phenotype" is used here as a convenient intermediate step to show the modification of the genotype by dominance or epistatic relations, which are usually considered genetic phenomena although really they are only consequences of the chemistry and physiology by which the genes produce their effects and are not involved in the mechanism by which genes are transmitted. By this device we can express separately the proportion ($h^2$) of the observed variance which is hereditary in the broader sense of that word (*i.e.*, including the effects of dominance and epistasis as "hereditary") and the proportion ($g^2$) of the hereditary (in that broad sense) variance which can be expressed by this additive scheme for representing genotype. The proportion of the observed variance which is hereditary in the narrow sense of that word (*i.e.*, which would contribute directly to parent-offspring correlations) would be $h^2g^2$.

Non-additive effects of gene combinations are separated into those due to dominance (D) and all others (I). This is done merely because the statistical consequences of dominance have been explored (2, 3, 12) and can be stated in definite formulae. The statistical consequences of epistatic interactions are not so well known and must be quite complex in some cases, but on the whole probably are much the same as those of dominance, *i.e.*, to lower the correlation between genotype and outward characteristic and hence to lower the observed correlations between relatives, but to affect the correlations between some kinds of relatives more than those between other kinds. So far as dominance and epistasis merely lower the parent-offspring correlation (*i.e.*, diminish the value of g) their effects nearly cancel out of the present problem since they weaken the correlation between daughter average and dam's genotype almost as much as they do the correlation between the dam's average and her genotype. However so far as dominance and epistasis lower the correlation between sisters less than they do that between parent and offspring, they actually prevent the progeny test from becoming relatively as accurate an indicator of the dam's genotype as it would if there were no dominance or epistasis.

The importance of epistasis in practical breeding problems is still a matter of conjecture, with some workers thinking it very unimportant and others speculating that it may often be highly important, especially in herds already improved to a point considerably above the average level of the breed. Wherever the parent-offspring correlation is high under nearly random mating (after having been corrected for the effects of common environment), it seems likely that epistasis is unimportant. However a low parent-offspring correlation under those circumstances does not prove that epistasis is important, since such a low correlation might also result from h being small, *i.e.*, from random variations in environment having important effects on the records.

The importance of dominance depends primarily on whether complete dominance is the rule, how abundant the dominant genes are in contrast to their recessive allelomorphs, and somewhat on the deviations from random mating. In a population mating at random with dominance complete and with no epistasis, $g^2$ for the effects of each gene pair $= \dfrac{2(1-q)}{2-q}$ or conversely $d^2 = \dfrac{q}{2-q} =$ the portion of variance in H due to dominance. (The letter q here represents the frequency of the dominant genes among all the genes occupying that locus in the whole breed.) This $d^2$ which Fisher calls "the dominance ratio" (3) is thought by him to be typically about ⅓ for most characteristics. Wright (12, pages 137–139) for other reasons, which appear to him more convincing, believes that this ratio is more typically about ⅕, even if complete dominance is the almost universal rule among all genes. Moreover he questions (13) whether such complete dominance is the rule, especially among modifying genes each of which has a comparatively small effect. Such modifying genes probably furnish much of the variance in quantitative characteristics which, like milk production, are vitally necessary in some degree but which may vary widely among different individuals in the breed.

Whether complete dominance is the rule or not, it is obvious that dominance will not often be the major obstacle in estimating an animal's genotype from its own appearance or performance. Even if dominance is universal and Fisher's estimate of $d^2$ is correct, g will generally have a value of about .82, while if Wright's view is correct g will usually be nearer .90. Both these figures are based on the assumption that epistasis is too small to be important. If epistasis is important, $d^2$ and $g^2$ would still retain about the same relation to each other, though both would be smaller. There will be occasional characteristics (of which lethals are notable examples) where dominance is complete and the recessive gene is so rare that the "dominance ratio" approaches close to unity. In those cases dominance may be the major source of the discrepancies between genotype and phenotype. However in such traits the recessive individuals are rare and each such trait will contribute but little to the total variance of the population. For example, in the United States in black breeds of cattle like the Angus, Galloway, or Holstein-Friesian, the gene for black must be around ten to fifteen times as frequent as its recessive allelomorph. Hence the "dominance ratio" for this particular trait will be well above .80. But the gene for red is so rare that less than 1% of the purebred calves of these breeds are born red. Hence this trait is a comparatively unimportant source of loss to breeders. Dominance has been discussed here more than its importance justifies, largely because it has received more attention than it deserves in the writings on applied genetics.

### SOLUTION UNDER SIMPLEST CONDITIONS

Figure 1 shows the relations and correlations as they would be under random mating. The correlation between one record of the dam and the dam's genotype will be $h_p g_p$ while the correlation between one record of one daughter and the dam's genotype will be hgab. When m unselected records of the dam are averaged together the correlation between that average and the dam's genotype becomes

$$h_p g_p \sqrt{\frac{m}{1 + (m-1)(h_p^2 + e_p^2 r_{E_p E_p})}}$$

which approaches

$$g_p \sqrt{\frac{h_p^2}{h_p^2 + e_p^2 r_{E_p E_p}}}.$$

as a limit as m becomes indefinitely large. If $e_p^2 r_{E_p E_p} = $ zero (the most favorable condition for high values of $r_{MG}$), this approaches $r_{MG} = g_p$ as a limit. Increasing the number of records observed thus tends to eliminate errors arising from random uncorrected variations in environment but does not tend to eliminate errors arising from dominance or epistasis or consistently biased and uncorrected environment.

On the other hand, the average record of n daughters each tested once is correlated with the dam's genotype thus:

$$hgab\sqrt{\frac{n}{1 + (n-1)(h^2g^2a^2b^2 + e^2r_{EE})}}$$

which approaches

$$\sqrt{\frac{h^2g^2a^2b^2}{h^2g^2a^2b^2 + e^2r_{EE}}}$$

as n becomes indefinitely large. This approaches unity when $e^2r_{EE}$ is zero. In other words, increasing n tends to cancel errors arising from the imperfections of both h and g, whereas increasing m tends to cancel the imperfections of $h_p$ but does nothing to correct for the fact that $g_p$ is less than unity. However the progeny test involves ab which arises from the intervening generation of Mendelian segregation and appears not at all in the correlation between dam's record and dam's genotype.

Thus a daughter's record usually is a less accurate (on account of ab) indicator of her dam's genotype than the dam's own record is, but the errors in daughters' records tend to be more completely cancelled by the averaging process than is the case with errors in the dam's record. Hence if g and $g_p$ were quite small, it *might* be possible for n to be so large that its cancellation of errors arising from dominance or epistasis would more than make up for the Mendelian sampling errors which affect the progeny average but not the dam's own records.

The problem of the relative accuracy of progeny test and dam's own record as indicators of the dam's genotype therefore becomes the quantitative one of balancing the errors introduced by ab against the greater effectiveness of the progeny average in eliminating errors introduced by dominance and epistasis. Unfortunately for the simplicity of the answer, $e^2r_{EE}$ and $e_p^2r_{E_pE_p}$ cannot safely be considered zero and their existence sets serious limits on the effectiveness of averaging as a means of eliminating errors, either from the daughter average or from the average of the dam's own records.

The value of ab will be nearly .5 even with moderately large departures from random mating (11, pages 118–119). Substituting that value for ab, the ratio (B) of $r_{AG}$ to $r_{MG}$ becomes

$$B = \frac{hg}{h_pg_p}\sqrt{\frac{n}{m}}\sqrt{\frac{1 + (m-1)(h_p^2 + e_p^2r_{E_pE_p})}{4 + (n-1)(h^2g^2 + 4e^2r_{EE})}} \quad (1)$$

If dam and daughters are produced by the same kind of a breeding system $g = g_p$ and if dams and daughters are equally typical samples of their generations and if the records used to represent the productiveness of dam and of daughters are chosen by a method equally fair to both, $h = h_p$ and the first part of this expression cancels. Assuming (as the case most favorable to the progeny test) that $m = 1$ and squaring to simplify the expression we have

$$B^2 = \frac{n}{4 + (n-1)(h^2g^2 + 4e^2r_{EE})} \quad\quad (2)$$

For the progeny test to be more accurate than the dam's own record $B^2$ must be larger than 1.0. This condition is satisfied when $n > 4 + (n-1)(h^2g^2 + 4e^2r_{EE})$. For this to be true, n must be larger than 4, even when $h^2g^2$ and $e^2r_{EE}$ are extremely small. If $e^2r_{EE}$ is as large as .25, n cannot possibly be large enough to make the progeny test average as reliable as the dam's own performance.

[*Editor's Note:* The material omitted at this point includes graphs in which equation (2) is evaluated to show the relative accuracy of progeny test and the individual's own performance for a range of values of heritability and environmental correlations among sibs. A discussion, "Conditions Which May Affect the Validity of the Simple Solution," is also omitted. In this, Lush reviews problems of nonrandom mating, the case where the sire's genotype is known accurately, and where there has been selection of dams. His conclusions from this analysis are summarized at the beginning of the section, "General Considerations and Discussion."]

## GENERAL CONSIDERATIONS AND DISCUSSION

The general conclusion to be drawn from all these considerations is that only under rare and unlikely combinations of conditions would a progeny test based on as few as four-daughters-average in an unselected population as accurate an indicator of a dam's breeding value as the dam's own performance. If the trait is highly hereditary, in the narrow sense of that word, or if the progeny resemble each other very much for reasons of having been exposed to a common environment, many more than four daughters may be required or it may even be quite impossible for the progeny test to average more accurate in a whole population than the dam's own performance, no matter how many daughters there are.

The superiority of the progeny test is greatest for traits which are least hereditary in the narrow sense of that word. For traits which are very faintly hereditary and for which the offspring do not resemble each other for reasons of having been under common environment, the progeny test can become several times as accurate as the dam's own record, but for that to be true the trait must be so slightly hereditary that neither indicator can be highly accurate!

The progeny test is needed most where one sex cannot express the trait, as in milk and fat production in dairy cattle, egg production in poultry, prolificacy in swine, etc. Since the biometrical relations between sire's genotype and progeny average are the same as those between dam's genotype and progeny average (except for sex-linked inheritance which must be a small part of the total inheritance in mammals) the foregoing considerations apply also to the progeny test of the sire. However, for traits which the sire cannot express himself, there is nothing but a pedigree estimate of the sire against which to compare the accuracy of a progeny test of him. Since a progeny test surpasses even the best pedigree estimate when there are more than three progeny (provided the offspring do not resemble each other very much for any other reason except that they are by the same sire), the progeny test in such cases is much more useful and more urgently needed for the sire than for the dam. However such progeny tests will rarely tell as much about the sire as the available information correctly used will tell about a dam of nearly equal age. Enthusiasm over some such slogan as the current one that "The next best thing to a proved sire is the son of a proved sire" should not cause us to forget that such a son is one generation of Mendelian segregation away from this sire and that half of his inheritance (a little more than half if the probably small amount of sex-linked inheritance in mammals is also considered) comes from his dam whose breeding value may usually be estimated more closely than that of his proved sire, if the available information is fully and fairly used.

Although the progeny test will not often be *superior* to the dam's own record (where no differences of selection are involved), it will be noticed from equation 2 that $B = .5$ even when n and m are only one. Therefore even the most fragmentary progeny test is worth much as an indicator of the dam's genotype. Naturally, as in any other prediction, both indicators should be used where both are available. The principles of multiple correlation govern the amount of attention to be paid to each indicator where both are used. However the formulae for the relative amount of attention to be given to the two records are not only complex but also involve a term (which is usually indeterminable) for the degree to which the dam's record and her daughters' records resemble each other because of being made under common environment. The ratio between the standard regression coefficients will be in the same direction from unity as is the ratio between the primary correlations (B in the earlier algebraic discussion) but will be more extreme. Perhaps as good an approximate rule as any would be to give a daughter's production nearly half as much attention as the cow's own record where there is only one record for each, but to give between $\frac{2}{3}$ and $\frac{4}{5}$ as much attention to the daughter average as to the cow's own record where there are at least two or three offspring. This rule would be approximately correct if daughters and dams had been equally selected from their generations. Where the dams have already been highly selected, more importance than this should be given to the progeny test in making further selections but it does not seem possible to develop any simple general rule for this, since such a rule would include terms for the intensity of selection among dams and among daughters and also a term for the extent to which the trait is really hereditary.

In actual practice selection will also be based in part on pedigree. Chronologically pedigree becomes available first, then the individual's own performance and the progeny test comes latest. As some selection is practiced on each basis, the possibilities of further gains by additional selections on the same basis rapidly diminish unless new information becomes available. The sampling nature of Mendelian inheritance sets a lower limit on the usefulness of pedigree than is inherently necessary for either of the other two bases of selection. Hence the very early selections on the basis of pedigree come near exhausting the possibilities in that direction, although increased knowledge of the performance of parents and other ancestors or collateral relatives does make it profitable occasionally to revise an earlier pedigree estimate. On the other hand, knowledge of the individual's own performance can continue to increase at a practically important rate as long as it lives and knowledge of its progeny can increase as long as previously untested progeny come on test and (at a slower rate) as long as progeny already tested continue to be tested at later ages.

This leads to the general picture that pedigree occupies first place only for such selections as must be made before the other two criteria are available. It is very distinctly in third place after a few such early selections have been made. Individual performance occupies first place after it becomes available and until considerable use has been made of it. Then progeny test is the most useful basis for further selections but when it is much used for such selections, the possibilities in that tend to be exhausted and the individual's own performance might again assume first place. Among mature animals under practical conditions individual performance and progeny test therefore will vary or alternate as most useful for further selections, according to the amount of new knowledge becoming available about each and according to how nearly the possibilities for selection on the basis of the existing knowledge about each have already been exhausted.

## SUMMARY

1. The biometrical relations governing the relative accuracy of progeny test and of the parent's own performance as indicators of the parent's breeding value are presented and discussed.

2. A solution under the simplest conditions is presented algebraically and graphically. Under those conditions there must be at least five offspring before the progeny test in a whole population will usually be a more accurate indicator of the parent's breeding value than the parent's own performance.

3. Most deviations from those simplest conditions have only slight effects on the solution. However any general resemblance between the offspring for any other reason than that they are half-sibs through the parent in question sets serious limits on the accuracy of the progeny test. On the other hand if the parents or the records used to represent them are more highly selected than the offspring or their records, the progeny test may become relatively more accurate than under the simple conditions for which the algebraic solution is given.

4. The progeny test is needed most for traits which cannot be expressed in one sex and for traits which are but slightly hereditary.

5. The bases for estimating breeding value are pedigree, own performance, and progeny test. As fast as some selection is practiced on one of these bases, the possibilities for further progress by additional selection on the same basis rapidly diminish and correspondingly increased attention should be given to one of the other bases.

## REFERENCES

(1) COPELAND, LYNN. Pedigree analysis as a basis of selecting bull calves. JOURNAL OF DAIRY SCIENCE 17: 93–102. 1934.

(2) FISHER, R. A. The correlation between relatives on the supposition of Mendelian inheritance. Trans. Roy. Soc., Edinburgh 52 part 2: 399–433. 1918.
(3) FISHER, R. A. On the dominance ratio. Proc. Roy. Soc., Edinburgh 42: 321–341. 1922.
(4) FISHER, R. A. The genetical theory of natural selection. 272 pp. Oxford. 1930.
(5) HOGBEN, LANCELOT. Nature and nurture. 143 pp. New York. 1933.
(6) JULL, MORLEY A. Progeny testing in poultry breeding as a means of evaluating the breeding potentiality of an individual. American Naturalist 67: 500–514. 1933.
(7) JULL, MORLEY A. Limited value of ancestor's egg production. Jour. of Heredity 25: 61–64. 1934.
(8) LUSH, JAY L. The number of daughters necessary to prove a sire. JOURNAL OF DAIRY SCIENCE 14: 209–220. 1931.
(9) TURNER, C. W. The mode of inheritance of yearly butterfat production. Missouri Agr. Exp. Station, Research Bul. No. 112. (Pages 42 to 50.) 1927.
(10) WRIGHT, SEWALL. Correlation and causation. Jour. Agr. Res. 20: 557–585. 1921.
(11) WRIGHT, SEWALL. Systems of mating. I. The biometric relations between parent and offspring. Genetics 6: 111–123. 1921.
(12) WRIGHT, SEWALL. Evolution in Mendelian populations. Genetics 16: 97–159. 1931.
(13) WRIGHT, SEWALL. Physiological and evolutionary theories of dominance. American Naturalist 68: 24–53. 1934.

# 6

Copyright © 1943 by the Genetics Society of America
Reprinted from Genetics **28**:476-490 (1943)

## THE GENETIC BASIS FOR CONSTRUCTING SELECTION INDEXES[1]

L. N. HAZEL[2]

*Iowa State College, Ames, Iowa*

"The key is man's power of accumulative selection: nature gives successive variations; man adds them up in certain directions useful to him."—Darwin, p. 35, sixth edition of *The Origin of Species.* 1920.

### INTRODUCTION

THE idea of a yardstick or selection index for measuring the net merit of breeding animals is probably almost as old as the art of animal breeding itself. In practice several or many traits influence an animal's practical value, although they do so in varying degrees. The information regarding different traits may vary widely, some coming from an animal's relatives and some from the animal's own performance for traits which are expressed once or repeatedly during its lifetime. LUSH (1935) emphasized that permanent improvement from phenotypic selection is proportional to the additively genetic (heritable) fraction of the observed variance and that this varies for different traits. DOBZHANSKY (1937) suggested "that most, and possibly all, genes have manifold effects." These factors make wise selection a complicated and uncertain procedure; in addition fluctuating, vague, and sometimes erroneous ideals often cause the improvement resulting from selection to be much less than could be achieved if these obstacles were overcome.

In the initial stages of breeding investigations conducted by the REGIONAL SWINE BREEDING LABORATORY and cooperating state experiment stations, an arbitrary method of selecting breeding animals had to be adopted. In the meantime the theoretical aspects of the problem were investigated while data were being collected. While many fundamental genetic problems are still incompletely solved (particularly as regards the prevalence of dominance, epistasis and pleiotropic effects of genes in quantitative inheritance, the nature of heterosis, and the interaction of genotype and environment), the accumulation of genetic knowledge justifies an exploration of this problem. According to formulas presented by HAZEL and LUSH (1943), selection for an index which gives proper weight to each trait is more efficient than selection for one trait at a time or for several traits with an independent culling level for each trait. The principles of constructing and using selection indexes which permit the attainment of maximum genetic progress are given in the present paper. Three selection indexes were constructed (and compared as to efficiency) from data taken on the Iowa Station swine herd from the fall of 1937 through the spring of 1940. Such indexes are subject to revision when the accuracy of

---

[1] Journal paper No. J-1121 of the IOWA AGRICULTURAL EXPERIMENT STATION, Ames, Iowa. Project No. 32.

This research was done in cooperation with the REGIONAL SWINE BREEDING LABORATORY, BUREAU OF ANIMAL INDUSTRY, U. S. DEPARTMENT OF AGRICULTURE.

[2] Now at the WESTERN SHEEP BREEDING LABORATORY, Dubois, Idaho.

the statistics upon which they are based can be increased by analyzing additional data.

Selection indexes, constructed with attention to the genetic and economic bases for the various traits, should be valuable in livestock breeding programs. GALTON's "Law of Regression," presented before the Mendelian nature of inheritance was clear, represents an early step in index construction. The sire index, widely used in selecting sires for butterfat and milk production, is a practical example of an index based on one trait but using information about several relatives. SMITH (1937) developed an index designed for the selection of plant lines, using FISHER's concept of discriminant functions to derive a linear equation based on observable characteristics as the best available guide to the genetic value of each line.

## ANALYTICAL METHOD

The net genetic improvement which can be brought about by selecting among a group of animals is the sum of the genetic gains made for the several traits which have economic importance. It is logical to weight the gain made for each trait ($\overline{G}_i$) by the relative economic value of that trait ($a_i$). Thus the average genetic superiority of a selected group over the group from which it was chosen is

(1) $$\overline{H} = a_1\overline{G}_1 + a_2\overline{G}_2 + \cdots + a_n\overline{G}_n.$$

The relative economic value for each trait depends upon the amount by which profit may be expected to increase for each unit of improvement in that trait. Good approximations to relative economic values often can be obtained from long-time price averages and cost-of-production figures. As an example, WINTERS (1940) found that the average price of wool per pound was 3.4 times that of lamb per pound. If additional feed or labor costs are associated with increased production for either trait, the increased cost per unit should be discounted when calculating relative economic values. These values may vary from breed to breed or from region to region within the same breed. They may change, even while a breeding program is in progress, if permanent shifts in market demand occur.

Animals vary in breeding value, as in phenotype, for each of the several traits. The aggregate value of an animal is the sum of its several genotypes (assuming a distinct genotype for each economic trait), each genotype being weighted according to the relative economic value of that trait. An animal's genotype for a given trait may be defined as the sum of the average (strictly additive) effects of its genes which influence that trait. Thus the aggregate genotype of an animal is

(2) $$H = a_1G_1 + a_2G_2 + \cdots + a_nG_n.$$

Environmental factors, dominance, and epistasis may make phenotypic performance unlike the genotype for that trait; hence animals having the highest values for H cannot be recognized directly with perfect accuracy. Selection for improved breeding value therefore must be practiced indirectly by selecting directly for a correlated variable (I) based on the phenotypic performance of each animal for the several traits. When selection is practiced on a large population, the genetic average of the selected group minus that of the original population represents the genetic gain from selection ($\overline{H}$ in equation 1), being

$$(3) \qquad \overline{H} = (\overline{\imath})R_{IH}\sigma_H.$$

Here $\overline{\imath}$ is the average superiority of the indexes (in standard deviation units) for the selected group as compared to the whole group, $R_{IH}$ is the correlation between H and I, while $\sigma_H$ is the standard deviation of H in the whole group.

The selection differential ($\overline{\imath}$) depends upon the proportion which can be culled, being limited by the percentage needed for replacements for a particular species. The standard deviation of breeding value,

$$\sigma_H = \sqrt{a_1^2\sigma_{G_1}^2 + a_2^2\sigma_{G_2}^2 + \cdots + 2a_1a_2\sigma_{G_1}\sigma_{G_2}r_{G_1G_2} + \cdots},$$

depends upon gene frequency and to some extent upon the mating system. These can be changed only a little by the breeder. Consequently the opportunity for increasing the progress expected from selection lies in making $R_{IH}$ as large as possible. Accordingly I is defined as

$$(4) \qquad I = b_1X_1 + b_2X_2 + \cdots + b_nX_n,$$

where the X's represent phenotypic performance for the several traits and the b's are multiple regression coefficients chosen so as to make $R_{IH}$ as large as possible. These regression coefficients may be calculated from n simultaneous equations

$$(5) \quad \begin{aligned} \beta_1 \phantom{r_{X_1X_2}} + \beta_2r_{X_1X_2} + \cdots + \beta_nr_{X_1X_n} &= r_{X_1H} \\ \beta_1r_{X_1X_2} + \beta_2 \phantom{r_{X_1X_2}} + \cdots + \beta_nr_{X_2X_n} &= r_{X_2H} \\ \cdots \cdots \cdots \cdots \cdots \cdots \cdots \cdots \cdots \cdots & \\ \beta_1r_{X_1X_n} + \beta_2r_{X_2X_n} + \cdots + \beta_n \phantom{r_{X_1X_2}} &= r_{X_nH} \end{aligned}$$

where $\beta_i = b_i \dfrac{\sigma_{X_i}}{\sigma_H}$ and $r_{X_iH}$ is the correlation between H and the i-th phenotypic measurement.

The simultaneous equations can be solved only if estimates of the various correlations can be calculated. The usual methods of interclass and intraclass correlation are generally sufficient to calculate the phenotypic correlations ($r_{X_iX_j}$). WRIGHT's (1934) method of path coefficients is convenient for calculating the more complex correlations between H and phenotypic performance ($r_{X_iH}$). The path coefficient diagram (fig. 1) indicates the various relations between H and the phenotypic measurement for each trait ($X_i$). This correlation is the sum of the various paths from $X_i$ to H, as follows,

$$(6) \qquad r_{X_iH} = r_{G_iX_i}\{d_1r_{G_1G_i} + d_2r_{G_2G_i} + \cdots + d_nr_{G_nG_i}\},$$

where
$$d_i = a_i \frac{\sigma_{G_i}}{\sigma_H}.$$

Therefore we must have estimates of genetic variability for each trait ($\sigma_{G_i}$), the correlation between genotype and phenotypic performance for each trait ($r_{X_i G_i}$), and the correlation between genotypes for different traits ($r_{G_i G_j}$) to solve the simultaneous equations.

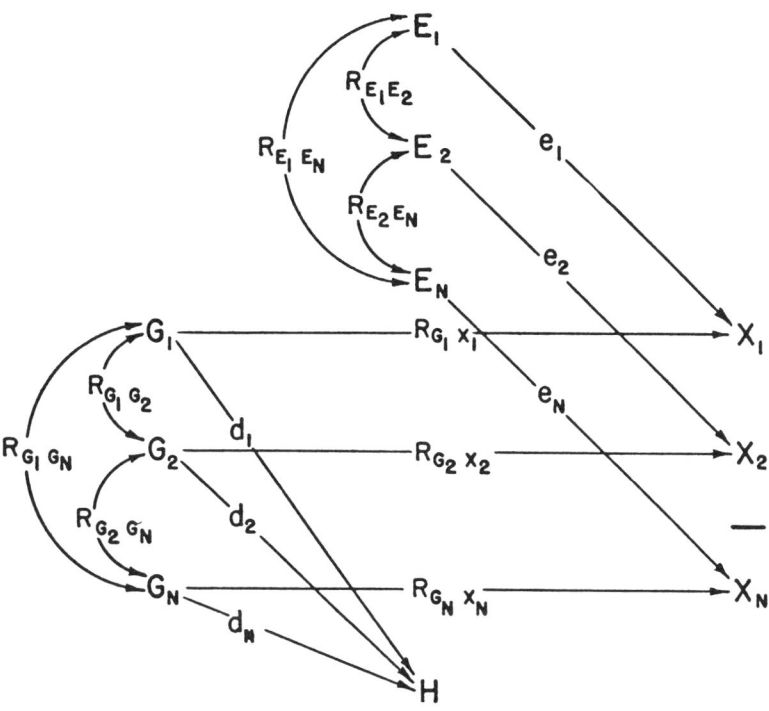

FIGURE 1. Path coefficient diagram showing the relation between phenotypic measurements ($X_i$) and the aggregate genotype (H). For further explanation of symbols see text.

Let an observed trait (X) be the sum of the average effects of the genes (G)[3] which an animal inherits, plus the combined effects of environment, dominance and epistasis (E); thus

(7) $$X = G + E.$$

Then the observed variance is

(8) $$\sigma_X^2 = \sigma_G^2 + \sigma_E^2,$$

[3] We have defined the genotype as the sum of the average effects of genes because WRIGHT (1921, 1935) has shown that only the average (additive) effects contribute to permanent gain from selection. The effects of environment, dominance, and epistasis may logically be grouped together since they act similarly to mask the genotype.

if G and E are uncorrelated, as would be the case unless particular efforts were made to give the better genotypes better or worse than average treatment. The correlation between the genotype and phenotype for the same trait is

$$r_{G_i X_i} = \frac{\sigma_{G_i}}{\sigma_{X_i}} = g_i, \tag{9}$$

where that trait is measured on the animal itself. The term $g_i$ is the square root of the heritable (additively genetic) fraction of the observed variance.

DARWIN observed the importance of correlated variation, as evidenced by the statement, "Hence if man goes on selecting, and thus augmenting, any peculiarity, he will almost certainly modify unintentionally other parts of the structure, owing to the mysterious laws of correlation." The genetic basis for this statement is evident if genes have manifold effects. Many cases of pleiotropic genes have been reported for laboratory animals, although they have received little attention in domestic animals. Linkage and non-random mating systems may also cause correlated variation; however, their effects would be less permanent and consequently less important in selection. Repeated crossing over ultimately makes the coupling and repulsion heterozygotes equally numerous, while the transient nature of breeding herds prevents sustained departure from random mating in most cases. An additional and usually much more important cause of correlated variation within an interbreeding population lies in the environmental circumstances peculiar to each animal, particularly for traits which develop during the same, or in adjacent, periods of time. Thus if trait I is correlated with trait J in the same animal, I may serve as an indicator either of the animal's genotype for J or of the environmental circumstances to which that animal was exposed when traits I and J were both being developed.

Statistically an observed phenotypic correlation may be analyzed into its constituents, a genetic correlation ($r_{G_i G_j}$) and an environmental correlation ($r_{E_i E_j}$), as indicated in figure 1. The observed phenotypic correlation between two traits measured on the same animal is

$$r_{X_i X_j} = g_i g_j r_{G_i G_j} + e_i e_j r_{E_i E_j}, \quad \text{where} \quad e_i = \frac{\sigma_{E_i}}{\sigma_{X_i}}. \tag{10}$$

Sex-limited traits and traits such as carcass merit cannot be measured directly on all breeding animals. Sometimes selection must be practiced before each animal's performance for every trait is known. For these reasons there is a possibility of increasing genetic progress in a breeding program by using information about the performance of relatives. Examples of how $r_{X_i H}$ may be calculated are given in the subsequent section where $X_i$ is (1) the individual's own performance, (2) the performance of the individual's dam, and (3) the average perfromance of a group of relatives of which the individual is a member.

## APPLICATION

Three selection indexes for young boars and gilts were constructed from data taken on the Iowa Station swine herd from the fall of 1937 through the spring of 1940. The history of the herd was given by BYWATERS (1937) and more recently by WHATLEY (1942). LUSH (1940) showed that the economic return from the swine enterprise depended largely upon three traits—growth rate, market suitability, and size of litter—the relative economic values of which were $\frac{1}{3}$, 1, and 2, respectively, in the units used in this study. Growth rate was measured by weight at 180 days of age. Market suitability was measured by a numerical score given by each of several judges. The details of these measurements were given by WHATLEY for 180-day weight and by STONAKER and LUSH (1942) for score. The measure of litter size was that suggested by LUSH and MOLLN (1942),

$$P = n_0 + n_{21} + n_{56} + W_{21}/10 + W_{56}/30,$$

the n's and W's referring to the number of pigs and weight of the litter, respectively, at the ages designated by the subscripts. Young boars and gilts are not old enough to have produced litters themselves at the time first selections must be made; consequently the estimate of productivity for each young animal is based upon the litter in which it was born.

The aggregate breeding value of an animal is therefore defined as

$$H = \tfrac{1}{3}G_W + G_S + 2G_P,$$

$G_W$, $G_S$ and $G_P$ referring to the genotypes for 180-day weight, market score and productivity, respectively. The following statistics are needed to construct one or more of the three indexes:

A. Phenotypic constants
  1. Standard deviation for each of the three traits
  2. Phenotypic correlation between each pair of traits
  3. Phenotypic correlations between the traits of relatives

B. Genetic constants
  1. Heritable fraction of the variance in each trait
  2. Genetic correlation between each pair of traits.

Differences due to season and line of breeding were eliminated in calculating these statistics, as selection is usually practiced between animals born in the same season and within the same line or other interbreeding population. The phenotypic constants shown in table 1 were calculated by the usual methods of variance and covariance analysis (SNEDECOR 1940).

Several procedures were outlined by LUSH (1941) for estimating heritability ($g_i^2$) in farm livestock. Intrasire regressions of offspring on dam were considered preferable to correlation coefficients in the present case because of the rather intense selection which had been practiced in the Iowa Station swine

TABLE 1

*Phenotypic constants necessary to construct one or more of the three indexes.*

| TRAIT | STANDARD DEVIATION | DEGREES OF FREEDOM |
|---|---|---|
| 180-day weight (in pounds) | 31.86 | 1513 |
| Market score (in points) | 4.78 | 1513 |
| Productivity (in points) | 9.72 | 282 |

| CORRELATION BETWEEN TRAITS | CORRELATION COEFFICIENT | STANDARD ERROR |
|---|---|---|
| Pig's own weight and score | 0.614 | 0.016 |
| Pig's own weight and productivity of its dam | −.024 | .059 |
| Pig's own score and productivity of its dam | −.081 | .059 |
| Weight of litter mates | .314 | .027 |
| Score of litter mates | .204 | .026 |
| Weight and score of litter mates | .186 | .026 |

herd for the traits under consideration. The intensity of selection was evidenced by the higher averages and lower variability for the dams as compared to their offspring. The intrasire regressions of offspring on dam are shown in table 2.

TABLE 2

*Regression of each trait for the offspring on each of the three traits for the dam.*

| OFFSPRING'S TRAIT | DAM'S TRAIT | SYMBOL* | REGRESSION COEFFICIENT | DEGREES OF FREEDOM |
|---|---|---|---|---|
| Weight | Weight | $b_{W_2 W_1}$ | 0.149 | 232 |
| Score | Score | $b_{S_2 S_1}$ | .049 | 189 |
| Productivity | Productivity | $b_{P_2 P_1}$ | .081 | 82 |
| Weight | Score | $b_{W_2 S_1}$ | .153 | 189 |
| Score | Weight | $b_{S_2 W_1}$ | .013 | 232 |
| Weight | Productivity | $b_{W_2 P_1}$ | −.067 | 204 |
| Productivity | Weight | $b_{P_2 W_1}$ | .004 | 52 |
| Score | Productivity | $b_{S_2 P_1}$ | −.024 | 204 |
| Productivity | Score | $b_{P_2 S_1}$ | .041 | 52 |

* The sub-subscripts 1 and 2 refer to dam and offspring, respectively, the regressions being those of offspring's trait on dam's trait in each case.

Values of $g_i$, calculated from the formula,[4] $g_i = \sqrt{2 b_{i_2 i_1}}$, are given in table 3 for the three traits.

[4] This formula was derived from the relation,
$$b_{i_2 i_1} = g_i^2 t + e_i^2 r_{EE}.$$
The genetic relationship between dam and offspring (t) was taken as 0.5, and the environmental correlation ($r_{EE}$) was assumed zero, since no particular efforts had been made to give parent and offspring similar or dissimilar treatment.

## TABLE 3

*Genetic constants derived from intrasire regressions of offspring on dam.*

| STATISTIC | SYMBOL | ESTIMATE |
|---|---|---|
| $\sqrt{}$ heritability of weight | $g_W$ | 0.546 |
| $\sqrt{}$ heritability of score | $g_S$ | .313 |
| $\sqrt{}$ heritability of productivity | $g_P$ | .402 |
| Genetic correlation between weight and score | $r_{G_W G_S}$ | .519 |
| Genetic correlation between weight and productivity | $r_{G_W G_P}$ | .0± |
| Genetic correlation between score and productivity | $r_{G_S G_P}$ | .0± |

Procedures for estimating the genetic correlations ($r_{G_i G_j}$) have not been developed previously. Equation 10 is not suitable for this purpose because no direct method is available for separating the genetic and environmental correlations for two traits measured upon the same animal. For example, any environmental accident such as differential exposure to parasites or infection which affected growth and plumpness would either raise or lower the growth rate and score of the same animal. Thus the actually observed correlation between the two traits on the same animal conceivably could be due wholly to such environmental circumstances, or to genes which affect both traits, or to a mixture of the two causes in any proportion. To measure the genetic correlations by themselves it was necessary to correlate one trait in one animal with the other in a relative. The formula,[5]

$$r_{G_i G_j} = \sqrt{\frac{b_{i_2 j_1} \cdot b_{j_2 i_1}}{b_{i_2 i_1} \cdot b_{j_2 j_1}}} = \sqrt{\frac{(\text{cov } I_2 J_1)(\text{cov } J_2 I_1)}{(\text{cov } I_2 I_1)(\text{cov } J_2 J_1)}},$$

was adopted because it appeared to be unbiased by selection and to utilize most of the available information. Estimates of the genetic correlations between each pair of the three traits are given in table 3. Consistent estimates could not be derived for $r_{G_W G_P}$ and $r_{G_S G_P}$, since in each case the two regressions required in the numerator differed in sign. Since these regressions are small and not significantly different from zero, it seemed more accurate to assign both $r_{G_W G_P}$ and $r_{G_S G_P}$ values of zero than to attempt an alternative method of calculating them.

The genetic constants probably include a small fraction of the epistatic deviations and exclude the average differences between groups of dams to

---

[5] Selection for the independent variable does not bias the corresponding regression coefficient (EISENHART 1939). The quantity $\sqrt{b_{i_2 j_1} \cdot b_{j_2 i_1}}$ consequently can be taken as an estimate of the correlation between two traits of dam and offspring in an unselected population. The biometric relation,

$$r_{i_2 j_1} = r_{j_2 i_1} = g_i g_j t r_{G_i G_j},$$

permits the derivation of the formula given above for $r_{G_i G_j}$. The two regressions (or correlations) actually provide two independent estimates of $r_{G_i G_j}$ in unselected populations and an arithmetic average of the two estimates would be less biased by sampling errors than the geometric average used here.

which sires were mated. Since some inbreeding had occurred, they may be slightly nearer zero than would be expected in a non-inbred population. If they are biased, and this bias is equal and in the same direction for the different genetic constants, the index will be no less accurate for selecting breeding animals.

Weight at 180 days of age and market score are available for each animal before breeding age. The first index constructed was based on these two traits. Values for $r_{WH}$ and $r_{SH}$ were calculated by substituting the statistics from tables 1 and 3 into formula 6 as follows:

$$r_{WH} = \frac{0.546}{\sigma_H} \{5.803(1) + (1.496)(.519)\} = 3.593/\sigma_H$$

$$r_{SH} = \frac{0.313}{\sigma_H} \{5.803(.519) + 1.496(1)\} = 1.409/\sigma_H.$$

The two simultaneous equations are:

$$\beta_W + 0.614\beta_S = 3.593/\sigma_H$$
$$0.614\beta_W + \beta_S = 1.409/\sigma_H.$$

Solving, we get $\beta_W = 4.381/\sigma_H$ and $\beta_S = -1.282/\sigma_H$.
Since $\beta_i = b_i \sigma_{Xi}/\sigma_H$, the first index is

$$I_1 = 0.137W - 0.268S,$$

where W and S represent the pig's own 180-day weight and market score, respectively. If the index can thus be made more convenient for use or for keeping records, it may be multiplied by any constant or any constant may be added to it without influencing its relative accuracy, since such procedure does not change $R_{IH}$.

A second index was constructed by using the productivity of the dam as a measure of each pig's productivity in the index, the lapse of one generation being compensated by multiplying $r_{PH}$ by one-half as follows:

$$r_{PH} = \frac{0.402}{2\sigma_H} (7.802) = 1.566/\sigma_H.$$

The three simultaneous equations were solved as before, the second index being

$$I_2 = 0.136W - 0.232S + 0.164P.$$

The third index was designed to include information about the average weight and score of the litter in which each pig was born, in addition to the three traits in the second index. The procedure used was to consider the average weight ($\overline{W}$) and score ($\overline{S}$) of the litter as a fourth and fifth variable, expressing the necessary correlations in terms of the correlations given in table 1 and as a function of the number of pigs per litter (k). This latter step was necessary because the number upon which the average is based influences the

GENETIC BASIS FOR SELECTION INDEXES 485

variability of the average and the amount of information in the average concerning each pig's breeding value. The additional phenotypic correlations in terms of k and the correlations in table 1 are

$$r_{w\bar{w}} = \sqrt{\frac{1 + (k-1).314}{k}},$$

$$r_{s\bar{s}} = \sqrt{\frac{1 + (k-1).204}{k}},$$

$$r_{w\bar{s}} = \frac{.614 + (k-1).186}{\sqrt{k[1 + (k-1).204]}},$$

$$r_{s\bar{w}} = \frac{.614 + (k-1).186}{\sqrt{k[1 + (k-1).314]}},$$

$$r_{P\bar{w}} = -.024\sqrt{\frac{k}{1 + (k-1).314}}$$

$$r_{P\bar{s}} = -.081\sqrt{\frac{k}{1 + (k-1).204}}, \quad \text{and}$$

$$r_{\bar{w}\bar{s}} = \frac{.614 + (k-1).186}{\sqrt{[1 + (k-1).314][1 + (k-1).204]}}.$$

The additional correlations between H and the litter averages are

$$r_{\bar{w}H} = \frac{3.593[1 + (k-1).5]}{\sigma_H\sqrt{k[1 + (k-1).314]}} \quad \text{and}$$

$$r_{\bar{s}H} = \frac{1.409[1 + (k-1).5]}{\sigma_H\sqrt{k[1 + (k-1).204]}}.$$

The five simultaneous equations were solved as before, giving the following partial regression coefficients:

$$b_W = .098$$

$$b_S = -.165$$

$$b_P = \frac{0.358\{0.164(k-1) + 0.016(k-1)^2\}}{2.18\{0.998(k-1) + 0.899(k-1)^2\}}$$

$$b_{\bar{w}} = \frac{k\{0.270 + (0.004)(k-1)\}}{7.14\{3.273(k-1) + 0.324(k-1)^2\}}$$

$$b_{\bar{s}} = \frac{-k\{0.070 + (0.02)(k-1)\}}{1.07\{0.491(k-1) + 0.049(k-1)^2\}}.$$

The absolute values of the three latter coefficients are given in table 4 for different values of k. The number of pigs in the litter has so little effect on

## TABLE 4

*Partial regression coefficients for dam's productivity, average weight of the litter and average score of the litter for different numbers of pigs in the litter, in the third index.*

| NUMBER OF PIGS IN THE LITTER (k) | $b_{\bar{P}}$ | $b_{\bar{W}}$ | $b_{\bar{S}}$ |
|---|---|---|---|
| 1  | 0.164 | 0.038 | −0.067 |
| 2  | .165  | .059  | − .113 |
| 3  | .165  | .072  | − .148 |
| 4  | .166  | .081  | − .175 |
| 5  | .166  | .088  | − .197 |
| 6  | .167  | .093  | − .215 |
| 7  | .167  | .098  | − .230 |
| 8  | .167  | .101  | − .243 |
| 9  | .168  | .104  | − .254 |
| 10 | .168  | .106  | − .264 |
| 11 | .168  | .109  | − .273 |
| 12 | .169  | .110  | − .280 |
| 13 | .169  | .112  | − .287 |
| 14 | .169  | .113  | − .293 |
| 15 | .169  | .115  | − .299 |

$b_{\bar{P}}$ that this could be taken as a constant (0.166 or 0.167) with very little error.

The amount of genetic progress expected when a given index is used in making selections is proportional to $R_{IH}$ (see formula 3). Hence these values provide a basis for choosing an index which is easy and simple to use yet which is of nearly maximum accuracy. For example the three indexes previously constructed may be compared as follows:

$$R_{I_1H} = \sqrt{\beta_W r_{WH} + \beta_S r_{SH}} = 0.363$$

$$R_{I_2H} = 0.395 \text{ and}$$

$$R_{I_3H} = 0.404, \text{ for } k = 5.$$

The second and third indexes are 8.8 and 11.3 percent, respectively, more efficient than the first. Since the time and effort expended in keeping records is but a small fraction of the total labor connected with a breeding program, the second index would almost certainly be preferable to the first. The third might also be chosen over the second, since genetic progress could be increased still further, and the extra labor would be only that of computing and using the litter averages from data already taken.

### DISCUSSION

FISHER'S (1930) "fundamental theorem of natural selection" and WRIGHT'S (1931) emphasis that genetic change depends upon genetic variability and selection intensity indicate that the factors which are important in natural selection also hold for selection as practiced by man. In equation 3, $\sigma_H$ is a measure of genetic variability and $\bar{\imath}$ a measure of selection intensity. The breeder in applying artificial selection to an animal population has the opportunity of increasing the accuracy of his selections (increasing $R_{IH}$) considerably over what may hold in the "trial-and-error" methods of natural selection. The breeder has additional opportunities of increasing $\bar{\imath}$ (within limits) and of increasing genetic variability by the intentional control of population size (inbreeding) and migration (outcrossing) which have not been considered in the present study. WRIGHT (1940) has examined the conditions under which the supplementary use of these latter methods may be advantageous.

For the special case where the traits are uncorrelated, $R_{IH}$ is a maximum when each regression coefficient is equal to (or proportional to) the product of the relative economic value and heritability for each trait ($b_i = a_i g_i^2$). The correlations which may exist between traits complicate the calculation of the partial regression coefficients, just as correlations between the independent variables do in any multiple regression analysis. When the phenotypic correlation is large as compared to the genetic correlation, the regression coefficient for a trait with little economic importance or slight heritable variation may be negative, because its function in the index then becomes mainly that of indicating the environment for a more important and more highly heritable trait. An example of this was seen in the negative regression coefficients for score in the three indexes calculated previously. However, selection for the animals having the highest indexes would create some improvement for score because of the positive genetic correlation between score and weight.

An index constructed from data taken on a herd in one locality may not be widely applicable. The reasons for this are:

1. Relative economic values for a trait may vary with the particular locality or nature of the enterprise.

2. The genetic constitution of herds may differ, especially where they are under distinctly non-random mating systems such as intense inbreeding.

3. Different managerial practices may cause the standard deviations for the traits to vary in different herds. The standard deviations for subjective traits such as market conformation measured by judging or by scores may vary because different judges will vary the range over which they spread their scores.

4. Few herds are large enough to provide data sufficient to make the sampling errors of the genetic constants small.

The best way to test whether or not selection indexes can be standardized and recommended for general use seems to be to compare several indexes constructed from data taken on different herds.

The data in the present study were sufficiently numerous to provide accurate estimates of the phenotypic constants in table 1. They were less satisfactory with regard to the reliability of the genetic constants given in table 3. Some idea of the general accuracy of these figures can be obtained by comparing them to similar estimates by other investigators. WHATLEY (1942) used several methods to estimate heritability of 180-day weight in the Iowa Station swine herd through 1938, concluding that "at least 30 percent and possibly more than 40 percent of the individual variance" was hereditary. WHATLEY and NELSON (1942) estimated that 180-day weight in the Oklahoma Duroc herd was 23 percent heritable, while BAKER and coworkers (1943) found a figure of 25 percent for 168-day weight in the Nebraska (North Platte substation) Duroc herd. These estimates help to substantiate the figure of 30 percent for the heritability of 180-day weight found in this study. WHATLEY and NELSON also estimated that market score was about 33 percent heritable, while STONAKER and LUSH (1942) obtained an estimate of 20 percent for the Iowa Station herd from data which included that used in this study. These estimates indicate that our estimate of 10 percent for the heritability of score may be too small. Previous reports of the heritability of productivity have not been made, but LUSH and MOLLN (1942) found that the correlation between litters by the same sow was between 0.15 and 0.20 for the items which are included in productivity. They quoted a number of investigations which substantiated their findings. In general these results substantiate our figure of 16 percent if most of the permanent differences in these items between sows are hereditary.

Other attempts to estimate genetic correlations have not been made. The genetic correlation of 0.52 between 180-day weight and score indicates that about half of the genes which influence one trait also influence the other (assuming equal gene frequency and equal effects of the genes). However most of the observed correlation of 0.61 was due to an environmental correlation. Although an effort was made to score all pigs at a constant weight of 225 pounds, part of this correlation was probably due to a subjective tendency on the part of the judges to assign scores in accordance with differences in age or weights of the pigs. The genetic correlations between productivity and weight or score may have been either positive or negative so far as the evidence from the present data indicate. They appeared to be small; hence assigning them a value of zero is unlikely to have caused serious errors in the indexes.

From the studies of heritability which have been made for economic traits in different farm animals, it seems that the best indexes which can be constructed will be far from perfect. The confusing effects of environment, dominance, and epistasis in masking genotypes cause the progress in the present case to be less than half of what might be made if genotypes could be recognized precisely. Thus the indexes constructed for swine permit from 36.1 to 40.4 percent as much gain as could be made with a perfect index (where $R_{IH} = 1$), which is the limit of what could be achieved if the exact Mendelian composition of every animal were known. These indexes could be improved somewhat by more perfect control of the environment, by the wise use of corrections for known environmental circumstances, by more accurately measuring differences in phenotypes, and by including the performance of additional relatives in the index; however, the use of these methods is limited by practical considerations. Although $R_{IH}$ is likely to increase with the age of the animal (as more becomes known about its phenotype and as its progeny becomes observable) so that more gain can be made from selecting within a group of older animals, the length of generation will also increase. DICKERSON and HAZEL (1942) have shown that the interval between generations in some cases is increased by progeny testing more than enough to offset the increased accuracy of selection, the net result of more emphasis on the progeny test then being a decrease in the *annual* rate of genetic improvement. While these considerations do not indicate much possibility of phenomenally rapid improvement in animal populations from selection alone, the progress which can be made with properly constructed indexes is considerably greater than can be expected when the ideals toward which selection is directed are confused or erroneous.

## CONCLUSIONS

The genetic gain which can be made by selecting for several traits simultaneously within a group of animals is the product of (1) the selection differential, (2) the multiple correlation between aggregate breeding value and the selection index, and (3) genetic variability. The first of these may be very small due to the breeder's carelessness, procrastination, etc., and is limited by the rate of reproduction for each species, while the third is relatively beyond man's control; hence the greatest opportunity of increasing the progress from selection is by insuring that the second is as large as possible.

A multiple correlation method of constructing selection indexes having maximum accuracy was presented. The following constants must be known in order to solve the simultaneous equations:

1. Relative economic values for the different traits
2. Phenotypic constants
    a. Standard deviations for each trait
    b. Correlation between each pair of traits
3. Genetic constants
    a. Heritability of each trait
    b. Genetic correlations between each pair of traits

Examples of the construction of selection indexes for young boars and gilts were presented from data taken on the Iowa Station swine herd using (1) 180-day weight and market score of the individual animal, (2) the two previous traits and productivity of the dam, and (3) the three previous traits and the average weight and score of the litter of which each pig is a member.

The progress which can be made by using the above indexes varied from 36 to 40 percent of that which could be made with a perfect index. The loss is due to the confusing effects of environmental circumstances, dominance, and epistasis, all of which can make phenotypes unlike genotypes.

## ACKNOWLEDGMENT

A formal statement of appreciation is scarcely indicative of the author's indebtedness to DR. JAY L. LUSH for the use of many of his unpublished notes, his inspiration and guidance. The author is also indebted to PROF. W. G. COCHRAN, whose suggestions regarding statistical procedure were invaluable.

## LITERATURE CITED

BAKER, M. L., L. N. HAZEL, and C. F. REINMILLER, 1943 The relative importance of heredity and environment in the growth of pigs at different ages. J. Anim. Sci. **2**: 1–13.

BYWATERS, J. H., 1937 The hereditary and environmental portions of the variance in weaning weights of Poland-China pigs. Genetics **22**: 457–468.

DICKERSON, G. E., and L. N. HAZEL, 1942 Effect of progeny testing on improvement expected within closed herds. (In process of publication.) Abstract in J. Anim. Sci. **1**: 342.

DOBZHANSKY, TH., 1937 Genetics and the origin of species. New York: Columbia University Press.

EISENHART, C., 1939 The interpretation of certain regression methods and their use in biological and industrial research. Ann. Math. Statist. **10**: 162.

FISHER, R. A., 1930 Genetical theory of natural selection. London: Oxford University Press.

HAZEL, L. N., and J. L. LUSH, 1943 The efficiency of three methods of selection. J. Hered. **33**: 393–399.

LUSH, J. L., 1935 The inheritance of productivity in farm livestock. Pt. V. Discussion of preceding contributions. Emp. J. Exp. Agric. **3**: 25–30.

—— 1940 Relative importance of dam's productivity, pig's own market score and pig's own growth rate, in selecting young boars and gilts for breeding. Research Item No. 18 of the Regional Swine Breeding Laboratory. Mimeographed. Ames, Iowa.

—— 1941 Intra-sire correlations or regressions of offspring on dam as a method of estimating heritability of characteristics. Proc. Amer. Soc. Anim. Prod. for **1940**: 293–301.

LUSH, J. L., and A. E. MOLLN, 1942 Litter size and weight as permanent characteristics of sows. Tech. Bull. U. S. Dept. Agric. No. **836**.

SMITH, F., 1937 A discriminant function for plant selection. Ann. Eugen. **7**: 240–250.

SNEDECOR, G. W., 1940 Statistical methods. Third edition. Ames, Iowa: Iowa State College Press.

STONAKER, H. H., and J. L. LUSH, 1942 Heritability of conformation in Poland-China swine as measured by scoring. J. Anim. Sci. **1**: 99–105.

WHATLEY, J. A., JR., 1942 Influence of heredity and other factors on 180-day weight in Poland-China swine. J. Agric. Res. **65**: 249–264.

WHATLEY, J. A., JR., and R. H. NELSON, 1942 Heritability of differences in 180-day weight and market score of Duroc swine. (Abstract.) J. Anim. Sci. **1**: 70.

WINTERS, L. M., 1940 Records of performance for meat animals. Emp. J. Exp. Agric. **8**: 259–268.

WRIGHT, S., 1921 Systems of mating. Genetics **6**: 111–178.

—— 1931 Evolution in Mendelian populations. Genetics **16**: 97–159.

—— 1934 The method of path coefficients. Ann. Math. Statist. **5**: 161–215.

—— 1935 The analysis of variance and the correlation between relatives with respect to deviations from an optimum. J. Genet. **30**: 243–256.

—— 1940 Genetic principles governing the rate of progress of livestock breeding. Proc. Amer. Soc. Anim. Prod. for **1939**: 18–26.

# 7

Copyright © 1952 by The University of Chicago
Reprinted from Am. Nat. **86**:293-298 (1952)

## THE PROBLEM OF ENVIRONMENT AND SELECTION

### D. S. FALCONER

The choice of the environmental conditions under which to practice selective breeding presents a problem of practical importance in many branches of livestock improvement, but for which genetical theory has not yet provided a satisfactory solution. The problem is whether the best results will be achieved when selection is carried out under the conditions in which the improved breed will eventually be required to live; or whether better results may be attained under some other conditions, for example under conditions more favorable for the expression of the desired character. On the one hand, one may argue with Hammond (1947) that an environment favorable to the expression of the desired character will allow more rapid progress under selection; and that if the improved breed is then transferred to less favorable conditions it will have attained a higher level of performance than could have been attained by the same amount of selection under the less favorable conditions. On the other hand, most geneticists would probably argue that performance (in respect of milk production, growth rate or any other character) in a favorable environment has a different genetic basis from performance in an unfavorable environment: a superior genotype in one environment could not be expected to be superior in a different environment. The genetic situation would thus be regarded as a case of genotype-environment interaction; and if the environments differed much, as for example temperate and tropical climates, or high and low planes of nutrition, an interaction large enough to vitiate Hammond's argument would be expected. It would therefore generally be recommended that selection should be carried out under the environmental conditions in which the improved breed is destined to live.

The possible existence of genotype-environment interactions has been widely recognized and often discussed: Haldane (1946) has specified a number of different forms which the interaction may take, and has discussed their bearing on livestock breeding; and Lerner (1950) has given a general discussion of the problem in relation to poultry breeding. But the concept of genotype-environment interaction does not seem to lead easily to a solution of problems connected with selection; differences of heritability cannot be taken into account, and a precise evaluation of Hammond's argument in genetic terms cannot be arrived at.

It is the purpose of the present paper to point out that, if only two different environments are considered, the interaction may be expressed as a genetic correlation. When so formulated the genetic aspect of the situation becomes clear, and a quantitative evaluation of the efficacy of different methods of selection may be easily obtained by the procedures already devised for dealing with genetic correlations. A similar treatment of genotype-environment interactions was derived independently by Reeve and Robertson, working on body size in Drosophila reared at different temperatures, and using progeny tests to estimate the importance of genotype-environment interactions (Reeve, in press). A more extended mathematical treatment of the relation between the analysis of variance and correlation techniques is given by Reeve and Robertson (in press).

Suppose that a character, for example growth rate, is measured in two different environments, for example high and low planes of nutrition. The two measurements are then not to be regarded as representing a single character, but two different characters; and the two environments are to be regarded as "treatments" which have to be applied in order that the characters may be measured. The solution of the problem now follows the method of analysis widely used in dealing with the behavior of correlated characters under selection (see for example Hazel, 1943, and Dempster and Lerner, 1947.

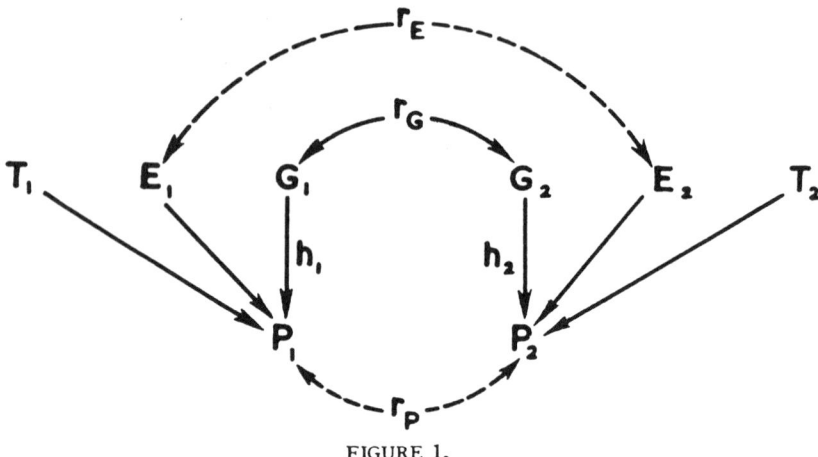

FIGURE 1.

The situation may be formally represented in a diagram (fig. 1), in which $P_1$ and $P_2$ represent the phenotypes of the two characters as measured in the two environments. There are three sources of variation affecting each of the two phenotypes: the two genotypes, $G_1$ and $G_2$; environmental differences within each treatment, $T_1$ and $T_2$; and environmental differences not associated with the treatments, $E_1$ and $E_2$. The two phenotypes are correlated through a genetic and, in theory at least, through an environmental path. The two genotypes, $G_1$ and $G_2$, are connected by a correlation, $r_G$. This genetic correlation, together with the path coefficients, $h_1$ and $h_2$ relating genotype to phenotype (which are the square roots of the two heritabilities) are all that is needed for the solution of the problem. But the completion of the diagram, showing all the sources of variation influencing the two phenotypes, makes for greater clarity. Variations within one treatment are not correlated with variations within the other, so no correlation path connects $T_1$ with $T_2$. But environmental differences not associated with the treatment are correlated, and $E_1$ is connected with $E_2$ by a correlation, $r_E$. To illustrate this point one may think of the treatment as being the plane of nutrition on which the animals are reared from weaning to maturity. Then the environmental correlation will operate through such factors as temperature and pre-weaning maternal influences.

In situations of the sort under discussion the two phenotypes will not be measurable in one and the same individual, and the environmental and phenotypic correlations will therefore have no reality; but we may regard them as being at least potentially existent in every individual. The genetic correlation is, of course, actually present in every individual since all individuals possess genes corresponding with both genotypes, even though only one genotype can find expression in a phenotype in any one individual. The complete path diagram is now exactly like one illustrating a simple case of correlated characters, except that the environmental source of variation has been split into two, one part associated with the treatment and the other not, and that the environmental and phenotypic correlations have no real existence. It is necessary to make the assumptions discussed by Lerner (1950, p. 83), that environmental variations, whether associated with the treatment or not, are uncorrelated with variations of genotype; and that the environmental variations within each treatment are small in comparison with the difference between the treatments, for if this were not so the interaction between genotype and environment which operates between the treatments would become important also within the treatments. In other words, it is assumed that the complete determination of the phenotype is specified by the formula $\sigma_P^2 = \sigma_G^2 + \sigma_E^2 + \sigma_T^2$.

Now suppose a population to be divided into two strains, and selection to be made for the phenotype $P_1$ in one strain and for $P_2$ in the other, the technique and intensity of selection being the same in the two strains. Then after some generations of selection, some animals of each strain are subjected to the treatment appropriate to the other strain, so that both phenotypes are measured in both strains. Suppose, further, that we are interested in the character $P_1$, which may be thought of as growth rate on a low plane of nutrition. The problem is, then, to find the amount of improvement in $P_1$ obtained as a correlated response to selection for $P_2$, and to compare this with the improvement of $P_1$ obtained as a direct response to selection for $P_1$ itself. By an application of the rules of path coefficients, the magnitude of the direct response may be written

(1) $$\Delta G_1 = \bar{i} h_1 \sigma_{G_1}$$

(see Hazel, 1934); and the magnitude of the correlated response may be written

(2) $$\Delta' G_1 = \bar{i} h_2 r_G \sigma_{G_1}$$

(This formula may also be readily derived from the formula given by Lerner (1950, p. 236) by rearrangement and substitution for the direct response,

$\Delta G_2$, as in (1) above.) The value of $\bar{\imath}$, the selection differential in standard measure, will be the same in the two cases since the manner of selection is assumed to be identical in the two strains. The ratio of the correlated to the direct response therefore becomes

$$(3) \quad \frac{\text{correlated response}}{\text{direct response}} = \frac{\Delta' G_1}{\Delta G_1} = \frac{h_2}{h_1} r_G$$

Thus we can evaluate the correlated response relative to the direct response simply in terms of the two heritabilities and the genetic correlation; and we can see clearly what conditions must be fulfilled if the correlated response is to be greater than the direct response. Some special terms are needed for clarity. Let the "primary environment" be the conditions under which the improved breed is destined to live, e.g. low plane of nutrition. The "desired character" is then the phenotype, e.g. growthrate, measured in the primary environment. Let the "secondary environment" designate the conditions under which selection is made, but under which the improved breed is not required subsequently to live, e.g. high plane of nutrition. The "direct response" is then the improvement in the desired character attained by selection in the primary environment, and the "correlated response" is the improvement in the desired character attained by selection in the secondary environment. The quantitative solution of the problem then leads to the following conclusions.

(1) An advantage of selection in the secondary environment would accrue only through an increase of heritability; but the increase of heritability would have to be great enough to offset the loss of efficiency through selection being made for a character that has not exactly the same genetic basis as the desired character. In fact the product, $h_2 r_G$, must be greater than $h_1$ where h is the square root of the heritability, and 2 refers to the secondary environment. (2) No general statement about the relative merits of the two methods of selection can be made: each case must be treated individually. The full solution requires the estimation of three genetic parameters, the two heritabilities and the genetic correlation. The determination of the genetic correlation could be made by methods similar to those used by Hazel (1943) and Dempster and Lerner (1947), but it might present practical difficulties. (3) Though each case requires its own solution, it seems probable that there will be few cases when direct selection will not yield the better result. If the two environments are very dissimilar, e.g. temperate and tropical climates, the genetic correlation will be low, and the difference in heritability would have to be very great for the correlated response to be greater than the direct. The expectation of a much higher heritability in the secondary environment would be the only justification for favoring selection in the secondary environment.

The common situation in which the experimenter or breeder seeks to improve heritability by the control of the environmental variation will be seen to be a special case of the general problem. If the nature of the environment is not altered, but merely its variability reduced, the genetic correlation will be unity, and the rate of progress becomes proportional to the square root of the heritability, as is already well known. The insertion of the genetic correlation into the formula enables one to allow for any effect that the control of the environment might have on the genetic basis of the character measured.

An experiment with mice has recently been described (Falconer and Latyszewski, in press) in which two strains were selected for body-weight, one on a high plane of nutrition and the other on a low. After some generations of selection the performance of each strain on the other diet was measured. It was found that neither of the two correlated responses was as great as the direct response; but when compared with the corresponding direct response, the correlated response following selection on the low plane was much greater than the correlated response following selection on the high plane. It was found, further, that heritability was higher on the low plane than on the high. The causal connection between these last two findings was not realized when the experiment was described, but it becomes quite clear when the problem is treated as one of genetic correlation and correlated responses. Neither the heritability nor the responses were very exactly determined, but they may serve for illustrating the method of treatment.

The observed heritabilities were 0.2 on the high plane and 0.3 on the low, and the ratio of the correlated to the direct response for weight on high plane was about 0.8. Therefore, substituting in equation (3), where 2 refers to low plane and 1 to high plane,

$$0.8 = \frac{\sqrt{0.3}}{\sqrt{0.2}} r_G,$$

whence $r_G = 0.65$.

The other correlated response, that is the change of weight on low plane following selection on high plane, was thought to be zero. This can now, however, be seen to be impossible. Taking the value found for $r_G$ and substituting again in equation (3), where 2 now refers to high plane and 1 to low, we find the expected ratio of the correlated to the direct response of weight on low plane to be about 0.5. This correlated response was very inexactly determined, and the observations were actually not inconsistent with this expected value.

An experiment designed to provide reliable estimates of the two heritabilities and of the two direct and correlated responses, would yield two separate estimates of the genetic correlation, and would therefore provide a check on the validity of the treatment of the problem outlined in this paper. Such an experiment is now being made with mice. The first stages of a similar experiment with pigs, in which selection for growth rate is being made on high and low planes of nutrition, are reported by Brugman (1950). The outcome of this experiment will be awaited with interest.

## SUMMARY

Situations involving an interaction between genotype and environment may be treated by the methods of genetic correlation, if only two different environments are considered. Formulation of the genotype-environment interaction in terms of a genetic correlation leads easily to a solution of problems connected with selection. In this way a precise answer can be given to the question whether it is better to carry out selection in the environment in which the improved breed is required eventually to live, or in some other environment more favorable to the expression of the desired character. Performance in the two environments is regarded as two different characters which are genetically correlated. Selection for one character will then bring about a correlated response of the other character. The magnitude of this correlated response may then be compared with that of the direct response to selection for the desired character itself. The ratio of the correlated to the direct response may be expressed in a simple formula involving the square roots of the two heritabilities and the genetic correlation. It is possible for the correlated response to be greater than the direct response, but it seems probable that this will seldom happen. The expectation of a great increase of heritability would be the only justification for favoring selection in an environment other than the one in which the improved breed is required to live.

## ACKNOWLEDGEMENTS

The writer gratefully acknowledges the benefit of discussions with Drs. E. C. R. Reeve, A. Robertson and F. W. Robertson.

## LITERATURE CITED

Brugman, H. H., 1950, The effect of the plane of nutrition on the carcass quality of a line of swine based on a Chester White and Danish Landrace cross. J. Animal Sci. 9: 602–607.

Dempster, E. R., and I. M. Lerner, 1947, The optimum structure of breeding flocks. II. Methods of determination. Genetics 32: 567–579.

Falconer, D.'S. and M. Latyszewski (in press), The environment in relation to selection for size in mice. J. Genet.

Haldane, J. B. S., 1946, The interaction of nature and nurture. Ann. Eugen. 13: 197–205.

Hammond, J., 1947, Animal breeding in relation to nutrition and environmental conditions. Biol. Rev. 22: 195–213.

Hazel, L. N., 1943, The genetic basis for constructing selection indexes. Genetics 28: 476–490.

Lerner, I. M., 1950, Population genetics and animal improvement. Cambridge.

Reeve, E. C. R. (in press), Expression of genes affecting a quantitative character in two environments. (Abstr.) Heredity.

Reeve, E. C. R. and F. W. Robertson (in press)

# 8

Reprinted from *J. Agric. Res.* **69**:459–476 (1944)

## EFFECTIVENESS OF SELECTION ON PROGENY PERFORMANCE AS A SUPPLEMENT TO EARLIER CULLING IN LIVESTOCK [1]

By G. E. Dickerson, *associate geneticist, Regional Swine Breeding Laboratory,* and L. N. Hazel, *associate animal husbandman, Western Sheep Breeding Laboratory, Bureau of Animal Industry, Agricultural Research Administration, United States Department of Agriculture* [2]

### INTRODUCTION

The extensive literature on the use of the progeny test in selecting breeding animals deals almost exclusively with its accuracy, as compared with that obtained from the use of pedigree, individual performance, or averages of collateral relatives, as an indicator of transmitting ability. Although several investigators, particularly Wright (*15*)[3] and Lush (*3, 4, 5*), have emphasized difficulties in the practical use of the progeny test, its accuracy under properly controlled conditions is unquestioned. However, from the standpoint of genetic progress expected from selection in a given period of time, the usefulness of the progeny test is greatly influenced by factors other than its relative accuracy. The most important of these are the age at which progeny tests can be obtained and the rate of reproduction. The longer interval between generations that results from use of the progeny test in selection tends to offset the advantage of more accurate selection and may actually reduce the rate of improvement obtained.

The purpose of this study is to examine the effectiveness of selection based on the progeny test when it is used to supplement earlier selection. The criterion of effectiveness is the average genetic improvement expected yearly from early selection alone as compared with that expected when use is made of the progeny test. The examples have been chosen to include economic traits in farm livestock for which the basis of earlier culling is restricted to individual performance, pedigree, or average performance of collateral relatives.

### ANNUAL IMPROVEMENT EXPECTED FROM SELECTION IN CLOSED POPULATIONS

The two factors that determine annual improvement from selection in any closed population are (1) the average genetic superiority of those animals selected to become parents over the group from which they were chosen ($\Delta P$) and (2) the average age of parents when their offspring are born or the average interval between generations ($T$). These averages are weighted according to the proportion of offspring from parents of different sex and age groups. Since $\Delta P$ represents the average genetic gain in $T$ years, the average annual gain is $\Delta G = \dfrac{\Delta P}{T}$.

---

[1] Received for publication April 28, 1943.
[2] The authors are indebted to Prof. W. G. Cochran of Iowa State College for suggestions on statistical procedure.
[3] Italic numbers in parentheses refer to Literature Cited, p. 475.

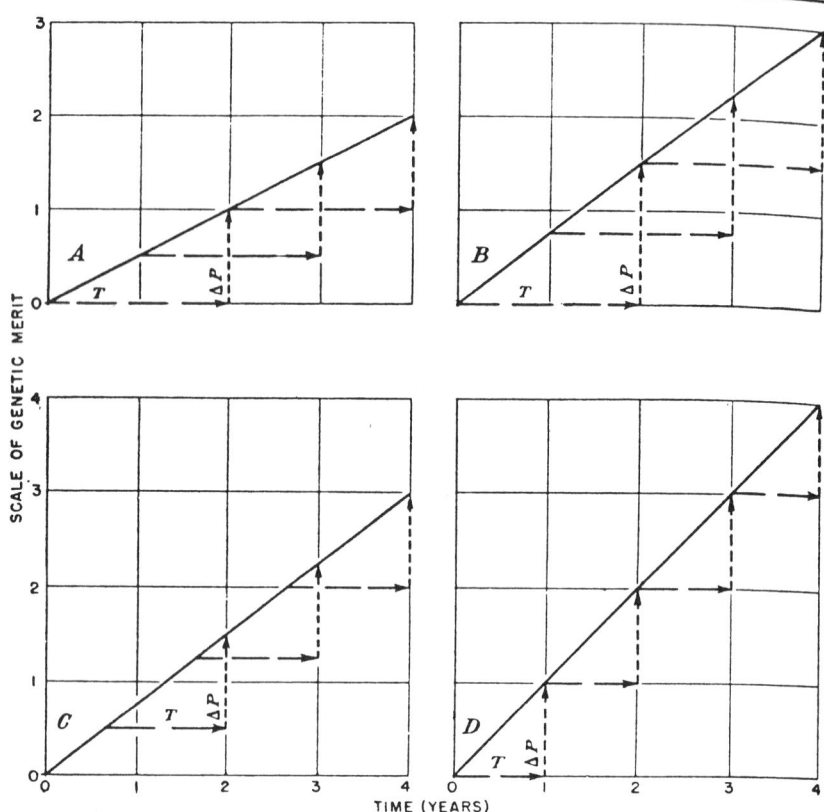

FIGURE 1.—Effect of interval between generations ($T$) and average genetic superiority of parents ($\triangle P$) on annual progress from selection ($\triangle G$). (Annual progress=$\triangle G=\frac{\triangle P}{T}$). $A$, $\frac{\triangle P}{T}=\frac{1}{2}=0.50$; $B$, $\frac{\triangle P}{T}=\frac{1.5}{2}=0.75$; $C$, $\frac{\triangle P}{T}=\frac{1}{1.33}=0.75$; $D$, $\frac{\triangle P}{T}=\frac{1}{1}=1.0$.

The most effective plan of making selections is the one that produces most improvement per unit of time. One plan may be more effective than another because it (1) increases $\triangle P$ and/or decreases $T$, (2) increases $\triangle P$ relatively more than $T$, or (3) decreases $\triangle P$ relatively less than $T$. The effect of changing $\triangle P$ and $T$ is illustrated in figure 1.

The parents of the animals born in any one year differ in age and in the intensity of the selection applied to the different age groups and sexes. When the different parents within each age and sex group have equal opportunity to produce offspring, the average yearly genetic progress expected in two successive cullings of sires and of dams is [4]

$$\triangle G = \frac{\overbrace{N_1 \triangle S_1 + N_2(\triangle S_1 + \triangle S_2)}^{\triangle P \text{ for sires}} + \overbrace{M_1 \triangle D_1 + M_2(\triangle D_1 + \triangle D_2)}^{\triangle P \text{ for dams}}}{\underbrace{N_1 Y_1 + N_2 Y_2}_{T \text{ for sires}} + \underbrace{M_1 Z_1 + M_2 Z_2}_{T \text{ for dams}}} \qquad (1)$$

[4] This formula is rigorously proved algebraically from the fact that the average breeding value of an unselected group of offspring tends to be the same as that of the parents and may be extended to any number of successive cullings. The assumption is made that the average difference between offspring born in successive years ($\triangle G$) is constant, as would be expected for polygenic traits in a closed population where a regular breeding plan was in use.

in which

$\Delta S_1$ = average genetic superiority of young sires retained in the first selection

$N_1$ = proportion of the offspring that are from young sires

$Y_1$ = average age of young sires when their offspring are born

$\Delta S_2$ = additional genetic superiority of sires obtained from the second culling of sires retained in the first selection

$N_2$ = proportion of the offspring produced by sires retained in the second selection

$Y_2$ = average age of sires saved in the second culling when their subsequent offspring are born

$\Delta D_1$, $M_1$, $Z_1$, $\Delta D_2$, $M_2$, and $Z_2$ have corresponding meanings for dams retained in the first and second cullings.

The general principles that govern progress from selection are the same for a whole breed as for a single closed herd. It is impractical to consider here the many forms that herd differences (genetic and environmental) may take, and they are ignored in the formulas for calculating $\Delta S_1$, $\Delta S_2$, etc. This procedure favors the progeny test, since the use of progeny averages helps to minimize errors in selection from random environmental variation within a herd but does not lessen those from environmental differences between herds unless a sire has progeny in more than one herd.

### GENETIC SUPERIORITY FROM FIRST CULLING ($\Delta S_1$ and $\Delta D_1$)

In estimating the selection differential or average gain in apparent merit of a selected group as compared with the group from which it was chosen ($\bar{\imath}$), it is assumed that the basis of selection ($I$) is normally distributed and that all individuals below a given level are culled. The expected size of the differential or apparent gain in either sex depends on the proportion saved ($p$), as illustrated in figure 2. In

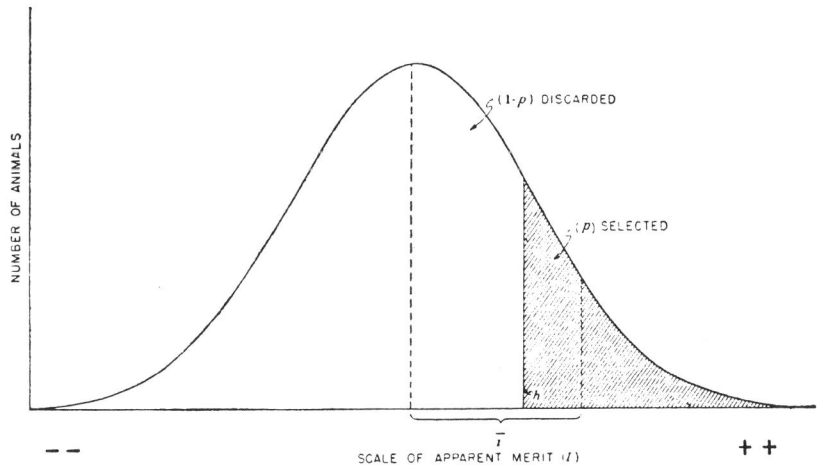

FIGURE 2.—A normal distribution showing how a population may be sharply divided at a point ($h$) into a selected ($p$) and a discarded (1-$p$) fraction. The average superiority in apparent merit of the selected fraction is ($\bar{\imath}$).

livestock breeding, $p$ (and consequently $\bar{i}$) is largely determined by such factors as rate of reproduction, longevity, and age at puberty, peculiar to each type of livestock. Values of $\bar{i}$ in standard deviation units for different values of $p$ are given for normally distributed populations of infinite size by Pearson (*10*) and may be calculated from Fisher and Yates (*1*) for smaller populations (from 2 to 50).

Since the selected group is chosen because of its superiority for some trait ($X$) or index ($I_1$) that is never perfectly correlated with transmitting ability ($G$), the average genetic superiority expected from the first selection is

$$\Delta S_1 \text{ or } \Delta D_1 = (\bar{i_1}) b_{GI_1} \sigma_{I_1} = (\bar{i_1}) r_{GI_1} \sigma_G. \tag{2}$$

Here $\bar{i_1}$ represents the selection differential in standard deviation units, $b_{GI_1}$ the regression of transmitting ability on apparent merit, $r_{GI_1}$ the corresponding correlation, and $\sigma_G$ the standard deviation of transmitting abilities. It is convenient to calculate $r_{GI_1}$ and $\sigma_G$ in terms of the hereditary and environmental portions of the observed variance of the population; for example, when the first selection is made on some phenotypic trait ($X$),

$$r_{GX} = \sqrt{\frac{G}{E+G}} = \sqrt{G}, \text{ and } \sigma_G = \sigma_X \sqrt{G},$$

where $G$ is the heritability or fraction of the observed variance caused by individual differences in transmitting ability $\left(\frac{\sigma_G^2}{\sigma_X^2}\right)$ and $E$ is the remaining fraction attributed to environment, dominance, and epistasis or gene interaction $\left(\frac{\sigma_E^2}{\sigma_X^2}\right)$.

## ADDITIONAL GENETIC SUPERIORITY FROM SECOND CULLING ($\Delta S_2$ AND $\Delta D_2$)

All the culling possible may be done on the basis of the first information available, in which case $\Delta S_2$, $N_2$, and $Y_2$ (or $\Delta D_2$, $M_2$, and $Z_2$) become zero in formula (1). If the number of animals retained in the first culling permits a second culling after additional information ($O$), such as the progeny test, becomes available, the maximum additional genetic superiority from the second culling is

$$\Delta S_2 \text{ or } \Delta D_2 = (\bar{i_2}) R'_{G \cdot I_1 O} \sigma'_G. \tag{3}$$

Distinction is made between the multiple correlation ($R'_{G \cdot I_1 O}$) and standard deviation of transmitting abilities ($\sigma'_G$) among animals retained in the first culling, as compared with those in an unselected group ($R_{G \cdot I_1 O}$ and $\sigma_G$), because it is mathematically convenient to calculate the former in terms of the latter.[5]

The selection differential for the second culling ($\bar{i}_2$) can be calculated from table 20 of Fisher and Yates (1) since it is not affected materially by the slight skewness expected in the distribution of $I_2$.[6]

## EFFECTIVENESS OF SELECTION FOR IMPORTANT TRAITS OF FARM ANIMALS

A regular plan of progeny testing may be effective in increasing rate of progress in one kind of animal but not in another even for similar traits, or for one kind of trait but not for another in the same population. The effectiveness depends largely on the age of parents when progeny-test information becomes available, but to some extent also on the rate of reproduction and the relative accuracy of information used for the first culling. Obviously these factors differ for particular kinds of animals and traits.

Several examples that illustrate the influence of these factors on the effectiveness of the progeny test have been chosen. These examples were selected because of their economic importance and because of the extensive breeding research that is being directed toward their improvement. They illustrate the effectiveness of the progeny test when used in conjunction with earlier selection based on pedigree, on individual performances, and on performance of collateral relatives, each of these plans being peculiarly fitted to making early selections for a different kind of trait.

### TRAITS MEASURED IN BOTH SEXES BEFORE BREEDING AGE

Many important traits, such as growth rate, economy of feed utilization, market conformation, fleece weight, and fleece length, can be

---

[5] The standard deviation of transmitting abilities among the group saved in the first culling is

$$\sigma'_G = \sigma_G \sqrt{1 - r_{GI}^2 (1 - \sigma_s^2)}$$

where $\sigma_s^2$ is the fraction of the original variance of $I_1$ that remains in the selected group. Values of $\sigma_s^2$ (hereafter designated as $\sigma_{ss}^2$ for sires and $\sigma_{sd}^2$ for dams) may be calculated for large populations from the formula $\sigma_s^2 = 1 - \bar{i}_1(\bar{i}_1 - h)$, suggested by Professor Cochran, where $h$ is the plus or minus deviation from the mean of the unselected population at the point of truncation of the normal curve (fig. 2).

The multiple correlation of $G$ with $O$ and $I_1$ among animals retained in the first culling is

$$R'_{G \cdot O I_1} = \sqrt{\frac{r_{GI_1}^{2'} + r_{GO}^{2'} - 2 r_{GI_1}^{'} r_{GO}^{'}}{1 - r_{GI_1}^{2'} r_{GO}^{2'}}}$$

The correlation between $G$ and $I_1$ among those selected in the first culling is

$$r'_{GI_1} = r_{GI_1} \sqrt{\frac{\sigma_s^2}{1 - r_{GI_1}^2 (1 - \sigma_s^2)}}$$

whereas that between $G$ and $O$ is

$$r'_{G_s O_s} = r_{G_s O_s} \sqrt{\frac{1 - r_{GI_1}^2 (1 - \sigma_{ss}^2)}{1 - r_{GI_1}^2 r_{G_s O_s}^2 (1 - \sigma_{ss}^2) - r_{GI_1}^2 r_{G_d O_s}^2 (1 - \sigma_{sd}^2)}}$$

for sires, and

$$r'_{G_d O_d} = r_{G_d O_d} \sqrt{\frac{1 - r_{GI_1}^2 (1 - \sigma_{sd}^2)}{1 - r_{GI_1}^2 r_{G_s O_d}^2 (1 - \sigma_{ss}^2) - r_{GI_1}^2 r_{G_d O_d}^2 (1 - \sigma_{sd}^2)}}$$

for dams.

[6] The exact selection differential expected in the second culling ($\bar{i}_2$) of a population of infinite size was calculated by a method, suggested by Professor Cochran, for varying proportions retained in the first ($p_1$) and second ($p_2$) culling and for different degrees of correlation between $I_1$ and $I_2$ in the unselected population. Even when $r_{I_1 I_2}$ is as large as 0.8, the exact value of $\bar{i}_2$ expected does not differ appreciably, because of skewness, from that expected for a normal distribution unless $p_2$ is much larger or smaller than 0.5. For example, when $p_1 = 0.2$, $p_2 = 0.1$, and $r_{I_1 I_2} = 0.8$, the exact expectancy for $\bar{i}_2$ is only 3 percent higher than for a normal distribution. In the examples that follow, $p_2$ is never larger than 0.5 nor smaller than 0.1 and $r_{I_1 I_2}$ does not exceed about 0.3.

measured on both males and females before they reach breeding age. The annual progress expected from selection based on individual performance alone in swine and sheep as compared with that expected from the supplementary use of the progeny test is indicated in the examples that follow.

When first selections are based on individual performance $(X)$ alone, the genetic superiority expected from the first selection is

$$\Delta S_1 \text{ or } \Delta D_1 = (\bar{i}_1) G \sigma_X. \qquad (4)$$

When second selections are based on the optimum combination of individual performance and progeny test (formula 3), the additional genetic superiority expected is

$$\Delta S_2 = (\bar{i}_2) G \sigma_X \sqrt{\sigma_{ss}^2 + \frac{nd(1-G)^2}{4(A+nB)+ndG(1-G)-nG^2(1-\sigma_{sd}^2)}} \qquad (5)$$

for sires, and

$$\Delta D_2 = (\bar{i}_2) G \sigma_X \sqrt{\sigma_{sd}^2 + \frac{n(1-G)^2}{4(A+nB)+nG(1-G)-nG^2(1-\sigma_{ss}^2)}} \qquad (6)$$

for dams, each of which produces one litter of $n$ progeny. The symbols and their interpretations in terms of the hereditary and environmental fractions of the variance are as follows:

$L$ = fraction of total variance due to differences in environment and in gene interaction that are alike for members of the same litter

$E$ = fraction that behaves as random environmental variation between litter mates

$A$ = fraction due to differences between litter mates = $E + G/2$

$B$ = fraction due to differences between paternal half-sibs, less $A = L + G/4$

$C$ = fraction due to differences between nonsibs, less $A + B = G/4$, so that $G + L + E = A + B + C = 1$

In addition,

$n$ = number of offspring per litter, and

$d$ = number of litters per sire.

Weight at 180 days may be used as an example for swine, in which $G = 0.30$, $L = 0.20$, $E = 0.50$, and $\sigma_x = 32$ pounds, in accord with values found by Whatley (14) and Hazel (2). We shall consider a 20-sow herd in which breeding stock are saved from the spring farrow only and $n = 5$. The age when offspring are born is 1 year for young boars and gilts and 2 years for the tested boars and sows. The selection differentials for formulas (4), (5), and (6) are calculated from table 20 of Fisher and Yates (1). For example, if 3 young sires are saved annually, $p = 3/50$, so that, for sires, $(\bar{i}_1) = 1.91$. When the best 1 of the 3 is kept after testing, $p = 1/3$ and $(\bar{i}_2) = 0.85$. The curves for annual progress in figure 3 were constructed from formula 1. Solid lines show progress expected from the use of 2, 3, and 4 young sires and 1 boar tested the year before, when no tested sows are used $(M_2 = 0)$, but the proportion of the litters by the tested boar $(N_2)$ varies from 0 to 0.8. Broken lines show progress expected as the proportion of litters farrowed by 2-year-old sows ($M_2$) varies from 0 to 0.5, when all litters are sired by 2 young boars ($N_2=0$).

Figure 3, $A$, (solid lines) shows that $\Delta G$ is maximum when 2 young sires are used each year on all sows. The use of 3 or 4 young sires decreases progress, as does increasing the proportion of offspring by tested sires. The use of 1 young sire instead of 2 would not increase $\Delta G$ since all offspring would then be paternal half-sibs, and the reduced genetic variability and heritability would more than cancel the slightly larger selection differential. Even a plan of testing the optimum number of 6 to 8 sires in an auxiliary herd of 20 sows and using the 2 best ones in the closed herd does not increase $\Delta G$, as shown at the extreme right of figure 3, $A$. This indicates that the progeny test for sires is not effective for this or similar traits in swine, regardless of herd size.

Figure 3, $A$, (broken lines), also shows that $\Delta G$ is about the same when the optimum proportion of the litters (10 to 20 percent) is from

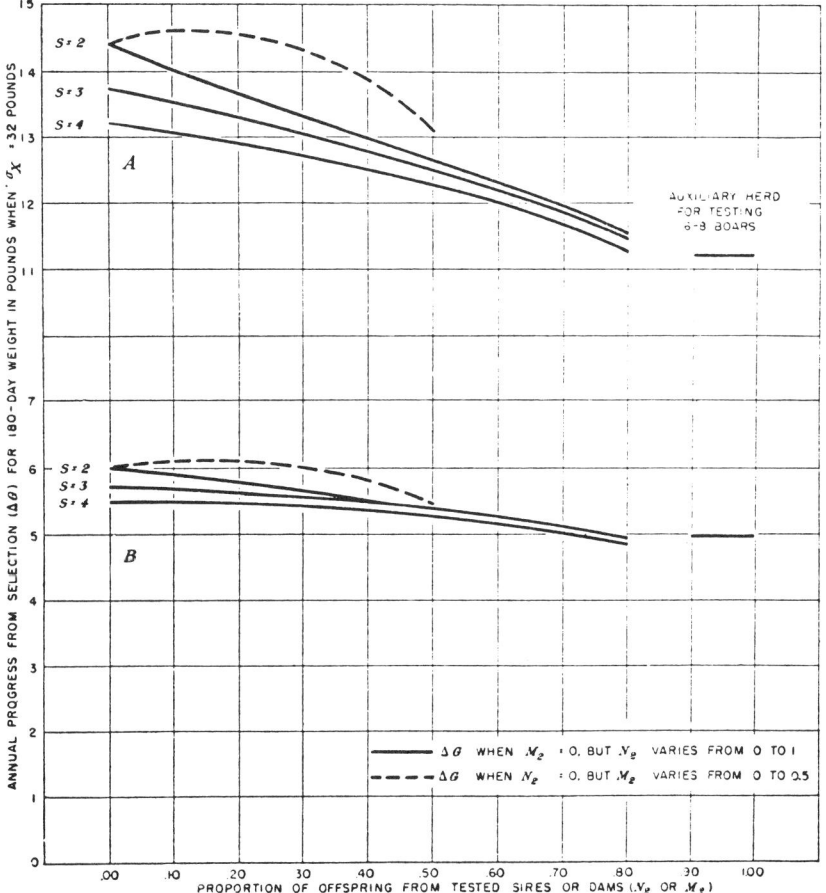

FIGURE 3.—Effect of progeny testing on genetic progress from selection for 180-day weight in a closed 20-sow herd of swine, when early culling is based on individual weights. $A$, When $G=0.30$, $L=0.20$; $B$, when $G=0.125$, $L=0.25$. $s$=number of sires tested each year.

tested sows as when only gilts are used. As $M_2$ is increased, the first selection of gilts becomes more effective, the second selection of sows becomes less effective, and the average age of dams increases. Up to $M_2=0.2$ the opposing influences nearly cancel, but as $M_2$ is increased further, progress declines.

It is apparent from formulas (5) and (6) that progeny testing is more likely to increase progress for traits of lower heritability ($G$ smaller, $E$ larger) that are unaffected by litter environment ($L=0$, $B=G/4$). As a population becomes more uniform genetically owing to inbreeding ($f$), heritability declines (that is, $G \cong \frac{G_o(1-f)}{1-G_o f}$). For example, as $f$ rises from 0 to 0.67, the heritability ($G$) of growth rate in swine would change roughly from 0.30 to 0.125, $L$ from 0.20 to 0.25, and $E$ from 0.50 to 0.625. Figure 3, $B$, shows that progeny testing of boars or sows does not increase $\Delta G$ even at this lower level of heritability, although the reduction in $\Delta G$ from the use of a tested boar is less marked. $G$ and $L$ appear to be a little smaller for conformation score at market weight, as shown by Stonaker and Lush (*12*), than for growth rate of swine, but not enough so to make use of progeny-tested sires advantageous.

The influence that rate of reproduction has on the effectiveness of the progeny test may be illustrated by comparing selection for body weight or fleece length in yearling sheep with that for growth rate in swine. The influence of the time required to obtain progeny tests of sires is shown by comparing selection for yearling traits with that for weanling traits in sheep. Table 1 shows the age distribution, fertility, and average age of dams expected in a flock of 100 ewes if all voluntary culling of females were done before breeding age. In the first selection the best 44 percent of the 50 ewe lambs or yearling ewes and the best 2, 3, or 4 of the 50 ram lambs or yearling rams are chosen. Formulas (4) and (5) may be used in calculating $\Delta D_1$, $\Delta S_1$, and $\Delta S_2$ ($\Delta D_2=0$, $n=1$, and $A+nB=1-G/4$).

TABLE 1.—*Age distribution, fertility, and average age of dams at lambing time in a flock of 100 ewes, when all voluntary culling is done before breeding age* [1]

| Age of females (years) | Ewes of each age per 100 breeding ewes | Lambs weaned per ewe | Fraction of all lambs weaned = $\frac{(2)\times(3)}{100}$ | $T$ for ewes $[(1)\times(4)]$ |
|---|---|---|---|---|
| (1) | (2) | (3) | (4) | (5) |
| | *Number* | *Number* | | *Years* |
| 0 | 22.0 | | | |
| 1 | 21.4 | | | |
| 2 | 20.9 | 0.62 | 0.129 | 0.259 |
| 3 | 19.7 | .93 | .184 | .552 |
| 4 | 18.3 | 1.22 | .222 | .893 |
| 5 | 16.4 | 1.17 | .192 | .960 |
| 6 | 13.8 | 1.12 | .153 | .924 |
| 7 | 10.9 | 1.10 | .120 | .842 |
| Total | 143.4 | | 1.000 | 4.430 |

[1] These data were taken from a report by Terrill (*13*) on the Rambouillet flock of the United States Sheep Experiment Station and Western Sheep Breeding Laboratory, Dubois, Idaho. They have been adjusted for voluntary culling after breeding age and for a 100-percent lamb crop at weaning age. The figures for ewe lambs and yearling ewes represent those necessary for replacements.

Figure 4, $A$, shows yearly progress for weanling (solid lines) and for yearling (broken lines) traits when heritability is 0.30 and rams are used first as yearlings. Use of the best ram tested the year before on the optimum proportion (0.6 to 0.7) of the ewes increases progress by about 4 percent for weanling traits but reduces progress for yearling traits, as compared with the use of only the 2 best yearling rams each year. The difference occurs solely because selection of rams on progeny performance can be made a year earlier for weanling than for yearling traits. As shown at the extreme right of the figure a still greater increase in progress (6 percent) could be obtained for weanling traits by testing the optimum number (7) of yearling rams each year in an auxiliary flock of 100 ewes and using the 2 best ones

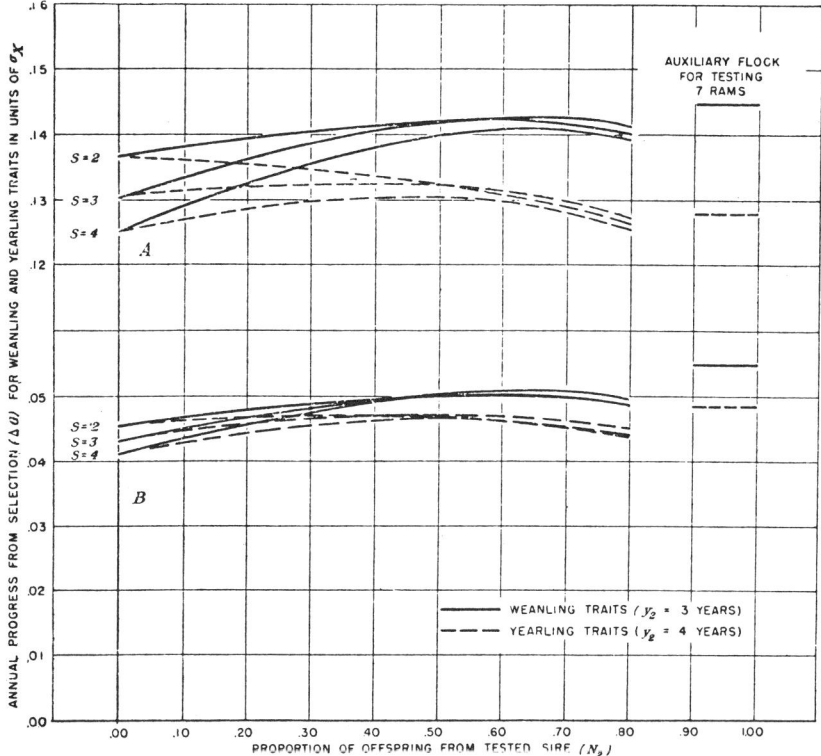

FIGURE 4.—Influence of progeny testing of rams on genetic progress from selection for weanling and yearling traits in a closed 100-ewe flock of sheep, when early culling is based on individual performance. $A$, When $G=0.30$, $L=0$; $B$, when $G=0.10$, $L=0$. $s$ = number of sires tested each year.

in the main flock the following year as 2-year-olds. This plan is not effective for yearling traits, however.

When heritability is only 0.10 (fig. 4, $B$) use of the best progeny-tested ram is expected to increase progress about 11 percent for weanling traits and 3 percent for yearling traits. By using the auxiliary flock for testing rams, the increase in progress would be raised to 20 percent for weanling traits and 5 percent for yearling traits.

Progress for weanling traits could be increased still further by testing each of the best ram lambs on a small number of ewes and then using the best tested rams as yearlings instead of as 2-year-olds.

For example, suppose the 4 best ram lambs were tested on 40 ewes and the remaining 60 ewes were mated to the best yearling ram tested the year before. Under this plan, yearly progress expected would be 20 and 28 percent greater, for heritabilities of 0.3 and 0.1, respectively, than if the 2 best untested yearling rams were used on the entire flock, whereas the advantage of testing yearling rams and using the best tested 2-year-old is only 4 and 11 percent, respectively. The increase in progress from testing rams in the auxiliary flock of 100 ewes is also greater if the tested rams can be used in the main flock as yearlings instead of as 2-year-olds (22 and 37 percent, respectively, for heritabilities of 0.3 and 0.1, instead of 6 and 20 percent). These maximum estimates of the gain in rate of improvement from using progeny-tested rams are far below the 500 percent gain claimed by McMahon ($9$) for heritability at 0.10. The discrepancy appears to be due largely to McMahon's assumptions that (1) the average of 7 progeny is perfectly correlated with the sire's genotype and (2) the interval between generations would not be lengthened by using tested rams. Actually, the correlation of a ram's genotype with the average of 7 progeny would be little larger ($r_{GO}=0.39$) than with the ram's own phenotype ($r_{GX}=\sqrt{0.10}=0.32$), in an unselected population and for heritability at 0.10. For weanling traits, the generation interval could actually be shortened a little, compared with the use of untested yearling rams, by testing ram lambs on part of the flock and using tested rams as yearlings on the rest of the ewes. However, progeny tests for yearling traits on rams used as lambs or yearlings would be obtained only in time for use of selected tested rams as 2- or 3-year-olds, and use of tested rams would lengthen the generation interval by 6 months or 1 year compared with the use of untested yearling rams.

### TRAITS MEASURED ONLY AFTER SLAUGHTER

Information on collateral relatives may not be of sufficient importance for traits measured in both sexes before breeding age to be considered in making selections. For traits measurable only in the carcass, there may be no other basis for making early selections. In a closed herd of swine, for example, several pigs from each litter may be slaughtered at market weight, the information being used in selecting collateral relatives and for progeny-testing the previous group of young sires.

The genetic superiority for a carcass trait ($X$) of boars or gilts selected in the first culling based on the average of $n$ litter mates is

$$\Delta S_1 \text{ or } \Delta D_1 = (\bar{i}_1)\frac{G}{2}\sigma_X \sqrt{\frac{n}{A+n(B+C)}}. \qquad (7)$$

The additional genetic superiority from the second culling based on the optimum combination of the average of $nd$ progeny and of $n$ litter mates is

$$\Delta S_2 = (\bar{i}_2)\frac{G}{2}\sigma_X \cdot$$

$$\sqrt{\frac{n}{A+n(B+C)}\left\{\sigma_{ss}^2+\frac{d(A+nB)^2}{(A+nB+ndC)[A+n(B+C)]-n^2C^2(d+1-\sigma_{ss}^2)}\right\}} \qquad (8)$$

for sires, and

$$\Delta D_2 = (\bar{i}_2)\frac{G}{2}\sigma_X \sqrt{\frac{n}{A+n(B+C)}\left\{\sigma_{sd}^2 + \frac{(A+nB)^2}{[A+n(B+C)]^2 - nC^2(2-\sigma_{ss}^2)}\right\}} \quad (9)$$

for dams having only one litter. The symbols used are those defined for formulas (4) to (6).

The curves in figure 5 are for a closed 20-sow herd in which carcass traits are observed on 2 pigs from each litter and when the $s$ young boars and the gilts used for breeding each year are chosen from 40

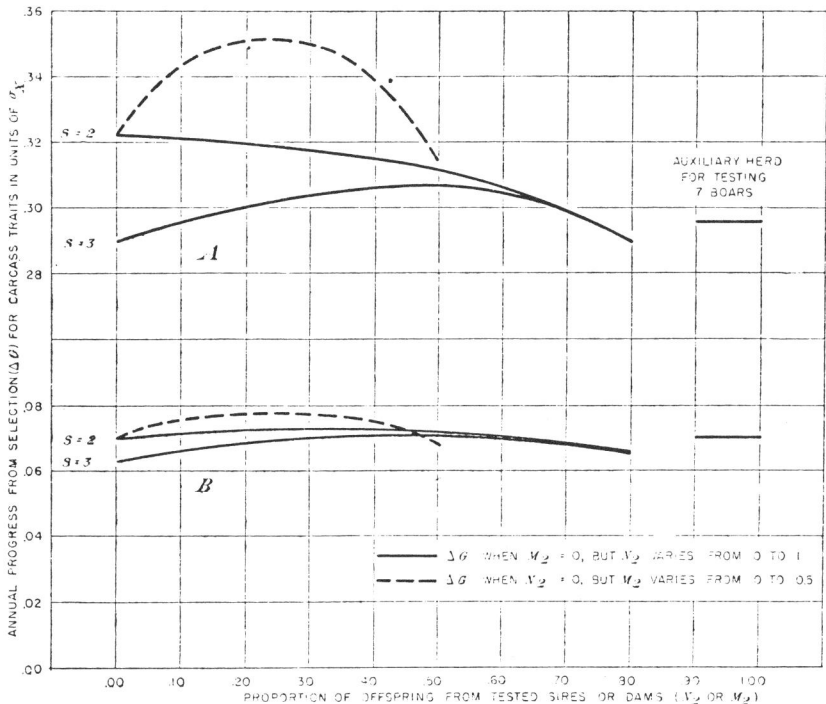

FIGURE 5.—Influence of progeny testing on genetic progress from selection for carcass traits in a closed 20-sow herd of swine, when the first culling is based on averages for 2 litter mates. $A$, When $G=0.50$, $L=0$; $B$, when $G=0.10$, $L=0$. $s=$ number of sires tested per year.

gilts in 20 litters and 10 boars in 10 litters. Use of the best sire tested the preceding year reduces progress expected when heritability is 0.50 (fig. 5, $A$), although progress is increased about 10 percent if optimum proportion of the litters (about 0.2) is from the best sows tested the year before. When heritability is as low as 0.10 (fig. 5, $B$), the progeny test is slightly effective. As shown at the extreme right of figure 5, progress is not increased by testing the optimum number (7) of boars in an auxiliary herd of 20 sows and using the best 2 in the main herd the next year.

Thus it seems unlikely that progeny testing can increase progress for carcass traits in swine appreciably if earlier culling of breeding animals can be based on the average performance of several litter mates. However, progeny tests are much more likely to be helpful in selecting for such traits in beef cattle or sheep, where the reproductive rate is lower and earlier culling must be based on the performance of half-sibs rather than full sibs.

### REPEATABLE TRAITS EXPRESSED ONLY IN THE FEMALE

Progeny testing of sires might be presumed to have special usefulness in selecting for such traits as butterfat production in dairy cattle or prolificacy in swine, since early culling must be based largely on pedigree. Because progeny tests have received so much attention both in the literature and in the practice of dairy-cattle improvement, butterfat production has been used in the example that follows.

The genetic superiority for butterfat production $(X)$ expected for bull or heifer calves selected in a first culling based on an average of $k''$ records for each of the dams is

$$\Delta S_1 \text{ or } \Delta D_1 = (\bar{i}_1)\frac{G}{2}\sigma_X\sqrt{\frac{k''}{E_r + k''[E_p + G]}}, \quad (10)$$

where the intraherd variance in butterfat production $(\sigma_X^2)$ is subdivided into the following fractions:

$G =$ heritability or fraction due to differences in transmitting ability

$E_p =$ fraction due to permanent differences in environment, to deviations from transmitting ability due to dominance and epistasis, and

$E_r =$ fraction due to random variation in environment between different records of the same cow, after adjustment for age.

The additional genetic superiority expected for sires selected in a second culling based on the best combination $(I_2)$ of the average production for $d$ daughters with $k$ records each and the average of $k''$ records of each sire's dam is (from formula 3)

$$\Delta S_2 = \frac{(\bar{i}_2) G \sigma_X}{2\sqrt{E_r + k''(E_p + G)}} \cdot$$

$$\sqrt{k''\sigma_{sd}^2 + \frac{kd[4(E_r + k''E_p) + 3k''G]^2}{4[E_r + k''(E_p + G)][4(E_r + kE_p) + kG(d+3)] - kk''G^2(d+1-\sigma^2_{sd})}}. \quad (11)$$

The additional genetic superiority expected for dams selected in a second culling based on the best combination of each cow's own average for $k'$ records and her dam's average for $k''$ records is

$$\Delta D_2 = \frac{(\bar{i}_2) G \sigma_X}{2\sqrt{E_r + k''(E_p + G)}} \cdot$$

$$\sqrt{k''\sigma_{sd}^2 + \frac{k'[4(E_r + k''E_p) + 3k''G]^2}{4[E_r + k''(E_p + G)][E_r + k'(E_p + G)] - k'k''G^2}}. \quad (12)$$

The results to be expected from progeny testing of dairy sires are shown in figure 6 for a closed herd of 120 cows. Heritability $(G)$ is 0.25 in figure 6, $A$, and 0.10 in $B$, but repeatability $(G + E_p)$ is 0.35 in both. These are roughly the upper and lower limits of heritability indicated in such studies as those of Lush, Norton, and Arnold (7) and Lush and Straus (8). In this example, one-fourth of the cows are replaced each year; only calves from three-fourths of the cows that have completed one or more records are considered in selecting breeding animals; and 90 percent of the cows raise calves each year. Three-fourths of the heifer calves (30/40) are saved in the first culling and kept for two lactations, after which one-half (15/30) are retained in the second culling for an average of 4 more lactations. Thus the average age of dams when the calves from which breeding stock are chosen (second record and later) are born is about $4\frac{2}{3}$ years; that is, $\frac{3 \text{ years}}{3} + \frac{2}{3}$ (5.5 years). The average number of records per dam is $k''=2\frac{2}{3}$. Also, young sires selected for progeny testing and used for 1 year when from 15 to 27 months of age are about $2\frac{1}{2}$ years old when their calves are born (that is, $Y=2\frac{1}{2}$ years). When the 1 best sire tested over each 2-year period (on the basis of $d$ daughters with $k=1$ record each) is used again on part of the herd for a 2-year period, his average age when his second group of calves is born will be about 8 years (that is, $Y_2=8$ years). Under these conditions, the annual progress expected is

$$\Delta G_t = \frac{N_1(\Delta S_1) + N_2(\Delta S_1 + \Delta S_2) + \frac{1}{3}(\Delta D_1) + \frac{2}{3}(\Delta D_1 + \Delta D_2)}{2.5N_1 + 8N_2 + 4\frac{2}{3}}.$$

Figure 6 shows that the progeny test is not effective under the conditions stated, $\Delta G_t$ decreasing as $N_2$ increases.

The total annual progress ($\Delta G_t$) is subdivided in figure 6 into that from the two successive cullings among females ($\Delta G_d$) and that from the two cullings among males ($\Delta G_s$). As Seath (11) has shown, much of the culling among females is for disease, breeding failure, and similar causes, so that $\Delta G_d$ for butterfat production actually may be considerably less than indicated in figure 6. This would make the curves for $\Delta G_t$ more nearly like those for $\Delta G_s$, which show the progeny test in a more favorable light. If no progress is made from selection of dams ($\Delta G_d=0$), progress from sire selection ($\Delta G_s$) is optimum when one-third to one-half of the cows are bred to the best 1 of 4 sires tested during the 2 preceding years, and $\Delta G_s$ is increased about twice as much (10 percent as compared with 5 percent) by progeny testing when heritability is 0.10 as when heritability is 0.25.

Use of progeny-tested dairy sires would be a little more likely to increase the rate of improvement if, instead of the average production of the daughters alone being used, that of the dams to which each sire was mated were also considered properly. This procedure, according to Lush (6), would make progeny tests in a population of many herds about 1.12 to 1.20 times as accurate as the use of the average production of the daughters alone. The exact amount depends largely on the correlation between the average production of the sire's daughters and that of their dams, and would, therefore, be less for comparisons between sires tested in the same herd and during the same years. In the above example (when selection of dams is presumed ineffective), increasing the accuracy of progeny tests 1.10 times would change the percentage increase in yearly improvement resulting from optimum use of tested sires only, from 10 to 13 if heritability is 0.10 and from 5 to 8 if heritability is 0.25.

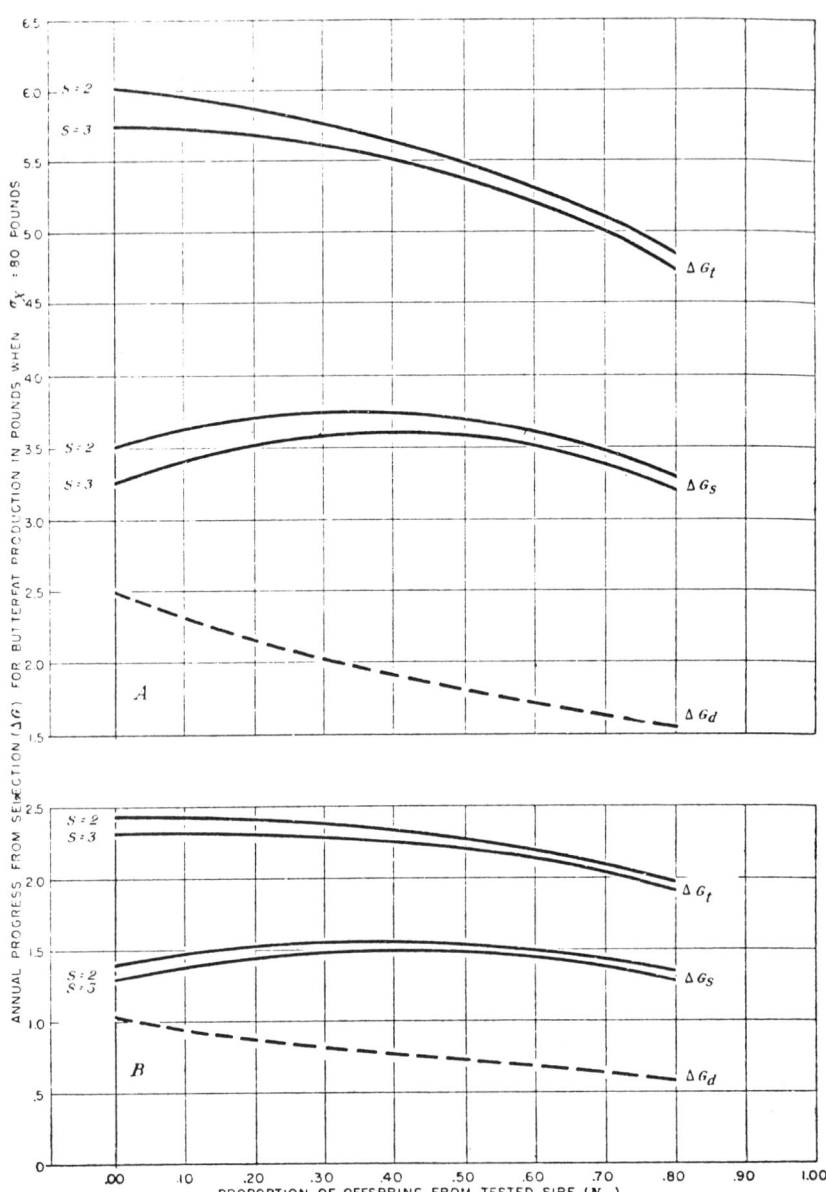

FIGURE 6.—Effect of using progeny-tested sires on genetic progress from selection for butterfat production in a closed 120-cow herd, when early culling i based on the dam's average production alone. A, When $G=0.25$, $E_p=0.10$; B, when $G=0.10$, $E_p=0.25$. $s=$ number of sires tested each year.

## DISCUSSION

The foregoing examples indicate that the possibilities of increasing progress by a regular plan for use of progeny-tested sires are limited to certain kinds of livestock and to certain traits. The reasons for this limitation may be illustrated by comparing the effectiveness of progeny testing in the different examples. First, the less the interval between generations is increased by progeny testing the more likely it is that progeny testing will increase progress. This is shown by contrasting the results of selecting for weanling and yearling traits in sheep (fig. 4). The only difference in these examples is that 1 year is required to obtain progeny-test information on weanling traits, whereas 2 years are required for yearling traits. Second, when the rate of reproduction is low, progeny testing of sires is more likely to increase progress. The resulting increase in genetic superiority of parents ($\Delta P$) tends to be larger, relative to that in the age of parents ($T$), when there is less opportunity for early culling, particularly among females. This is the reason that progeny testing affects progress more favorably for yearling traits in sheep (fig. 4) than for growth rate in swine (fig. 3), and when little voluntary culling of females is possible for butterfat production in dairy cattle ($\Delta G_s$ in fig. 6). Third, if the basis for making first selections is relatively inaccurate, the progeny test is more likely to be effective, because there is more environmental variation to be discounted by the progeny test and more of the genetic variation remains among animals tested. This is illustrated by the contrast in the effectiveness of the progeny test for high and low heritability (figs. 3, 4, 5, and 6).

Thus a combination of circumstances, largely beyond the breeder's control, operates to make the use of a regular plan of progeny testing a wise or unwise procedure. In many cases when the progeny test is most easily applied it may actually reduce genetic progress (fig. 3). Even when the circumstances indicate the use of progeny-tested sires, there is danger that its full effectiveness will not be realized in practice because of unwise judgment. Too many or too few young sires may be tested on too many or too few females, so that the optimum use of young and tested sires is not attained (fig. 4). Although little attention has been given to this point, it becomes important once a breeder decides to use a regular plan of progeny testing.

The progeny test is not likely to be more effective in increasing genetic progress in actual practice than under the conditions assumed in the foregoing examples. Where assumptions had to be made, these generally favored the progeny test. For example, the average of a sire's offspring was assumed to be unbiased because of special treatment or selection among the offspring before the trait ($X$) was measured. Although the sharp truncation assumed here is not likely to exist in actual practice, particularly for any single trait, this seemingly overestimates the selection differential for all age groups in both sexes. This favors the older animals, since more culling for age, sterility, disease, and other factors and more deaths would occur among them.

It was necessary to assume that the genetic gain from the first and the second selections ($\Delta S_1$ and $\Delta S_2$ or $\Delta D_1$ and $\Delta D_2$) was constant from year to year. Of course, this would not be exactly true, particularly for small herds. Nevertheless, the estimates of $\Delta G$ in the foregoing examples represent the average expectancy for any of the plans for regular use of progeny-tested sires. The effectiveness of progeny testing would be somewhat greater than indicated in these examples if a tested sire were used only when one of exceptional merit was found, or if the proportion of the herd mated to the tested sire were varied according to his apparent superiority. Although there are some notable cases in which the progeny test has identified animals of exceptional transmitting ability, these cases are rare in breed histories and offer uncertain possibilities to the individual breeder who can test only a limited number of sires. Breeders who regularly use the best young sires extensively would be better able to recognize and make use of the occasional outstanding sire, if still available, than breeders who regularly use a tested sire on most of the herd and use young sires sparingly.

For simplicity, selection for one trait at a time was considered in the examples. Methods have been developed by Hazel (2) for expressing net merit as a linear function of several traits, including the performance of relatives. The relative effectiveness of the progeny test in selecting for an index based on several traits would be much the same as that for a single trait. In fact, the formulas developed herein for individual traits could be applied directly to more complicated cases by considering the index as a single trait.

The effect of progeny testing on the genetic progress in an entire breed is much the same as in a closed herd. A breed is just a much larger closed herd, in which another source of variation, herd differences, must be considered and in which inbreeding may be a negligible factor. Thus the same general conclusions for selection within closed herds apply for the much larger closed population of an entire breed. The regular use of progeny tests does not increase, and may decrease, the rate of progress unless the progeny-test information can be obtained early, the reproductive rate is low, and there is little or no basis for earlier culling.

These conclusions do not conflict in any way with the fact that unbiased progeny-test information always increases the accuracy of selection for traits that are influenced much by dominance, epistasis, or environmental variations. They simply mean that in the time required to carry out the progeny test the genetic progress from selection based on pedigree, individual merit, or family averages may be more than that obtained from selection on the progeny test. For the improvement of most traits, in most kinds of livestock, these conclusions point unmistakably toward the fuller use of pedigree, individual merit, and family averages for early culling in order to keep the interval between generations short and progress maximum.

The technique of artificial insemination may increase the advantage of using selected progeny-tested sires if the population is sufficiently large and if the reproductive rate of males is increased markedly thereby, as in sheep and cattle. If fewer sires are needed, more progress is expected from the more intense selection of young sires. However, mating each young sire to larger numbers of females also is likely to increase the accuracy of selecting between tested sires more than enough to offset the increased intensity selection of young sires, particularly for traits low in heritability.

In the literature, progeny testing does not always refer to use of progeny-tested individuals. The progeny-test breeding so successfully practiced on laboratory animals, poultry, and plants is often based on selection between and within the progenies themselves rather than between parents on the basis of the progeny test, and consequently does not increase the interval between generations. Actually, this is selection based on a combination of individual performance and family average. It differs from the methods indicated for farm livestock only in the greater emphasis on selection between families or progenies which a higher reproductive rate permits.

## SUMMARY AND CONCLUSIONS

Annual improvement from selection in a closed herd or breed is the ratio of the average genetic superiority of parents (compared with the unselected group from which they were chosen) to the average age of parents when offspring are born.

Examples of progress expected from selection based on pedigree, individual performance, or averages for collateral relatives, with and without the supplementary use of the progeny test, have been given for representative economic traits of farm animals.

A regular plan of progeny testing is unlikely to increase, and may reduce, progress unless (1) the progeny-test information becomes available early in the tested animal's lifetime, (2) the reproductive rate is low, and (3) the basis for making early selections is relatively inaccurate. These factors are largely beyond the breeder's control, being relatively unchangeable for a particular kind of animal and trait.

Opportunity for improvement from selection is nearly maximum for most traits when (1) culling is based on individual performance, family average, and pedigree and (2) the interval between generations is kept short. Possible exceptions are weanling traits in sheep and carcass traits in sheep and beef cattle.

## LITERATURE CITED

(1) FISHER, R. A., and YATES, F.
    1938. STATISTICAL TABLES FOR BIOLOGICAL, AGRICULTURAL AND MEDICAL RESEARCH. 90 pp. London and Edinburgh.
(2) HAZEL, L. N.
    1943. THE GENETIC BASIS FOR CONSTRUCTING SELECTION INDEXES. Genetics 28: 476–490, illus.
(3) LUSH, J. L.
    1933. THE BULL INDEX PROBLEM IN THE LIGHT OF MODERN GENETICS. Jour. Dairy Sci. 16: 501–522, illus.
(4) ———
    1935. PROGENY TEST AND INDIVIDUAL PERFORMANCE AS INDICATORS OF AN ANIMAL'S BREEDING VALUE. Jour. Dairy Sci. 18: 1–19, illus.
(5) ———
    1936. GENETIC ASPECTS OF THE DANISH SYSTEM OF PROGENY-TESTING SWINE. Iowa Agr. Expt. Sta. Res. Bul. 204, 196 pp., illus.
(6) ———
    1944. THE OPTIMUM EMPHASIS ON DAMS' RECORDS WHEN PROVING DAIRY SIRES. Jour. Dairy Sci. (In press.)
(7) ——— NORTON, H. W., III, and ARNOLD, F.
    1941. EFFECTS WHICH SELECTION OF DAMS MAY HAVE ON SIRE INDEXES. Jour. Dairy Sci. 24: 695–721, illus.
(8) ——— and STRAUS, F. S.
    1942. THE HERITABILITY OF BUTTERFAT PRODUCTION IN DAIRY CATTLE. Jour. Dairy Sci. 25: 975–982.
(9) MCMAHON, P. R.
    1940. INCREASED PROFITS FROM SHEEP THROUGH PROGENY TESTING AND CULLING FOR PRODUCTION. Massey Agr. Col., Palmerston, New Zeal. Sheep Farmers Proc. 9: 37–49, illus.
(10) PEARSON, K., ed.
    [1931]. TABLES FOR STATISTICIANS AND BIOMETRICIANS. Pt. 2, ed. 1, 262 pp., illus. Cambridge.
(11) SEATH, D. M.
    1940. THE INTENSITY AND KIND OF SELECTION ACTUALLY PRACTICED IN DAIRY HERDS. Jour. Dairy Sci. 23: 931–951, illus.

(12) STONAKER, H. H., and LUSH, J. L.
    1942. HERITABILITY OF CONFORMATION IN POLAND-CHINA SWINE AS EVALUATED BY SCORING. Jour. Anim. Sci. 1: [99]–105.
(13) TERRILL, C. E.
    1939. SELECTION OF RANGE RAMBOUILLET EWES. Amer. Soc. Anim. Prod. Proc. (1939) 32: 333–340.
(14) WHATLEY, J. A., Jr.
    1942. INFLUENCE OF HEREDITY AND OTHER FACTORS ON 180-DAY WEIGHT IN POLAND CHINA SWINE. Jour. Agr. Res. 65: 249–264, illus.
(15) WRIGHT, S.
    1932. ON THE EVALUATION OF DAIRY SIRES. Amer. Soc. Anim. Prod. Proc. (1931) 24: 71–78.

# 9

*Copyright © 1974 by the American Dairy Science Association*
*Reprinted from J. Dairy Science* **57**:963–972 (1974)

# GENERAL FLEXIBILITY OF LINEAR MODEL TECHNIQUES FOR SIRE EVALUATION

C. R. Henderson

**Abstract**

Linear model methods applied to the mixed model provide a flexible and powerful tool for sire evaluation under a wide variety of situations. These techniques combine the known, desirable properties of selection index and the capability of linear model methods to deal with large sets of data with unequal subclass numbers. The problem to be solved is to find the best evaluation of a sire that is regarded as a random individual from some specified subpopulation (group). Defining as best that evaluation (prediction) which, in the class of linear functions of the observations, is unbiased and has the smallest possible variance of prediction errors, the problem has been solved provided relative values of elements of the variance-covariance matrix of random variables in the model are known. The method derived to meet these specifications is Best Linear Unbiased Prediction. The computing technique is a simple modification of least squares.

**Introduction**

Recent developments in artificial breeding of dairy cattle have magnified difficulties of sire evaluation but also have increased potentialities for genetic improvement. Problems have been created by widespread use of frozen semen which has made possible nationwide and international distribution of the progeny of a sire and has caused great overlapping of generations. Varying selection practices among artificial insemination (AI) studs and different goals among dairymen, who have the opportunity to select any one of many hundreds of bulls, have destroyed the essentially random distribution across herds of bulls within the area of operation of certain studs that existed at one time. This past random distribution made sire evaluation, at least within regions, relatively accurate when simple methods were used. The situation now is different. Genetic trends and overlapping generations create difficulties in comparisons between younger and older sires. Differential use of sires by dairymen has caused the progeny of some bulls to be compared with herdmates that are better genetically than the herdmates of other bulls. In brief, certain assumptions inherent in the herdmate method are not tenable. These include: (1) all sires with tested daughters from a breed are a random sample from a single population and (2) the distribution of sires across herds is random.

As a consequence of these changes in the industry, relatively simple and unsophisticated methods like the herdmate method or the contemporary comparison method are no longer suitable for sire evaluation. But the statistical and computing problems in devising a suitable program are indeed formidable. As a minimum, an acceptable method should account for genetic trends of AI sires, including differential trends among studs and the natural service population of sires, for nonrandom distribution of sires across herds, for the fact that the sires to be evaluated are not a sample from a single population, for environmental trend, for seasonal differences, for herd environmental differences, and for herd differences in the average genetic merit of the dams.

A method that incorporates these features has existed since 1949. This is Henderson's mixed model method, originally called maximum likelihood. Particular applications of it were described in 1949 and 1952, and the general result was reported in 1950 (3, 4, 5). Some basic proofs were published in 1959 and 1963 (8, 6). The historical basis for the methods was discussed in the symposium honoring J. L. Lush (7). Although the method has been available for sire evaluation for nearly 25 yr, its application required a recognized need for it and advances in computer technology sufficient to make it economically feasible.

The mixed model method is most powerful and flexible, some features being:

1) The evaluations are unbiased. That is, the predictor (the evaluation) and the predictand (the unknown value to be evaluated) have the same expected values.

2) The evaluations have minimum variance of prediction errors.

3) The method is easy to learn by those familiar with least squares.

4) It is easy to modify if conditions change.

Received September 27, 1974.

5) Its properties are clearly defined and unambiguous.

6) It takes advantage of modern statistical computing techniques developed for linear models.

7) It yields variances of prediction errors.

8) It sometimes eliminates bias due to selection and culling and in all cases provides a mechanism for checking such bias.

The mixed model method is a combination of the best features of selection index techniques and least squares techniques. An explanation of this merger requires a brief description of both. A more detailed description is given by Henderson (7).

### Selection Index Principles

A vector of records, say $y$, is distributed jointly with some nonobservable random vector, say $w$. A sample $[y', w']$ vector is drawn, and from $y$ we wish to predict the value of the $i^{th}$ element of $w$. The selection index evaluation is

$$w_i = \alpha_i + b'_i (y - \theta) \quad [1]$$

where $\alpha_i$ = mean of $w_i$, $\theta$ = vector of means of $y$, and $b_i$ is a vector of weights obtained by solving

$$V b_i = c_i. \quad [2]$$

$V$ = variance-covariance matrix of $y$, and $c_i$ = $i^{th}$ column of $C$, the matrix of covariances between $y$ and $w$.

This method was known by both Lush and Wright as early as 1931 (10, 13). Smith (11) presented an application in plant breeding and Hazel (2) in animal breeding. Hazel also showed how to apply the method to multiple trait selection with account taken of economic values. Let $m'w$ be a weighting of genetic values according to the economic values represented by the elements of $m$. Then the index is

$$m'\alpha + b' (y - \theta) \quad [3]$$

where $b$ is the solution to

$$V b = C m. \quad [4]$$

Henderson showed many years ago (see Karam et al. (9) for an example) that the same solution could be obtained by finding the selection index evaluation of the appropriate elements of $w$ and then weighting these evaluations by the elements of $m$. A formal proof of this result was published in 1963 (6).

What desirable properties can be ascribed to selection index? Some of them are stated below for both the normal and nonnormal cases. Means of $y$ and $w$ must be known, a highly unlikely event.

1) Properties when $y$ and $w$ have a joint multivariate normal distribution.

a) Of all functions of $y$ (including nonlinear ones), truncation selection by the index maximizes the expected mean of the selected variables in the case of equal information.

b) Of all functions of $y$, the index maximizes the correlation between the predictor and the predictand.

c) Of all functions of $y$, the index minimizes the average squared errors of prediction.

d) In the unequal information case, and in the class of linear unbiased predictors, the index maximizes the probability of a correct pairwise ranking.

e) The predictor is unbiased even though this was not a condition imposed in its derivation.

f) The predictor is the conditional mean of $w$ given $y$.

2) The following properties are true regardless of whether the distribution is normal.

a) Of all linear functions of $y$, the index maximizes the correlation between the predictor and the predictand.

b) Of all linear functions of $y$, the index minimizes the average squared errors of prediction.

c) The predictor is unbiased.

In the normal case the index is called Best Prediction (BP), and in the nonnormal case it is called Best Linear Prediction (BLP).

### Selection Index Modified for Unknown Means

The index method requires that the following parameters of the joint distribution of $y$ and $w$ be known: mean of $y$, mean of $w$, variance of $y$, and covariance between $y$ and $w$. In some animal breeding applications, and especially in single trait selection, the required variances and covariances may have been estimated well enough to regard the estimates as parameter values. In most cases, however, we may have no knowledge of some of the parameters determining the means of $y$ and $w$. Consequently, the same data to be used in index evaluation must also be used for estimation of means. The developers of the index method either did not recognize this as a problem or did not know an optimum method for coping with the problem. What, in fact, users have done is to estimate the parameters by some method (regular least squares, for example) and then substitute these estimates for the corresponding parameter values. The question logically follows as to what are the best estimates of the means to substitute. It turns out logically that we should use the generalized least squares estimates described in equation [6] and not regular least squares as described in equation [8].

This result was proved by deriving the unbiased linear prediction that has the smallest average squared errors of prediction (6). It must be recognized that biased predictors exist that have smaller mean squared errors, but there is no way to discriminate between these and other biased predictors that will have larger mean squared errors than the unbiased predictors. Consequently, we restrict ourselves to unbiased evaluations. We also assume that we know the required variances and covariances apart from some scalar, $\sigma^2$.

A general statement of the form of the means of $y$ and $w$ can be written

mean of $y = X\beta$,
mean of $w = p\beta$,

where $x$ and $p$ are known matrices and $\beta$ is an unknown fixed vector.

Thus, the predictor of $w_i$ should, in order to be unbiased, have expectation

$p'_i \beta$

where $p'_i$ is the $i^{th}$ row of $p$.

Consequently, in the class of linear functions of $y$ having this expectation, we find the one that has minimum squared errors of prediction. It turns out that the result is

$$\hat{w}_i = p'_i \hat{\beta} + b'_i (y - X\hat{\beta}) \qquad [5]$$

where $b_i$ is the solution to the regular index equations [2] and $\hat{\beta}$ is any solution to

$$X'V^{-1}\hat{\beta} = X'V^{-1}y. \qquad [6]$$

Equations [6] can be recognized as the classical Aitken (1) generalized least squares equations. One warning is necessary, namely, $p'_i \beta$ must be estimable. That is, it must be possible to find some function of $y$ that has expectation $p'_i \beta$.

I have chosen to call the foregoing method Best Linear Unbiased Prediction (BLUP). The BLUP properties are true regardless of the form of the distribution, provided the necessary variances and covariances are known. When the distribution is multivariate normal, it has the following additional properties:

1) The predictor is the maximum likelihood estimator, the generalized least squares estimator, and the best linear unbiased estimator of the conditional mean of $w_i$ given the records, $y$.

2) If the mean of $w$ is a null vector, then BLUP, in the class of all linear unbiased predictors, maximizes the probability of a correct pairwise ranking of the elements of $w$.

Thus, BLUP has nearly all the properties desired in a prediction or evaluation procedure when the variances and covariances are known. However, the modified selection index of [5] and [6] above is not feasible computationally in any large-scale sire evaluation problem, requiring as it does inversion of $V$, the variance-covariance matrix of the records, $y$. The number of records in the northeastern sire evaluation totals nearly a half million even though we use first records of AI progeny exclusively. A national program using first and later records on both AI and NS progeny would involve several million records, but the cost of inverting matrices of order more than a few hundred is prohibitive on present computers. Fortunately, however, the Henderson Mixed Model Method does not require inverting $V$ and gives the same result as the modified selection index version of BLUP. Before presenting that method, we shall describe ordinary least squares since the mixed model method is similar computationally, although different conceptually.

**Least Squares Evaluation**

The application of least squares techniques to estimation and hypothesis testing in multiply classified data with unequal numbers has been known since the early 1930's, the first complete description having been presented by Yates (14). A large proportion of animal breeding research in the past two decades has made use of these techniques. Consequently, many researchers are familiar with the methodology.

The conventional model of least squares is

$$y = X\beta + \varepsilon \qquad [7]$$

where $y$ is the observation vector (the records); $X$ is a known matrix; $\beta$ is an unknown, fixed vector; and $\varepsilon$ is a nonobservable random vector with mean, a vector of zeros, and variance-covariance matrix, $V\sigma^2$, where $V$ is a known matrix and $\sigma^2$ is a scalar, possibly unknown.

In many applications, $V$ is an identity matrix. That is, the elements of $\varepsilon$ are assumed to have the same variance and to be uncorrelated. In the latter case, regular least squares is appropriate and the regular least squares equations are

$$X'X\hat{\beta} = X'y. \qquad [8]$$

Then, the best linear unbiased estimate of the estimable linear function, $k'\beta$ is

$$k'\hat{\beta}$$

where $\hat{\beta}$ is any solution to [8].

When $V \neq I$, generalized least squares is required for best linear unbiased estimations. Then the equations are as in [6]. Some authors use the terms weighted least squares and generalized least squares interchangeably. Others reserve the term weighted least squares for the case where $V$ is diagonal but not the identity matrix. This paper used the latter terminology.

The foregoing least squares models are called fixed models. In contrast, most genetic applications invoke a mixed model, which can be described as follows:

$$y = X\beta + Zu + e \qquad [9]$$

where $y$, $X$, and $\beta$ are defined the same as in [7]; $Z$ is a known matrix; $u$ is a nonobservable random vector (sire values, for example) with null means and variance-covariance matrix $G\sigma^2$; $e$ is a nonobservable random vector with null means and variance-covariance matrix $R\sigma^2$; $G$ and $R$ are known; the scalar quantity, $\sigma^2$, may be unknown; and $u$ and $e$ are regarded as uncorrelated.

Then the variance of $y$ is $(R + ZGZ')\sigma^2$. That is, $V = R + ZGZ'$.

In most applications $R = I$ or at least has a form that is easy to invert as compared to inversion of $V$.

Now suppose we use regular least squares as a method for evaluation. This is done by writing least squares equations as though $u$ were fixed rather than random. Then the regular least squares equations are

$$\begin{pmatrix} X'X & X'Z \\ Z'X & Z'Z \end{pmatrix} \begin{pmatrix} \tilde{\beta} \\ \tilde{u} \end{pmatrix} = \begin{pmatrix} X'y \\ Z'y \end{pmatrix}. \qquad [10]$$

The least squares prediction of $w_i$, if one exists, is

$$\tilde{w}_i = p'_i \tilde{\beta} + \tilde{u}_i$$

where $\beta$, $u$ are any solution to [10]. However, one of the problems in least squares evaluation is that $p'_i\beta + u_i$ must be estimable under a fixed $u$ model. That is, the expectation of $p'_i\tilde{\beta} + \tilde{u}_i$ must be $p'_i\beta + u_i$ under the fixed $u$ model. This likely is not true. It may be possible, however, to evaluate differences, e.g.,

$$(p'_i\beta + u_i) - (p'_j\beta + u_j).$$

If the predictand of interest is not estimable under a model of fixed $u$, the least squares equations have an infinity of solutions, and the predictor is not invariant to the solution, an obvious and most undesirable property of this method.

In any case, the least squares evaluation does not have minimum variance of prediction errors, and further it tends either to overevaluate or to underevaluate seriously animals with small amounts of information. Details of this problem are described by Henderson (7) in a section dealing with selection and culling bias.

**Mixed Model Methodology**

A simple modification of regular least squares [10] leads to a BLUP solution to the evaluation problem in situations of unequal numbers if $R = I$. We simply add $G^{-1}$ to a submatrix of the least squares coefficient matrix, thus

$$\begin{pmatrix} X'X & X'Z \\ Z'X & Z'Z + G^{-1} \end{pmatrix} \begin{pmatrix} \hat{\beta} \\ \hat{u} \end{pmatrix} = \begin{pmatrix} X'y \\ Z'y \end{pmatrix}. \qquad [11]$$

Henderson et al. (8) have proved that any solution to $\hat{\beta}$ in [11] is also a solution to $\hat{\beta}$ in [6] and also that the solution to $\hat{u}$ in [11] is the same as $b'_i(y - X\hat{\beta})$ of [5] [6]. Therefore, $p'_i\hat{\beta} + \hat{u}_i$ from [11] yields BLUP with its desirable properties but often with much easier computations. Note, for example, that $V^{-1}$ is not required. In many applications, $G$ is a diagonal matrix; that is, the elements of $u$ are uncorrelated. In that case, $\sigma_e^2 G^{-1}$ can be written as the diagonal matrix

$$\begin{pmatrix} \sigma_e^2/\sigma_{u_1}^2 & & 0 \\ & \sigma_e^2/\sigma_{u_2}^2 & \\ 0 & & \ddots \end{pmatrix}.$$

That is, the only change from least squares is the addition of certain variance ratios to some of the diagonals of the coefficient matrix.

If $R \neq I$, modification of [11] is required. In most applications, $R$ will be diagonal or, at the least, block diagonal:

$$\begin{pmatrix} X'R^{-1}X & X'R^{-1}Z \\ Z'R^{-1}X & Z'R^{-1}Z + G^{-1} \end{pmatrix} \begin{pmatrix} \hat{\beta} \\ \hat{u} \end{pmatrix} = \begin{pmatrix} X'R^{-1}y \\ Z'R^{-1}y \end{pmatrix}. \qquad [12]$$

If $G$ is a large nondiagonal matrix, its inversion can be avoided by a method presented by Henderson (7).

In many applications, equations [11] can be solved rapidly by the Gauss-Seidel iterative method (12). The addition of relatively large numbers to the diagonals of the u equations is the reason for rapid convergence. This is fortunate because if many sires are to be evaluated, the number of equations is too large for a conventional solution.

Now suppose that we wish to evaluate some animals whose genetic values do not enter into the model for [9]. Let these be called the vector, $u_o$. Two methods are available for handling this situation. The first is to modify [11] to obtain a direct solution to $u_o$. The equations for doing this are

$$\begin{pmatrix} X'X & X'Z & 0 \\ Z'X & Z'Z + T_{11} & T_{12} \\ 0 & T'_{12} & T_{22} \end{pmatrix} \begin{pmatrix} \hat{\beta} \\ \hat{u} \\ \hat{u}_o \end{pmatrix} = \begin{pmatrix} X'y \\ Z'y \\ 0 \end{pmatrix} \quad [13]$$

$T_{11}$, $T_{12}$, and $T_{22}$ are computed as follows. Let the variance-covariance matrix of $(u' \colon u'_o)$ be

$$\begin{pmatrix} G & G_c \\ G'_c & G_o \end{pmatrix} \sigma^2, \quad [14]$$

and

$$\begin{pmatrix} G & G_c \\ G'_c & G_o \end{pmatrix}^{-1} = \begin{pmatrix} T_{11} & T_{12} \\ T'_{12} & T_{22} \end{pmatrix} \quad [15]$$

A second method which yields the same result is

$$\hat{u}_o = G'_c G^{-1} \hat{u}. \quad [16]$$

Note that, if $u$ were observable, the selection index evaluation of $u_o$ would be $G'_c G^{-1} u$.

Variances of prediction errors can be computed directly from the mixed model coefficient matrix. Let a symmetric generalized inverse of the coefficient matrix of [11] or [12] be

$$\begin{pmatrix} C_{11} & C_{12} \\ C'_{12} & C_{22} \end{pmatrix}. \quad [17]$$

Then the variance-covariance matrix of $P\hat{\beta}$ is $PC_{11}P'\sigma^2$; Cov $(P\hat{\beta}, \hat{u}) =$ null matrix; Cov$(P\hat{\beta}, \hat{u} - u) = PC_{12}\sigma^2$; Var$(\hat{u}) = (G - C_{22})\sigma^2 =$ Cov$(\hat{u}, u)$; Var$(\hat{u} - u) = C_{22}\sigma^2$; and variance-covariance matrix of errors of prediction of $P\beta + u = (PC_{11}P' + PC_{12} + C'_{12}P' + C_{22})\sigma^2$.

In most practical situations, the matrix of coefficients is too large to invert by conventional methods. Nevertheless, iterative methods may be suitable.

Write [12] as
$$T\gamma = r$$
where $\gamma' = [\hat{\beta}' \colon \hat{u}']$ and $r' = [y'X \colon y'Z]$.

Suppose we want the variance of prediction errors of $k'\beta + m'u$, which we define as prediction errors of $p'\gamma$. Then solve iteratively for $q$ in
$$Tq = p.$$
Now the desired variance is $p'q\sigma^2$.

### Illustration of BLP, BLUP, and Mixed Model Method

We will illustrate BLP, BLUP, and the mixed model version of BLUP with a simple example in sire evaluation. Suppose that we have the following progeny data on five sires in two herds. Sires 1 and 2 are a sample of two from group 1, and sires 3, 4, and 5 are a sample of three from group 2. Groups and herds are fixed.

Number of progeny

|      | Group 1 | | Group 2 | | |      |
|------|---|---|---|----|---|------|
| Herd Sire | 1 | 2 | 3 | 4 | 5 | [18] |
| 1    |   | 5 | 4 | 0 | 10 | 1 |
| 2    |   | 0 | 2 | 5 | 1 | 5 |

Progeny totals

$$\begin{pmatrix} 51 & 36 & - & 74 & 9 \\ - & 25 & 31 & 10 & 54 \end{pmatrix} \quad [19]$$

A possible linear model is
$$y_{ijkl} = h_i + g_j + s_{jk} + e_{ijkl}.$$
Suppose Var$(s_{jk}) = .1\sigma^2$, and
$$\text{Var}(e_{ijkl}) = \sigma^2.$$
All variables are uncorrelated.

*BLP (selection index).* Suppose we want to predict the value of each sire when his progeny are distributed equally across the two herds. Then the predictand is
$$g_j + s_{jk} + (h_1 + h_2)/2.$$
The selection index for sire 2 as an example is
$$g_1 + (h_1 + h_2)/2 + b_1 (\bar{y}_{112.} - h_1 - g_1) + b_2 (\bar{y}_{212.} - h_2 - g_1)$$
where $b_1$ and $b_2$ are the solution to
$$\begin{pmatrix} .35 & .10 \\ .10 & .60 \end{pmatrix} \begin{pmatrix} b_1 \\ b_2 \end{pmatrix} = \begin{pmatrix} .10 \\ .10 \end{pmatrix};$$
$b_1 = .25$, and $b_2 = .125$.

The other subclass means are not used because they are uncorrelated with $s_2$ and uncorrelated with the subclasses that are correlated with $s_2$. See equations [1] and [2] for the general procedure.

*BLUP selection.* Suppose now that **g** and **h** are unknown. We first use the methods of [5] and [6]. The generalized least squares equations are, after considerable algebra,

$$\begin{bmatrix} 12.509 & -1.289 & 5.833 & 5.387 \\ -1.289 & 9.473 & 1.250 & 6.935 \\ 5.833 & 1.250 & 7.083 & 0 \\ 5.387 & 6.935 & 0 & 12.321 \end{bmatrix} \begin{bmatrix} \hat{h}_1 \\ \hat{h}_2 \\ \hat{g}_1 \\ \hat{g}_2 \end{bmatrix} =$$

$$\begin{bmatrix} 93.812 \\ 78.354 \\ 72.125 \\ 100.042 \end{bmatrix} \quad [20]$$

No unique solution exists because the sum of the first two equations is equal to the sum of the last two. A particular solution is $\hat{h}_1 = 7.139$, $\hat{h}_2 = 8.881$, $\hat{g}_1 = 2.736$, and $\hat{g}_2 = 0$.

Then the prediction of $g_1 + s_2 + (h_1 + h_2)/2$ is

2.736 + (7.139 + 8.881)/2 + .25(9 − 7.139 − 2.736) + .125(12.5 − 8.881 − 2.736) = 10.638.

*BLUP by mixed model method.* The mixed model equations, obtained by adding $10 = \sigma^2_e/\sigma^2_s$ to the diagonal coefficient of the sire equations in regular least squares are in [21]. A solution is $[\hat{h}_1 \ \hat{h}_2] = [7.139 \ 8.881]$, $[\hat{g}_1 \ \hat{g}_2] = [2.736 \ 0]$, and $[\hat{s}_{11} \ \hat{s}_{12} \ \hat{s}_{23} \ \hat{s}_{24} \ \hat{s}_{25}] = [.108 \ -.108 \ -.893 \ .178 \ .716]$. Then the prediction of $g_1 + s_2 + (h_1 + h_2)/2$ is 2.736 − .108

+ (7.139 + 8.881)/2 = 10.638 as before.

Note $g_j + (h_1 + h_2)/2$ is estimable. Consequently, the prediction is invariant to whatever solution is taken to [20] or [21].

### Applications of Mixed Model Methods to Sire Evaluation

Now to illustrate the generality and flexibility of the mixed model to sire evaluations, we will present applications beginning with simple assumptions and progressing through increasing complexity.

*Simple one-way classification.* Suppose that a random sample of sires from a single population is mated to a random sample of dams, and the progeny are subject to the same random environmental influences. Then the model would be

$$y_{ij} = \mu + s_i + e_{ij}$$

where $\mu$ is fixed and unknown, s and e are uncorrelated variables with means zero and variances $\sigma^2_s$ and $\sigma^2_e$.

Let the number of progeny on the $i^{th}$ sire be $n_i$. Then the mixed model equations are

$$\begin{bmatrix} n. & n_1 & n_2 & \cdots \\ n_1 & n_1 + \sigma^2_e/\sigma^2_s & 0 & \cdots \\ n_2 & 0 & n_2 + \sigma^2_e/\sigma^2_s & \cdots \\ \vdots & \vdots & \vdots & \end{bmatrix} \begin{bmatrix} \hat{\mu} \\ \hat{s}_1 \\ \hat{s}_2 \\ \vdots \end{bmatrix} = \begin{bmatrix} y.. \\ y_1. \\ y_2. \\ \vdots \end{bmatrix} \quad [22]$$

It is easy to verify by substitution in [22] that

$$\begin{bmatrix} 20 & 0 & 9 & 11 & 5 & 4 & 0 & 10 & 1 \\ 0 & 13 & 2 & 11 & 0 & 2 & 5 & 1 & 5 \\ 9 & 2 & 11 & 0 & 5 & 6 & 0 & 0 & 0 \\ 11 & 11 & 0 & 22 & 0 & 0 & 5 & 11 & 6 \\ 5 & 0 & 5 & 0 & 15 & 0 & 0 & 0 & 0 \\ 4 & 2 & 6 & 0 & 0 & 16 & 0 & 0 & 0 \\ 0 & 5 & 0 & 5 & 0 & 0 & 15 & 0 & 0 \\ 10 & 1 & 0 & 11 & 0 & 0 & 0 & 21 & 0 \\ 1 & 5 & 0 & 6 & 0 & 0 & 0 & 0 & 16 \end{bmatrix} \begin{bmatrix} \hat{h}_1 \\ \hat{h}_2 \\ \hat{g}_1 \\ \hat{g}_2 \\ \hat{s}_{11} \\ \hat{s}_{12} \\ \hat{s}_{23} \\ \hat{s}_{24} \\ \hat{s}_{25} \end{bmatrix} = \begin{bmatrix} 170 \\ 120 \\ 112 \\ 178 \\ 51 \\ 61 \\ 31 \\ 84 \\ 63 \end{bmatrix} \quad [21]$$

$$\hat{\mu} = \sum_i \frac{y_{i.}}{n_i + \sigma^2_e/\sigma^2_s} \bigg/ \sum_i \frac{n_i}{n_i + \sigma^2_e/\sigma^2_s}, \text{ and}$$

$$\hat{s}_i = \frac{n_i}{n_i + \sigma^2_e/\sigma^2_s} (\bar{y}_{i.} - \hat{\mu}).$$

If $\mu$ were known, the selection index (BLP) evaluation of the $i^{th}$ sire would be

$$\frac{n_i}{n_i + \sigma^2_e/\sigma^2_s} (\bar{y}_{i.} - \mu).$$

*Sires cross-classified with environments — no interaction.* A random sample of sires from the same population has daughters in different environments (herds, for example) in varying proportions. Suppose we regard these environments as fixed and assume no interaction between sires and environments. Then the model can be written as

$$y_{ijk} = \mu + h_i + s_j + e_{ijk} \quad [23]$$

where $\mu$ and $h$ are fixed, and $s$ and $e$ are uncorrelated variables with means zero and variances $\sigma^2_s$ and $\sigma^2_e$. The mixed model equations can be written as

$$\begin{bmatrix} C_{11} & C_{12} \\ C'_{12} & C_{22} \end{bmatrix} \begin{bmatrix} \hat{h} \\ \hat{s} \end{bmatrix} = \begin{bmatrix} r_1 \\ r_2 \end{bmatrix} \quad [24]$$

where $C_{11}$ is a diagonal matrix with elements $n_{1.}$, $n_{2.}$, etc.; $C_{12}$ is the table of $n_{ij}$; $C_{22}$ is a diagonal matrix with elements $n_{.1} + \sigma^2_e/\sigma^2_s$, $n_{.2} + \sigma^2_e/\sigma^2_s$, etc.; $r_1 = [y_{1..} \ y_{2..} \ldots]'$; and $r_2 = [y_{.1.} \ y_{.2.} \ldots]'$.

$\hat{\mu}$ is not included because the rank is one less than order. By deleting $\hat{\mu}$, the best linear unbiased estimator of $\mu + h_i$ is $\hat{h}_i$.

Storing the entire set of equations [24] in memory can be avoided by absorbing $\hat{h}_i$ into $\hat{s}$ at the end of data input for each herd. One can then build into memory only $C_{22} - C'_{12}C^{-1}_{11}C_{12}$ and $r_2 - C'_{12}C^{-1}_{11}r_1$ which then can be used to solve for $\hat{s}$. If this reduced matrix and right hand side are too large to store in memory, one can write preliminary information on tape or disc at the end of each herd and later sort and sum to obtain the desired equations.

Only a simple modification of equations [24] is required to handle the situation in which herds are uncorrelated variables with means zero and variance $\sigma^2_h$. The mixed model equations become

$$\begin{bmatrix} C_{00} & C_{01} & C_{02} \\ C'_{01} & C_{11} & C_{12} \\ C'_{02} & C'_{12} & C_{22} \end{bmatrix} \begin{bmatrix} \hat{\mu} \\ \hat{h} \\ \hat{s} \end{bmatrix} = \begin{bmatrix} r_0 \\ r_1 \\ r_2 \end{bmatrix} \quad [25]$$

where $C_{00} = n_{..}$; $C_{01} = [n_{1.} \ n_{2.} \ldots]$; $C_{02} = [n_{.1} \ n_{.2} \ldots]$; $C_{12}$, $C_{22}$, $r_1$, and $r_2$ are the same as in [24]; $C_{11}$ is the same as in [24] except that $\sigma^2_e/\sigma^2_h$ is added to the diagonals; and $r_0 = y_{...}$.

*Sires cross-classified with herds — interaction present.* Now the model for fixed $h$ becomes

$$y_{ijk} = \mu + h_i + s_j + \gamma_{ij} + e_{ijk}. \quad [26]$$

The elements of [26] are the same as in [23] except now $\gamma_{ij}$ are added. These are considered variables, uncorrelated with any others, and with variance, $\sigma^2_\gamma$. Now the mixed model equations are

$$\begin{bmatrix} C_{11} & C_{12} & C_{13} \\ C'_{12} & C_{22} & C_{23} \\ C'_{13} & C'_{23} & C_{33} \end{bmatrix} \begin{bmatrix} \hat{h} \\ \hat{s} \\ \hat{\gamma} \end{bmatrix} = \begin{bmatrix} r_1 \\ r_2 \\ r_3 \end{bmatrix} \quad [27]$$

where $C_{11}$, $C_{12}$, $C_{22}$, $r_1$, and $r_2$ are the same as in [24];

$$C_{13} = \begin{bmatrix} m'_1 & 0 & 0 & \cdots \\ 0 & m'_2 & 0 & \cdots \\ 0 & 0 & m'_3 & \cdots \\ \cdot & \cdot & \cdot & \\ \cdot & \cdot & \cdot & \\ \cdot & \cdot & \cdot & \end{bmatrix}$$

where $m'_i = [n_{i1} \ n_{i2} \ldots]$; $C_{23} = [D_1 : D_2 : \ldots]$ where $D_i$ is a diagonal matrix with elements $[n_{i1} \ n_{i2} \ldots]$; $C_{33}$ is a diagonal matrix with elements $n_{ij} + \sigma^2_e/\sigma^2_\gamma$; and $r_3 = [y_{11.} \ y_{12.} \ldots]'$.

Now all columns and rows of [27] pertaining to subclasses with $n_{ij} = 0$ can be deleted. $\hat{\gamma}$ is easy to absorb. Let

$$p_{ij} = \frac{\sigma^2_e}{\sigma^2_\gamma} \frac{n_{ij}}{n_{ij} + \sigma^2_e/\sigma^2_\gamma} \text{ and}$$

$$w_{ij} = \frac{\sigma^2_e}{\sigma^2_\gamma} \frac{y_{ij.}}{n_{ij} + \sigma^2_e/\sigma^2_\gamma}$$

Then proceed as in [24] except that $p_{ij}$ is substituted for $n_{ij}$ and $w_{ij}$ is substituted for $y_{ij.}$.

If herds are random, the equations are like those of [25] after absorbing $\hat{\gamma}$ except that we use $p_{ij}$ and $w_{ij}$ rather than $n_{ij}$ and $y_{ij.}$.

Two consequences of interaction between sire and herd are that predictions are less accurate and that relatively less emphasis is placed on subclass means with large $n_{ij}$ as compared to the no interaction model.

*Relationship between interaction and $c^2$.* Problems concerning $c^2$ can be solved by using interaction instead and with considerably easier computations, I believe. Consider the model just above: sires, herds, and interaction with herds random. The corresponding $c^2$ model is

$$y_{ijk} = \mu + h_i + s_j + \varepsilon_{ijk}$$

where h and s are uncorrelated variables.

The elements of $\varepsilon$ have certain nonzero covariances, namely $\text{Cov}(\varepsilon_{ijk}, \varepsilon_{ijk}') = t$, say, and all covariances between elements of $\varepsilon$ with either different sires or different herds = 0. Then the following relationships exist between the parameters of the two models:

| Interaction Model | | $c^2$ Model |
|---|---|---|
| $\sigma_h^2$ | = | $\sigma_h^2$ |
| $\sigma_s^2$ | = | $\sigma_s^2$ |
| $\sigma_e^2 + \sigma_\gamma^2$ | = | $\sigma_\epsilon^2$ |
| $\sigma_\gamma^2$ | = | t |

*Different populations of sires.* Suppose now the same assumptions as in one-way classification except that the sires come from different populations, say $q_1$ from population 1, $q_2$ from population 2, etc. Then the model can be written

$$y_{ijk} = g_i + s_{ij} + e_{ijk}.$$

Now the best prediction from mixed model methods can be obtained by setting up and solving a separate set of equations for each group, as illustrated by group i in [28].

Note that [28] is identical to [22] except that $g_i$ is substituted for $\mu$.

The evaluation of the $ij^{th}$ sire is $\hat{g}_i + \hat{s}_{ij}$.

*Different groups of sires in different environments—no interaction.* The model is now like that of [23] except that groups are added.

$$y_{ijkl} = h_i + g_j + s_{jk} + e_{ijkl}. \qquad [29]$$

An example of these mixed model equations was in [21].

The computing strategy to follow in this model, which is the Northeastern AI Sire Comparison Method when $h_i$ refers to fixed herd-year-season, is to proceed by these steps:

1) Accumulate the data for a single herd (or herd-year-season) and absorb herd into sires (regarded temporarily as fixed and ignoring groups).

2) When all data have been processed, the equations can be written as

$$\begin{bmatrix} C_{11} & C_{12} & C_{13} & \cdots \\ C'_{12} & C_{22} & C_{23} & \cdots \\ C'_{13} & C'_{23} & C_{33} & \cdots \\ \cdot & \cdot & \cdot & \\ \cdot & \cdot & \cdot & \\ \cdot & \cdot & \cdot & \end{bmatrix} \begin{bmatrix} s_1 \\ s_2 \\ s_3 \\ \cdot \\ \cdot \\ \cdot \end{bmatrix} = \begin{bmatrix} r_1 \\ r_2 \\ r_3 \\ \cdot \\ \cdot \\ \cdot \end{bmatrix} \qquad [30]$$

where $s_1$ refers to sires in group 1, etc.

3) Compute coefficients of groups in sire equations by post-multiplying the coefficient matrix of [30] by matrix [31],

$$\begin{bmatrix} 1 & 0 & 0 & \cdots \\ 0 & 1 & 0 & \cdots \\ 0 & 0 & 1 & \cdots \\ \cdot & \cdot & \cdot & \\ \cdot & \cdot & \cdot & \\ \cdot & \cdot & \cdot & \end{bmatrix} . \qquad [31]$$

Call the resulting product matrix $C_{sg}$. 1 refers to a column vector of 1's of length, respectively, the number of sires in the group. The transpose of this product becomes the coefficients of sires in the group equations.

4) Compute the coefficients of groups in the group equations by pre-multiplying $C_{sg}$

$$\begin{bmatrix} n_{i.} & n_{i1} & n_{i2} & \cdots \\ n_{i1} & n_{i1} + \sigma_e^2/\sigma_s^2 & 0 & \cdots \\ n_{i2} & 0 & n_{i2} + \sigma_e^2/\sigma_s^2 & \cdots \\ \cdot & \cdot & \cdot & \\ \cdot & \cdot & \cdot & \\ \cdot & \cdot & \cdot & \end{bmatrix} \begin{bmatrix} \hat{g}_i \\ \hat{s}_{i1} \\ \hat{s}_{i2} \\ \cdot \\ \cdot \\ \cdot \end{bmatrix} = \begin{bmatrix} y_{i..} \\ y_{i1.} \\ y_{i2.} \\ \cdot \\ \cdot \\ \cdot \end{bmatrix} . \qquad [28]$$

by the transpose of [31].

5) Compute the right-hand side of group equations by pre-multiplying

$$\begin{bmatrix} r_1 \\ r_2 \\ \cdot \\ \cdot \\ \cdot \end{bmatrix}$$

by the transpose of [31].

6) Add $\sigma^2_e/\sigma^2_s$ to the diagonal coefficient of each sire equation.

7) Solve the resulting equations, remembering that the rank is at least one less than the order. If there is no complete confounding between groups and herds, the rank is exactly one less.

8) Then the prediction of the difference between two sires is

$$\hat{g}_i + \hat{s}_{ij} - \hat{g}_{i'} - \hat{s}_{i'j'}.$$

*Accounting for relationships among sires.* A simple modification of all preceding methods will take into account the additive relationships among sires to be evaluated. Let $A$ = numerator relationship matrix for these sires. Then instead of adding $\sigma^2_e/\sigma^2_s$ to the diagonals of the sire equations, add $A^{-1}\sigma^2_e/\sigma^2_s$ to the submatrix of coefficients of $\hat{s}$ in the sire equations.

*Using repeated records.* It may be desirable to use all records on progeny rather than first records only, and this is particularly true when herd sizes are small. Assuming the simple repeatability model and using only repeated records made in the same herd, the linear model without interaction is

$$y_{ijklmn} = h_{ij} + g_k + s_{kl} + p_{iklm} + e_{ijklm} \quad [32]$$

where $h_{ij}$ refers to the $j^{th}$ year-season in the $i^{th}$ herd and is regarded as fixed in this discussion; $g_k + s_{kl} + p_{iklm}$ refers to the producing ability of the $m^{th}$ daughter of the $kl^{th}$ sire in the $i^{th}$ herd; and s, p, and e are regarded as uncorrelated variables with variances under a simple genetic model as follows:

$$\sigma^2_s = h^2\sigma^2_y/4,$$
$$\sigma^2_p = (r - h^2/4)\sigma^2_y, \text{ and}$$
$$\sigma^2_e = (1 - r)\sigma^2_y$$

where $h^2$ = heritability and $r$ = repeatability.

Efficient computing strategies are particularly important in models of increasing complexity like this one. The following would appear to be a good procedure:

1) Enter data sorted by cows within herd.
2) At the end of each cow's data absorb **p** into **h** and **s** (ignore **g** and regard **s** as fixed for the time being).
3) At the end of the data for the $i^{th}$ herd absorb the $h_{ij}$ into **s**.
4) At the end of all data input for a breed proceed as was described for end of data input in model [29].

**General Advice Regarding Use of Mixed Model Methods**

1) Try to account in the model for major sources of variance and of bias, but keep the model as simple as possible within these limitations.

2) Try to write the model with as many as possible of the sets of variables mutually uncorrelated as well as uncorrelated with other sets of variables. Examples given previously were use of interaction in place of $c^2$, and writing real producing abilities as sums of uncorrelated sire and cow components rather than real producing abilities with nonzero correlations between half-sibs. Remember that a model which will generate the actual variance-covariance matrix of records is equivalent to any other model that generates the same matrix. The choice among equivalent models is dictated by computational efficiency.

3) Use care that the rank of the coefficient matrix to be solved is not smaller than is assumed in the solution strategy. If it is smaller, most computer programs will come up with a solution even though it may be ridiculous. We have found that a good test to check the accuracy of the solution is to post-multiply the coefficient matrix by the solution and then check whether the resulting vector is approximately equal to the right-hand side of the equations. The most likely cause of a problem in rank would be the use of a group or set of groups of sires in herds that use no other groups. The better the design for progeny testing the less is the likelihood of problems with the solution and the smaller will be the errors of prediction. It is impossible to obtain a completely balanced design if we wish to compare different generations of sires. Consequently, methods like those described in this paper will be required.

4) Consider use of an iterative solution to the equations, usually after absorbing nuisance parameters and variables. Such situations are likely to converge rapidly because of the addi-

tion of relatively large quantities to some of the diagonal coefficients.

**References**

(1) Aitken, A. C. 1935. On least squares and linear combinations of observations. Proc. Roy. Soc. Edin. 55:42.
(2) Hazel, L. N. 1943. The genetic basis for constructing selection indexes. Genetics 28:476.
(3) Henderson, C. R. 1949. Estimation of changes in herd environment. J. Dairy Sci. 32:706. (Abstr.)
(4) Henderson, C. R. 1950. Estimation of genetic parameters. Ann. Math. Stat. 21:309.
(5) Henderson, C. R. 1952. Specific and general combining ability in Heterosis. J. W. Gowen, ed. Iowa State College Press, Ames.
(6) Henderson, C. R. 1963. Selection index and expected genetic advance. NAS-NRC:982.
(7) Henderson, C. R. 1973. Sire evaluation and genetic trends in Proceedings of the animal breeding and genetics symposium in honor of Dr. Jay L. Lush. ASAS and ADSA, Champaign, Illinois.
(8) Henderson, C. R., O. Kempthorne, S. R. Searle, and C. M. von Krosigk. 1959. The estimation of environmental and genetic trends from records subject to culling. Biometrics 15:192.
(9) Karam, H. A., A. B. Chapman, and A. L. Pope. 1953. Selecting lambs under flock conditions. J. Anim. Sci. 12:148.
(10) Lush, J. L. 1931. The number of daughters necessary to prove a sire. J. Dairy Sci. 14:209.
(11) Smith, H. F. 1936. A discriminant function for plant selection. Ann. Eug. 7:240.
(12) Van Norton, R. 1959. The solution of linear equations by the Gauss-Seidel method in Mathematical methods of digital computers. A. Ralston and H. S. Wilf, eds. John Wiley & Sons, Inc., New York.
(13) Wright, S. 1931. On the evaluation of dairy sires. Proc. Amer. Soc. Anim. Prod. 24th Annu. Meeting:71.
(14) Yates, F. 1934. The analysis of multiple classifications with unequal numbers in the different subclasses. J. Amer. Stat. Ass. 29:51.

# GENETICAL PREDICTIONS OF SELECTION RESPONSE

# Editor's Comments
# on Papers 10 Through 13

**10  FISHER**
Excerpt from *The Fundamental Theorem of Natural Selection*

**11  HALDANE**
*A Mathematical Theory of Natural and Artificial Selection. Part VII. Selection Intensity as a Function of Mortality Rate*

**12  COMSTOCK, ROBINSON, and HARVEY**
*A Breeding Procedure Designed to Make Maximum Use of Both General and Specific Combining Ability*

**13  ROBERTSON**
*A Theory of Limits in Artificial Selection*

Providing we are attempting to make predictions of selection response over one or very few generations, the methods used in the papers of the previous section can be used. When longer-term predictions are required, we need to take into account the actual effects and frequencies of genes influencing the trait. This section includes four papers that fit into this category. In reverse order, the last two, by Comstock et al. (Paper 12) and by A. Robertson (Paper 13) are certainly discussions of long-term response. Paper 11, by Haldane, deals with the rates of change of gene frequency from artificial selection per generation and is thus a prerequisite for consideration of response over succeeding generations. The first paper, by Fisher (Paper 10), is based on arguments at the individual locus level, but summarizes them at the level of variances.

R. A. (later Sir Ronald) Fisher's "Fundamental Theorem of Natural Selection" is published in his book, *The Genetical Theory of Natural Selection* (1930, 1958), and an eight-page section from the first edition (1930), in which the theorem is derived, has been reprinted as Paper 10. The first rather than second edition has been chosen partly because of precedence and partly because additional parts of the derivation put into the later edition are generalizations rather than simplifications. Fisher's (1918) benchmark paper on the correlation among relatives is included in Part I. For a biography, see Box (1978).

The theorem is simply stated: "The rate of increase in fitness of any organism at any time is equal to its genetic variance in fitness at that time" (p. 35). The derivation is, however, rather more obscure, and there has been much controversy about and discussion of the conditions under which the theorem applies; we can do little more here than give an outline and references. The extensive discussion is not surprising in view of the apparent generality of the theorem, which Fisher considered "bears some remarkable resemblances to the second law of thermodynamics . . . . Each requires the constant increase of a measurable quantity, in the one case the entropy of a physical system and in the other the fitness, measured by $m$, of a biological population," where $m$ is the Malthusian parameter of population increase (p. 36).

Price (1972) attempts to make clear what Fisher meant by the theorem and gives an entertaining account. In particular, he argues that the predicted change in mean fitness is only the component due to natural selection and not the total change, some of which comes from environmental change; he points out that dominance and epistasis were regarded as a matter of the environment. This implies a severe limitation of the theorem. More particularly, problems of its use apply with linked epistatic loci (Kimura, 1965; Nagylaki, 1976; and see Ewens, 1979, Ch. 6, for a full discussion). In such cases, it is possible for fitness to decline over a period of many generations. In a general model, the theorem as actually stated therefore almost never holds *exactly*.

While clearly lacking the power and generality that Fisher originally thought, the quantification of the relation between variance and response to natural selection has still been important in predicting rates of evolutionary change. In addition, there is also a so-called secondary law, due to Robertson (1966), which states that the rate of change in any metric trait equals its additive genetic covariance with fitness. This can be derived, as can the fundamental theorem itself, by using statistical arguments such as those employed in Section II for predicting rates of response (see, for example, Falconer, 1981, Ch. 20).

J. B. S. Haldane's important series of papers on "A Mathematical Theory of Natural and Artificial Selection" commenced in 1924. This volume includes Part VII of the series, entitled "Selection Intensity as a Function of Mortality Rate." It is better known to quantitative geneticists than to others, and is reprinted as Paper 11. Haldane made numerous other major contributions to population genetics, primarily on selection. His earlier work is summarized in *The Causes of Evolution* (Haldane, 1932). For a discussion of Haldane's life and work in its many diverse areas, see Dronamraju (1968).

*Editor's Comments on Papers 10 Through 13*

In Paper 11 Haldane computes the relative frequencies of different genotypes in a population undergoing truncation selection for a quantitative trait, and from this what we would call the selective value. This enables formulae obtained for models of natural selection, which are generally stated in terms of selective value, to be extended to artificial selection (e.g., Paper 13). Haldane deals with the special case of two types, A and B (e.g., haploids), in equal frequency, with means $+\lambda$ and $-\lambda$ standard deviation units. The proportion selected is $1/(1 + z)$, and the truncation point is X in standard deviation units; he finds that the relative proportions selected of the two types is $k = 2\lambda (z + 1)e^{-\frac{1}{2}X^2}/\sqrt{2\pi}$ for the case where the variability within each type is the same ($\mu = 0$). In modern notation (e.g., Falconer, 1981, p. 186), k is the selective value, s; $1/(1 + z)$ is the proportion selected, p; $\lambda$ is the effect of the gene in terms of the phenotypic standard deviation, $a/\sigma_p$; and $e^{-\frac{1}{2}X^2}/\sqrt{2\pi}$ is the ordinate of the standardized normal distribution at the truncation point, z, which equals ip where i is the standardized selection differential. Hence $s = 2ia/\sigma_p$.

Haldane generalized the result to differential variance in the two types ($\mu \neq 0$), but this has received little attention. For fuller analyses of the calculation of selective values and thus of change in gene frequency, see Griffing (1960) for discussions including epistasis, Latter (1965a) for the case where the assumption that $a/\sigma_p$(Haldane's $\lambda$) is small breaks down and second-order terms can not be ignored, and Falconer (1960, 1981, Ch. 11) for alternative derivations, in 1960 by regression methods. Formulae such as Haldane's are useful in predicting how quickly gene frequencies change under artificial selection for different models of gene numbers and effects, and thus how long equations based on statistical quantities such as heritability in the base population are likely to remain valid. However, such analyses are complicated by linkage disequilibrium (see Paper 23).

In the applications of quantitative genetics to animal and plant breeding discussed in the previous section, the objectives are solely to maximize the breeding values of selected animals—in other words, to utilize the additive genetic variability in the population. Yet many traits, notably yield in plants and reproductive rate in animals, show substantial heterosis. If this heterosis is due, at least in part, to loci at which heterozygotes are superior in performance to both homozygotes, then straightforward selection in a single population will not achieve maximum performance. This can be obtained only by producing crossbreds from lines in which the alternative complementary alleles are fixed. Several breeding procedures have been suggested and used in order to make best use of the heterosis, among them inbreeding and line crossing, as pioneered by East and Hayes (1912) in maize, recurrent selection to a tester, proposed by Hull (1945) where recurrent means testing and selection every generation, as opposed to the intermittent process of inbred line production and testing, and reciprocal recurrent selection (RRS), proposed in 1949 by Comstock, Robinson, and Harvey (Paper 12). The last of these makes the most use of quantitative genetic principles, was the most thoroughly analyzed, and was responsible for many experiments and practical applications.

R. E. Comstock and H. F. Robinson of North Carolina State College (now University), with P. H. Harvey and others, undertook a series of major investigations into the genetic variability and the degrees of dominance in populations deriving from crosses of inbred lines or open pollinated varieties of corn, and a paper reviewing their analytical methods appears in Part I (Comstock and Robinson, 1952). These analyses were aimed at providing data on which breeding decisions, such as whether to use RRS, could be made.

In the reciprocal recurrent selection (RRS) method proposed by Comstock et al., two non-inbred parental pure-bred populations are maintained. Crosses are made each generation between the lines, cross-bred performance is recorded, and individuals are chosen on the basis of their performance in crosses to breed pure to reproduce the parental populations. Essentially, the method involves selection on cross-bred progeny or sib record and can be accommodated within the index theory of Hazel (Paper 6), for example, simply by replacing genetic variances of purebreds by variances and covariances of pure and cross performance. This approach can predict current rates of response, but not long-term responses, for these depend on the degree of dominance, as mentioned previously. In Paper 12, the authors deal thoroughly with this issue, but point out that RRS is likely to be at least as effective as other cross-testing methods in the absence of overdominance. It is not, however, as efficient as simple mass selection for highly heritable nonheterotic traits, because progeny or sib testing is required for RRS.

Among the experiments stimulated by the proposals of Comstock et al. was that by Bell, Moore, and Warren, which is included in a later section as Paper 20.

It is notable in Paper 12 that Comstock et al. consider the limits that can be achieved by different breeding methods, *assuming* all favorable genes are fixed. In practice, populations undergoing artificial selection (and, for that matter, natural selection) are finite in size; for example, many millions of egg-laying poultry may be multiplied from a nucleus flock of twenty or so sires, and all the sires needed in a dairy cattle population can be bred each generation from just a single sire by using artificial insemination. The consequent drift or sampling effects mean that some favorable genes will be lost by chance from the population. What then is the selection limit, and what influences it when finite population size is taken into account?

This problem was solved by Alan Robertson (Paper 13) using results obtained from diffusion equations by Kimura (1957) for the probabilities of fixation of genes in natural populations, by replacing selective value *(s)* by a function of the gene effect following Haldane (Paper 11). The further reparametrization in terms of the product $Ns$, where $N$ is the effective population size, enabled the influence of change of population size and selection intensity to be evaluated, and, for example, gave the result that the limit is maximized by selecting one-half of the population—that is, intense selection for rapid early gains is likely to incur a penalty in the long term. Robertson was also able to suggest an upper limit to the total response of $2N$ times the response in the first generation, except for rare recessives, and showed that division of resources into sublines would not affect the ultimate limit for additive genes.

Robertson's paper broke new ground in the application of quantitative genetics to animal and plant improvement in quantifying the effects of drift and introducing the ideas of selection limits. It is notable, however, that the time scale of the predictions is very long—$N$ generations for dairy cattle might represent 1,000 years—so the practical applications of the results have been less obvious. Further, because of the dependence of the probabilities of fixation on initial frequencies and effects, there is no straightforward answer to the breeder's question, "What is the minimum safe size for my population?" Theoretical results were extended to the case of linkage at two loci (Latter, 1965b; Hill and Robertson, 1966), and then to many loci (Robertson, 1970), and to a variety of alternative breeding schemes. A number of selection experiments to check the theory have also been undertaken; these are referred to in Section IV.

A number of other nonmathematical studies by Robertson are included in this volume (Papers 19, 22, and 25), and a more theoretical paper is reprinted in Part I. These cover population and quantitative genetics theory and laboratory experiment. Robertson's breadth of work can be judged by his classical papers on dairy cattle breeding (for example Rendel and Robertson, 1950; Robertson and Rendel, 1950; Robertson, 1966).

Many other papers could have been reprinted in this section on theory; for a start Wright, Fisher and Haldane get inadequate coverage. A notable paper on selection in natural populations is that by Kimura

(1958), for artificial selection there are *inter alia* well-known papers by Griffing (1960) and Bohren et al. (1966), and Dempster (1955) obtained some results for long-term selection prior to Robertson's exposition (Paper 13).

## REFERENCES

Bohren, B. B., W. G. Hill, and A. Robertson, 1966, Some Observations on Asymmetrical Correlated Responses to Selection, *Genet. Res. (Cambridge)* **7**:44-57.

Box, J. F., 1978, *R. A. Fisher: The Life of a Scientist,* Wiley, New York.

Comstock, R. E., and H. F. Robinson, 1952, Estimation of Average Dominance of Genes, in *Heterosis,* J. W. Gowen, ed., Iowa State University Press, Ames, pp. 494-516.

Dempster, E. R., 1955, Genetic Models in Relation to Animal Breeding Problems, *Biometrics* **11**:535-536.

Dronamraju, K. R., ed., 1968, *Haldane and Modern Biology,* Johns Hopkins Press, Baltimore.

East, E. M., and H. K. Hayes, 1912, Heterozygosis in Evolution and in Plant Breeding, *Bureau of Plant Ind. Bull. 243,* U.S.D.A., pp. 1-58.

Ewens, W. J., 1979, *Mathematical Population Genetics. Biomathematics, vol. 9,* Springer-Verlag, Berlin.

Falconer, D. S., 1960, *Introduction to Quantitative Genetics,* Oliver and Boyd, Edinburgh.

Falconer, D. S., 1981, *Introduction to Quantitative Genetics,* 2nd ed., Longmans, London.

Fisher, R. A., 1918, The Correlation Between Relatives on the Supposition of Mendelian Inheritance, *R. Soc. (Edinburgh) Trans.* **52**:399-433.

Fisher, R. A., 1958, *The Genetical Theory of Natural Selection,* 2nd. rev. ed., Dover, New York.

Griffing, B., 1960, Theoretical Consequences of Truncation Selection Based on the Individual Phenotype, *Aust. J. Biol. Sci.* **13**:307-343.

Haldane, J. B. S., 1932, *The Causes of Evolution,* Longmans, Green, London.

Hill, W. G., and A. Robertson, 1966, The Effect of Linkage on Limits to Artificial Selection, *Genet. Res. (Cambridge)* **8**:269-294.

Hull, F. H., 1945, Recurrent Selection for Specific Combining Ability in Corn, *J. Am. Soc. Agron.* **37**:134-145.

Kimura, M., 1957, Some Problems of Stochastic Processes in Genetics, *Ann. Math. Stat.* **28**:882-901.

Kimura, M., 1958, On the Change of Population Fitness by Natural Selection, *Heredity* **12**:145-167.

Kimura, M., 1965, Attainment of Quasi Linkage Equilibrium When Gene Frequencies Are Changing by Natural Selection, *Genetics* **52**:875-890.

Latter, B. D. H., 1965a, The Response to Artificial Selection Due to Autosomal Genes of Large Effect. I. Changes in Gene Frequency at an Additive Locus, *Aust. J. Biol. Sci.* **18**:585-598.

Latter, B. D. H., 1965b, The Response to Artificial Selection Due to Autosomal Genes of Large Effect. II. The Effects of Linkage on Limits to Selection in Finite Populations, *Aust. J. Biol. Sci.* **18**:1009-1024.

Nagylaki, T., 1976, The Evolution of One- and Two- Locus Systems, *Genetics* **83:**583–600.

Price, G. R., 1972, Fisher's "Fundamental Theorem" Made Clear, *Ann. Hum. Genet.* **36:**129–140.

Rendel, J. M., and A. Robertson, 1950, Estimation of Genetic Gain in Milk Yield by Selection in a Closed Herd of Dairy Cattle, *J. Genet.* **50:**1–8.

Robertson, A., 1966, A Mathematical Model of the Culling Process in Dairy Cattle, *Anim. Prod.* **8:**95–108.

Robertson, A., 1970, A Theory of Limits in Artificial Selection With Many Linked Loci, in *Mathematical Topics in Population Genetics,* K. Kojima, ed., Springer-Verlag, Berlin, pp. 246–288.

Robertson, A., and J. M. Rendel, 1950, Use of Progeny Testing With Artificial Selection in Dairy Cattle, *J. Genet.* **50:**21–31.

# 10

Copyright © 1930 by Oxford University Press
Reprinted by permission from pages 30-37 of *The Genetical Theory of Natural Selection,* Clarendon Press, Oxford, 1930, 272p.

# THE FUNDAMENTAL THEOREM OF NATURAL SELECTION

## R. A. Fisher

### The genetic element in variance

Let us now consider the manner in which any quantitative individual measurement, such as human stature, may depend upon the individual genetic constitution. We may imagine, in respect of any pair of alternative genes, the population divided into two portions, each comprising one homozygous type together with half of the heterozygotes, which must be divided equally between the two portions. The difference in average stature between these two groups may then be termed the average excess (in stature) associated with the gene substitution in question. This difference need not be wholly due to the single gene, by which the groups are distinguished, but possibly also to other genes statistically associated with it, and having similar or opposite effects. This definition will appear the more appropriate if, as is necessary for precision, the population used to determine its value comprises, not merely the whole of a species in any one generation attaining maturity, but is conceived to contain all the genetic combinations possible, with frequencies appropriate to their actual probabilities of occurrence and survival, whatever these may be, and if the average is based upon the statures attained by all these genotypes in all possible environmental circumstances, with frequencies appropriate to the actual probabilities of encountering these circumstances. The statistical concept of the excess in stature of a given gene substitution will then be an exact one, not dependent upon chance as must be any practical estimate of it, but only upon the genetic nature and environmental circumstances of the species. The excess in a factor will usually be influenced by the actual frequency ratio $p:q$ of the alternative genes, and may also be influenced, by way of departures from random mating, by the varying reactions of the factor in question with other factors; it is for this reason that its value for the purpose of our argument is defined in the precise statistical manner chosen, rather than in terms of the average sizes of pure genotypes, as would be appropriate in specifying such a value in an experimental population, in which mating is under control, and in which the numbers of the different genotypes examined is at the choice of the experimenter.

For the same reasons it is also necessary to give a statistical definition of a second quantity, which may be easily confused with that just defined, and may often have a nearly equal value, yet which must be distinguished from it in an accurate argument; namely the average effect produced in the population as genetically constituted, by the substitution of the one type of gene for the other. By whatever rules mating, and consequently the frequency of different gene combinations, may be governed, the substitution of a small proportion of the genes of one kind by the genes of another will produce a definite proportional effect upon the average stature. The amount of the difference produced, on the average, in the total stature of the population, for each such gene substitution, may be termed the average effect of such substitution, in contra-distinction to the average excess as defined above. In human stature, for example, the correlation found between married persons is sufficient to ensure that each gene tending to increase the stature must be associated with other genes having a like effect, to an extent sufficient to make the average excess associated with each gene substitution exceed its average effect by about a quarter.

If $a$ is the magnitude of the average excess of any factor, and $\alpha$ the magnitude of the average effect on the chosen measurement, we shall now show that the contribution of that factor to the genetic variance is represented by the expression $pq\alpha a$.

The variable measurement will be represented by $x$, and the relation of the quantities $a$ to it may be made more clear by supposing that for any specific gene constitution we build up an 'expected' value, $X$, by adding together appropriate increments, positive or negative, according to the natures of the genes present. This expected value will not necessarily represent the real stature, though it may be a good approximation to it, but its statistical properties will be more intimately involved in the inheritance of real stature than the properties of that variate itself. Since we are only concerned with variation we may take as a primary ingredient of the value of $X$, the mean value of $x$ in the population, and adjust our positive and negative increments for each factor so that these balance each other when the whole population is considered. Since the increment for any one gene will appear $p$ times to that for its alternative gene $q$ times in the whole population, the two increments must be of opposite sign and in the ratio $q : (-p)$. Moreover, since their difference must be $a$, the actual values cannot but be $qa$ and $(-pa)$ respectively.

The value of the average excess $a$ of any gene substitution was obtained by comparing the average values of the measurement $x$ in two moieties into which the population can be divided. It is evident that the values of $a$ will only be properly determined if the same average difference is maintained in these moieties between the values of $X$, or in other words if in each such moiety the sum of the deviations, $x - X$, is zero. This supplies a criterion mathematically sufficient to determine the values of $a$, which represent in the population concerned the average effects of the gene substitutions. It follows that the sum for the whole population of the product $X(x - X)$ derived from each individual must be zero, for each entry $qa$ or $(-pa)$ in the first term will in the total be multiplied by a zero, and this will be true of the items contributed by every factor severally. It follows from this that if $X$ and $x$ are now each measured from the mean of the population, the variance of $X$, which is the mean value of $X^2$, is equal to the mean value of $Xx$. Now the mean value of $Xx$ will involve $a$ for each Mendelian factor; for $X$ will contain the item $qa$ in the $p$ individuals of one moiety and $(-pa)$ in the $q$ individuals of the other, and since the average values of $x$ in these two moieties differ by $a$, the mean value of $Xx$ must be the sum for all factors of the quantities $pqaa$. Thus the variance of $X$ is shown to be $W = \Sigma(pqaa)$ the summation being taken over all factors, and this quantity we may distinguish as the *genetic* variance in the chosen measurement $x$. That it is essentially positive, unless the effect of every gene severally is zero, is shown by its equality with the variance of $X$. An extension of this analysis, involving no difference of principle, leads to a similar expression for cases in which one or more factors have more than two different genes or allelomorphs present.

The appropriateness of the term genetic variance lies in the fact that the quantity $X$ is determined solely by the genes present in the individual, and is built up of the average effects of these genes. It therefore represents the genetic potentiality of the individual concerned, in the aggregate of the mating possibilities actually open to him, in the sense that the progeny averages (of $x$, as well as of $X$) of two males mated with an identical series of representative females will differ by exactly half as much as the genetic potentialities of their sires differ. Relative genetic values may therefore be determined experimentally by the diallel method, in which each animal tested is mated to the same series of animals of the opposite sex, provided that a large number of offspring can be obtained from each such mating. Without obtaining individual values, the genetic variance of the population may be derived from the correlations between relatives, provided these correlations are accurately obtained. For this purpose the square of the parental correlation divided by the grandparental correlation supplies a good estimate of the fraction, of the total observable variance of the measurement, which may be regarded as genetic variance.

It is clear that the actual measurements, $x$, obtained in individuals may differ from their genetic expectations by reason of fluctuations due to purely environmental circumstances. It should be noted that this is not the only cause of difference, for even if environmental fluctuations were entirely absent, and the actual measurements therefore determined exactly by the genetic composition, these measurements, which may be distinguished as *genotypic*, might still differ from the genetic values, $X$. A good example of this is afforded by dominance, for if dominance is complete the genotypic value of the heterozygote will be exactly the same as that of the corresponding dominant homozygote, and yet these genotypes differ by a gene substitution which may materially affect the genetic potentiality represented by $X$, and be reflected in the average measurement of the offspring. A similar cause of discrepancy occurs when gene substitutions in different factors are not exactly additive in their average effects. The genetic variance as here defined is only a portion of the variance determined genotypically, and this will differ from, and usually be somewhat less than, the total variance to be observed.

It is consequently not a superfluous refinement to define the purely genetic element in the variance as it exists objectively, as a statistical character of the population, different from the variance derived from the direct measurement of individuals.

## Natural Selection

The definitions given above may be applied to any characteristic whatever; it is of special interest to apply them to the special characteristic $m$ which measures the relative rate of increase or decrease. The two groups of individuals bearing alternative genes, and consequently the genes themselves, will necessarily either have equal or unequal rates of increase, and the difference between the appropriate values of $m$ will be represented by $a$, similarly the average effect upon $m$ of the gene substitution will be represented by $\alpha$. Since $m$ measures fitness to survive by the objective fact of representation in future generations, the quantity $pqa\alpha$ will represent the contribution of each factor to the genetic variance in fitness; the total genetic variance in fitness being the sum of these contributions, which is necessarily positive, or, in the limiting case, zero. Moreover, any increase $dp$ in the proportion of one type of gene at the expense of the other will be accompanied by an increase $\alpha dp$ in the average fitness of the species, where $\alpha$ may of course be negative; but the definition of $a$ requires that the ratio $p : q$ must be increasing in geometrical progression at a rate measured by $a$, or in mathematical notation that

$$\frac{d}{dt}\log\left(\frac{p}{q}\right) = a$$

which may be written

$$\left(\frac{1}{p} + \frac{1}{q}\right)dp = a\,dt,$$

or

$$dp = pqa\,dt$$

whence it follows that,

$$\alpha\,dp = pqa\alpha\,dt$$

# FUNDAMENTAL THEOREM OF NATURAL SELECTION

and, taking all factors into consideration, the total increase in fitness,

$$\Sigma(a\,dp) = \Sigma(pqaa)dt = Wdt.$$

If therefore the time element $dt$ is positive, the total change of fitness $Wdt$ is also positive, and indeed the rate of increase in fitness due to all changes in gene ratio is exactly equal to the genetic variance of fitness $W$ which the population exhibits. We may consequently state the fundamental theorem of Natural Selection in the form:

*The rate of increase in fitness of any organism at any time is equal to its genetic variance in fitness at that time.*

The rigour of the demonstration requires that the terms employed should be used strictly as defined; the ease of its interpretation may be increased by appropriate conventions of measurement. For example, the ratio $p:q$ should strictly be evaluated at any instant by the enumeration, not necessarily of the census population, but of all individuals having reproductive value, weighted according to the reproductive value of each.

Since the theorem is exact only for idealized populations, in which fortuitous fluctuations in genetic composition have been excluded, it is important to obtain an estimate of the magnitude of the effect of these fluctuations, or in other words to obtain a standard error appropriate to the calculated, or expected, rate of increase in fitness. It will be sufficient for this purpose to consider the special case of a population mating and reproducing at random. It is easy to see that if such chance fluctuations cause a difference $\delta p$ between the actual value of $p$ obtained in any generation and that expected, the variance of $\delta p$ will be

$$\frac{pq}{2n},$$

where $n$ represents the number breeding in each generation, and $2n$ therefore is the number of genes in the $n$ individuals which live to replace them. The variance of the increase in fitness, $a\delta p$, due to this cause, will therefore be

$$\frac{1}{2n}(pqa^2),$$

and since, with random mating, the chance fluctuation in the different gene ratios will be independent, and the values of $a$ and $\alpha$ are no longer distinct, it follows that, on this condition, the rate of increase of fitness, when measured over one generation, will have a standard error due to random survival equal to

$$\frac{1}{T}\sqrt{\frac{W}{2n}}$$

where $T$ is the time of a generation. It will usually be convenient for each organism to measure time in generations, and if this is done it will be apparent from the large factor $2n$ in the denominator, that the random fluctuations in $W$, even measured over only a single generation, may be expected to be very small compared to the average rate of progress. The regularity of the latter is in fact guaranteed by the same circumstance which makes a statistical assemblage of particles, such as a bubble of gas obey, without appreciable deviation, the laws of gases. A visible bubble will indeed contain several billions of molecules, and this would be a comparatively large number for an organic population, but the principle ensuring regularity is the same. Interpreted exactly, the formula shows that it is only when the rate of progress, $W$, when time is measured in generations, is itself so small as to be comparable to $1/n$, that the rate of progress achieved in successive generations is made to be irregular. Even if an equipoise of this order of exactitude, between the rates of death and reproduction of different genotypes, were established, it would be only the rate of progress for spans of a single generation that would be shown to be irregular, and the deviations from regularity over a span of 10,000 generations would be just a hundredfold less.

It will be noticed that the fundamental theorem proved above bears some remarkable resemblances to the second law of thermodynamics. Both are properties of populations, or aggregates, true irrespective of the nature of the units which compose them; both are statistical laws; each requires the constant increase of a measurable quantity, in the one case the entropy of a physical system and in the other the fitness, measured by $m$, of a biological population. As in the physical world we can conceive of theoretical systems in which dissipative forces are wholly absent, and in which the entropy consequently remains constant, so we can conceive, though we need not expect to find, biological populations in which the genetic variance is absolutely zero, and in which fitness does not increase. Professor Eddington has recently remarked that 'The law that entropy always increases—the second law of thermodynamics—holds, I think, the supreme position among the laws of nature'. It is not a little instructive that so similar a law should hold the supreme position among the biological sciences. While it is possible that both may ultimately be absorbed by some more general principle, for the present we should note that the laws as they stand present profound differences— (1) The systems considered in thermodynamics are permanent; species on the contrary are liable to extinction, although biological improvement must be expected to occur up to the end of their existence. (2) Fitness, although measured by a uniform method, is qualitatively different for every different organism, whereas entropy, like temperature, is taken to have the same meaning for all physical systems. (3) Fitness may be increased or decreased by changes in the environment, without reacting quantitatively upon that environment. (4) Entropy changes are exceptional in the physical world in being irreversible, while irreversible evolutionary changes form no exception among biological phenomena. Finally, (5) entropy changes lead to a progressive disorganization of the physical world, at least from the human standpoint of the utilization of energy, while evolutionary changes are generally recognized as producing progressively higher organization in the organic world.

The statement of the principle of Natural Selection in the form of a theorem determining the rate of progress of a species in fitness to survive (this term being used for a well-defined statistical attribute of the population), together with the relation between this rate of progress and its standard error, puts us in a position to judge of the validity of the objection which has been made, that the principle of Natural Selection depends on a succession of favourable chances. The objection is more in the nature of an innuendo than of a criticism, for it depends for its force upon the ambiguity of the word chance, in its popular uses. The income derived from a Casino by its proprietor may, in one sense, be said to depend upon a succession of favourable chances, although the phrase contains a suggestion of improbability more appropriate to the hopes of the patrons of his establishment. It is easy without any very profound logical analysis to perceive the difference between a succession of favourable deviations from the laws of chance, and on the other hand, the continuous and cumulative action of these laws. It is on the latter that the principle of Natural Selection relies.

*A Mathematical Theory of Natural and Artificial Selection. Part VII. Selection intensity as a function of mortality rate.* By Mr J. B. S. HALDANE, Trinity College.

[*Received* 12 November, *read* 8 December 1930.]

The assumption is often made that when competition is extremely intense at any stage in a life cycle, natural selection is bound to be intense also. This assumption will be examined quantitatively and it will be shown that the intensity of selection may diminish and become negative at high rates of elimination, while at its best its increase is extremely slow.

The intensity of competition is measured by the ratio, $z$, of organisms eliminated, to survivors. This may be small, e.g. $z = 0.1$ or less for the period between birth and maturity in civilised human societies. It may exceed $10^6$ in marine organisms producing many million eggs per year, or spermatozoa of which $10^9$ are ejaculated at a time. But in few cases can it exceed $10^{12}$.

Confining ourselves for the moment to a population consisting of two types $A$ and $B$, the intensity of selection is measured by the coefficient of selection $k$, where the ratio of $A$ to $B$ is increased $1 + k$ times as the result of selection. $k$ is taken to be small throughout the argument.

Consider a character whose measure $x$ is normally distributed, according to Gauss' law, in the $A$ and $B$ groups, the standard deviation being the same in each, and the differences between the means and the standard deviations being small in comparison with the standard deviations. For example, Johansen (1926) found the mean breadths of 8·091 and 8·152 mm. and standard deviations of ·400 and ·405 mm. in two lines, $BB$ and $GG$, of beans, the difference being clearly significant in the first case, doubtfully so in the second. If all individuals in which the variate $x$ falls below a certain value are eliminated by selection, we can readily calculate the proportion of the whole population eliminated, and the proportion of $A$ to $B$ among the survivors.

Conditions are not grossly dissimilar under natural selection. We may imagine a variate, to be called viability, which is normally distributed and such that only those individuals possessing more than a certain viability survive. The large size of its standard deviation compared to the difference of the mean values would signify the relatively large part played by chance in natural selection. The best studied case is that of pollen-tube growth, described by Buchholz and Blakeslee (1929). Here those tubes which arrive first at the ovules are selected. The distribution of growth rates is definitely skew, but the skewness is not likely to affect the general character of the result if the two types compared are sufficiently similar. Where viability depends on a greater variety of accidental causes, as is generally the case, the distribution is likely to be more normal.

Without loss of generality we can put the initial numbers of $A$ and $B$ equal, and take the mean value of $x$ as zero and its standard deviation as unity. We suppose the mean value of $x$ for the $A$ type to be $\lambda$ and its standard deviation to be $1+\mu$, the corresponding values for $B$ being $-\lambda$ and $1-\mu$. The ratio of the frequency of any value of $x$ in the population to that in a strictly normal population is $1-(1+x^2)(\lambda+\mu x)^2 +$ higher powers of $\lambda$ and $\mu$. Hence the mixed population is normal to the second order of small quantities provided that $\lambda x$ and $\mu x^2$ are small. $x$ will rarely exceed 7 even in a population of $10^{12}$.

Then provided that the population is numerous compared with both the numbers surviving and eliminated, the survivors will be those members for which $x > X$, $X$ being given by

$$\frac{1}{\sqrt{2\pi}} \int_X^\infty e^{-\frac{1}{2}x^2} dx = \frac{1}{z+1}.$$

The proportion of the $A$ type exceeding this value is

$$\frac{1}{\sqrt{2\pi}(1+\mu)} \int_X^\infty e^{\frac{-(x-\lambda)^2}{2(1+\mu)^2}} dx = \frac{1}{\sqrt{2\pi}} \int_{(X-\lambda)/(1+\mu)}^\infty e^{-\frac{1}{2}t^2} dt$$

$$= \frac{1}{\sqrt{2\pi}} \int_X^\infty e^{-\frac{1}{2}t^2} dt + \frac{1}{\sqrt{2\pi}} \int_{(X-\lambda)/(1+\mu)}^X e^{-\frac{1}{2}t^2} dt$$

$$= \frac{1}{z+1} + \frac{(\lambda+\mu X)}{\sqrt{2\pi}} e^{-\frac{1}{2}X^2}, \text{ approximately,}$$

so that
$$k = \frac{2(\lambda+\mu X)(z+1)}{\sqrt{2\pi}} e^{-\frac{1}{2}X^2},$$

the value of $X$ being found as above.

First consider the case when $\mu = 0$, i.e. the standard deviations are equal. The value of $q = k/2\lambda$ is plotted against $\log_{10} z$ in the figure (calculated from Pearson's (1924) tables). When $z = 1$, $q = (2/\pi)^{\frac{1}{2}} = \cdot 798$. When $z$ is large, we may put

$$\frac{1}{1+z} = \frac{e^{-\frac{1}{2}\lambda^2}}{\sqrt{2\pi} X} \text{ approximately,}$$

whence
$$q = \sqrt{\log_e \frac{z^2}{2\pi}} \text{ approximately.}$$

So the intensity of selection only increases extremely slowly with $z$. Thus $q$ is only doubled when $z$ increases from 1 to about 6·4, or from 10 to 1800, and only increased 9 times over the whole range from 1 to $10^{12}$. On the other hand when $z$ is small, $q$ approximates to $z\sqrt{-\log_e 2\pi z^2}$, a small quantity of the order of $z$, and is roughly proportional to $z$ over small ranges. For example, when $z = 10^{-4}$, $q = \cdot0004$, and when $z = 10^{-2}$, $q = \cdot03$. To sum up, the efficiency of selection increases very rapidly with $z$ until about 80 % of the population is eliminated, and thereafter very slowly.

The only experimental data known to me are those of Correns (1918) who measured the sex-ratio of *Melandrium* when pollinated with mixtures of male-producing and female-producing pollen, and used numbers of pollen-grains either less than that of ovules, so that $z = 0$, or greater, in various proportions. When $z$ approximated to 6, $k$ was 0·195; when $z$ was about 142, $k$ rose to ·710. The value of $k$ thus increased only 3·6 times while $z$ increased 24 times. According to Fig. 1 the increase of $q$ should be only 1·8 times. But the values of both $z$ and $k$ are very uncertain, thus the value ·195 of $k$ has a standard error of ·07. Figures well within the limit of experimental error would give complete agreement with the theory. Moreover $\mu$ is probably not zero nor is the distribution of growth rates normal. Certainly, however, $k$ does not increase anything like proportionally to $z$, even when $\lambda$ has the somewhat large value of 0·1, which as we shall see later will tend to exaggerate the rate of increase of $k$ with $z$.

When $\mu$ is not zero the case is rather more complicated. If $\lambda$ and $\mu$ have the same sign, i.e. the type with the largest mean has also the largest standard deviation, selection favours them unless $X$ is negative and less than $-\lambda/\mu$. In this case the group of lower average viability will be favoured when competition is very slight, but their selective advantage will be extremely small at best. For example, Johansen's bean line $GG$ had a mean breadth of 8·152 mm. with standard deviation ·415, while the corresponding figures for the line $MM$ were 7·976 and ·348. Hence $\lambda = \cdot101$ and $\mu = \cdot076$. Selection for greater breadth would favour line $MM$ slightly when $z$ was less than 0·1, while for higher values $GG$ would be considerably favoured, $k$ being ·159 when $z = 1$.

In many cases the coefficient of variation of the two groups is approximately equal, i.e. $\mu = \lambda$. In four of Johansen's pure lines the coefficient of variation for length only varied between 5·0 % and 4·4 %. In a family described on p. 136 two slightly impure genotypes had coefficients of 7·0 % and 6·8 %, the heterozygote a coefficient of 6·8 %. It thus seems likely that in a large number of cases $\lambda$ and $\mu$ will be very nearly equal. When this is the case $q$ is small and negative for low mortalities, attaining a minimum value of $-0\cdot070$ when $z = \cdot066$, i.e. with a mortality of 6·2 %, vanishes when $z = \cdot1886$, i.e. with a mortality of 15·9 %, and then increases, being ·798 when $z = 1$ (50 % mortality), 23·4 when $z = 10^6$, and 57·6 when $z = 10^{12}$. For large values of $z$, $k$ varies as $\log z$.

Even though $\lambda$ and $\mu$ are not quite equal, in a very large number, perhaps the majority of cases, $\lambda/\mu$ will lie between ·5 and 2, and the direction of selection will be reversed at a mortality of between 31 °/₀ and 2·3 °/₀. It is of interest to note that during the last fifty

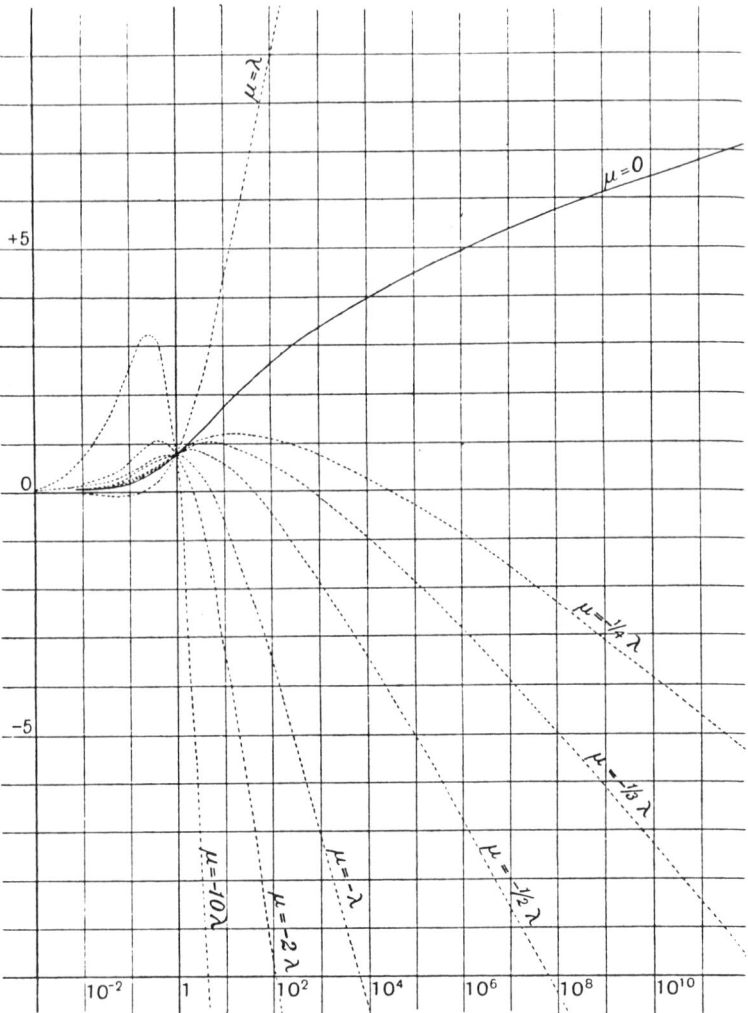

Fig. 1. Abscissa: $\text{Log}_{10} z$. Ordinate: $q$, which measures the intensity of selection.

years infantile mortality in most civilised countries has fallen from well above the critical value of 15·9 °/₀ to well below it. It seems probable, therefore, that the direction of selection for certain genes has been reversed.

*of natural and artificial selection* 135

If $\lambda$ and $\mu$ have opposite signs, the group of highest average viability will be favoured until $\bar{X}$ exceeds $-\lambda/\mu$, i.e. until

$$z > \frac{\sqrt{2\pi}}{\int_{-\lambda/\mu}^{\infty} e^{-\frac{1}{2}x^2} dx} - 1.$$

If $-\lambda/\mu$ is fairly large, say greater than 3, this becomes approximately

$$z > \sqrt{2\pi}\frac{\lambda}{\mu} e^{\lambda^2/\mu^2}.$$

In practice however since $z$ rarely exceeds $10^{12}$, the direction of selection is not likely to be reversed if $-\mu < \lambda/7$.

In the figure $q$ is plotted against $z$ when

$$\mu = 0, \lambda, -10\lambda, -2\lambda, -\tfrac{1}{2}\lambda, -\tfrac{1}{3}\lambda, -\tfrac{1}{4}\lambda.$$

The maximum intensity of selection is reached when $z$ is greater or less than unity according as $-\mu/\lambda$ is less or greater than unity.

But wherever it is not zero the results of slight and intense competition are in opposite directions, although the required competition may sometimes be too intense or the selection too slight to be of practical importance. This is in full accordance with the views of Bidder, who points out that, where "cataclasms" occasionally destroy the vast majority of a species, characters which are useless or worse under normal conditions may be selected. He specially mentions the case of a violent or erratic response of an animal by migration or otherwise to unfavourable environments, which would be likely to lower the average viability, but increase its dispersion.

It is easy to extend the above arguments to a population consisting of many genotypes. To take one example, suppose that $\mu = 0$, but $\lambda$ is normally distributed with a standard deviation $\sigma$. Then the new frequency of any value of $\lambda$ will be given by

$$df = \frac{1+q\lambda}{\sqrt{2\pi}\,\sigma} e^{-\frac{1}{2}\lambda^2/\sigma^2} d\lambda.$$

The new mean value of $\lambda$ will therefore be

$$\frac{\int_{-\infty}^{\infty} \lambda(1+q\lambda) e^{-\frac{1}{2}\lambda^2/\sigma^2} d\lambda}{\int_{-\infty}^{\infty} (1+q\lambda) e^{-\frac{1}{2}\lambda^2/\sigma^2} d\lambda}$$

$$= \frac{q}{\sqrt{2\pi}\,\sigma} \int_{-\infty}^{\infty} \lambda^2 e^{-\frac{1}{2}\lambda^2/\sigma^2} d\lambda, \text{ approximately,}$$

$$= q\sigma^2.$$

All the results here given apply only to the results of a single act of selection. The way in which the population will change depends on the way in which the mean viability and its dispersion are inherited, and on the system of mating. The effects of these have been considered in former papers of this series.

The theory can readily be extended to cover cases where $\lambda$ and $\mu$ are no longer small, but the results are no longer elegant or simple. In particular the proportion of types in the original population must be taken into account. When the difference of the means is large compared with the standard deviations, $A$ being more viable than $B$, it is convenient to take $\kappa$ as the intensity of selection, where $u$ is the ratio of $A$ to $B$ before, and $ue^\kappa$ after selection. $\kappa$ and $k$ are of course equal when both are small. It is clear that

$$\kappa = \log_e \left( \frac{1+z}{1+uz} \right),$$

approximately, so that the intensity of selection is proportional to $z$ when this is small, but becomes very large when $z = 1/u$. Such intense selection occasionally occurs in nature, for example between normal types and semi-lethal mutants, but its results as between competing types of organism would be very rapid, and it is not of much interest in a study of evolution. In general when $\lambda$ is not small, the value of $q$ for any value of $z$ will be increased.

## REFERENCES.

G. P. BIDDER (1930). "The importance of cataclasms in evolution." *Nature*, **125**, p. 783.

J. T. BUCHHOLZ and A. F. BLAKESLEE (1929). "Pollen-tube growth in crosses between balanced chromosomal types of *Datura Stramonium*." *Genetics*, **14**, p. 538.

C. CORRENS. "Fortsetzung der Versuche zur experimentellen Verschiebung des Geschlechtsverhältnisses." *Sitzber. K. Preuss. Akad. Wiss.* 1918, p. 1175.

W. JOHANSEN (1926). *Elemente der exakten Erblichkeitslehre*, p. 171.

K. PEARSON (1924). *Tables for Statisticians and Biometricians*.

## A Breeding Procedure Designed To Make Maximum Use of Both General and Specific Combining Ability[1]

R. E. COMSTOCK, H. F. ROBINSON, AND P. H. HARVEY[2]

UTILIZATION of heterosis between homozygous lines is best exemplified in corn breeding practice. Breeding programs directed at the development of improved commercial corn hybrids consist in general of two phases. The first involves the development of lines and choice of those to be tested in hybrid combinations; the second, the comparison of hybrids among the selected lines. In the development and selection of lines attention is commonly given to genetic diversity in origin, plant characters known to be important (including some attention to yield of inbreds), and general combining ability of the lines. In some programs general combining ability is measured only in the final stage prior to comparison of specific hybrids. However, selection among early generation inbreds on the basis of general combining ability was suggested by Jenkins (5)[3] and has been used extensively by other workers. It is not until the second phase of the process that attention is given specific combining ability, except for that inherent in the development of lines from several genetically divergent sources.

Sprague and Tatum (7) have demonstrated the importance of specific combining ability as a source of variance among single-crosses. Their findings are a logical consequence of the sorts of gene action that could be responsible for heterosis. In view of the role of specific combining ability in determining the performance of hybrids, it obviously would be desirable if procedures were available for guiding the development of lines in such a way that their combining ability in predetermined combinations would be enhanced.

East (3) has postulated a hereditary mechanism in corn involving sets of multiple alleles in which certain of the heterozygous genotypes possible at a given locus are superior to any of the possible homozygous genotypes. Hull (4) advanced reasons for believing that genes conditioning corn yield exhibit over-dominance. While the two proposals are not the same, they have in common the concept of heterozygosity superior to homozygosity of the best genes in a population. It is this aspect of both proposals that has special significance relative to breeding methods. Hull pointed out that with respect to loci at which over-dominance (heterozygote superior to best homozygote) is operative, improvement of hybrids resulting from selection among lines on the basis of either their own performance or their general combining ability will level off at a point below the potential maximum yield of hybrids. He proposed a procedure which he calls recurrent selection for specific combining ability that would be more effective with respect to improvement stemming from genes that exhibit over-dominance. He pointed out, however, that it would be inefficient relative to genes that are partially dominant.

Crow (2) concluded that heterosis much in excess of 5% of the mean of the random breeding population cannot be explained on the basis of simple dominance alone, but must rest on other sorts of genotype effects such as interactions of non-allelic genes or superiority of the heterozygote to either homozygote, i.e., over-dominance. Robinson et al. (6) reported an estimate of the average degree of dominance of genes conditioning grain yield in corn that was in the over-dominance range. The method which they employed for the estimate is described in detail by Comstock and Robinson (1). It is pointed out in both of these manuscripts that linkage could be responsible for genetic behaviour which on the basis of present evidence could not be distinguished from that which would result from over-dominance.

At present available evidence does not establish either the existence or nonexistence of over-dominance. Moreover, even though there were over-dominance at some loci, partial dominance could be the rule at others. In view of this situation the authors set out to discover a breeding and selection method that would be effective regardless of the level of dominance and which by giving attention to specific combining ability from the outset, might be more effective for genes showing complete or partial dominance than are current procedures. The authors are using the term recurrent reciprocal selection to designate the method described below. For brevity it will be referred to in what follows as simply reciprocal selection.

### Description of the Reciprocal Selection Method

Foundation material from two sources is used. The hybrid or hybrids to be developed will involve crossing material descended from these two sources, hence the sources should be as genetically divergent as possible. Two varieties, two synthetics, or the $F_2$ generation plants of the two single crosses involved in a successful double cross can serve as the source material.

---

[1] Contribution from the Institute of Statistics and Agronomy Department, North Carolina Experiment Station, Raleigh, N. C., in cooperation with the Division of Cereal Crops and Diseases, Bureau of Plant Industry, Soils, and Agricultural Engineering, U. S. Dept. of Agriculture, as Journal Paper 307. Presented in part at the annual meeting of the American Society of Agronomy in Fort Collins, Colo., August 24-27, 1948. Received for publication, February 21, 1949.
[2] Professor and associate professor, Institute of Statistics, and agronomist in the Division of Cereal Crops and Diseases and research professor in agronomy, North Carolina State College, respectively.
[3] Figures in parentheses refer to "Literature Cited", p. 367.

$s_0$ or $s_1$ plants from source A are self-pollinated and at the same time out-crossed to plants from source B. Selection is based on experimental comparison of testcross progenies and selected plants are interbred the third year using their selfed seed produced the first year. The cycle is reinitiated the fourth year. Source B plants are tested against source A plants in the same way. Outlined in greater detail the procedure would be as indicated below. (The numbers of plants and progenies indicated were arbitrarily chosen and could of course be varied. Attention has not been given to the question of optima).

*Year 1.*—Out-cross each of about 200 plants from source A with four or five plants taken at random from source B, and each of about 200 plants from source B with four or five plants from source A. Self-pollinate all plants used as pollen parent in these out-crosses.

*Year 2.*—Conduct two field trial comparisons of the progeny of crosses made in year 1. The one would involve progenies of source A plants as pollen parent; the other, progenies of source B plants as pollen parent. All seed from each of the four or five crosses involving a single pollen parent would be bulked to produce a single progeny from that parent. Thus there would be at most 200 entries in each of the two trials. The actual number would depend on success of pollinations in year 1 and the design chosen for the progeny comparisons. Perhaps 169 progenies of each group would be compared in 13 by 13 lattice designs.

*Year 3.*—Plant seed produced by self-fertilization in year 1 using seed from only those plants in each of the source groups (A and B) whose progenies were superior in the field trials of year 2. Within each source group make all or a large number of the possible single crosses between plants from which seed was planted.

*Years 4, 5, and 6.*—Repeat the procedures of years 1, 2, and 3 using as a starting point the group A and B seed produced in year 3.

By testing source A plants in crosses with source B plants, selection pressure for those specific genes which contribute most to the cross of material from the two sources is insured. The criteria for evaluation of the test cross progenies should be those on which commercial hybrids are evaluated for the geographical area involved. For example, progenies that exhibit lodging or disease susceptibility should not be considered even though their yield is high. In years favorable to their measurement selection might well be based completely or in large part on such agronomic characters.

Sufficient plants should be selected in each generation to hold inbreeding within the two source material groups at a low level since otherwise the within-group variability on which selection operates will be lost. For the same reason the plants selected in any generation should not all trace to one or two matings made in an earlier cycle of the program. To insure this, controlled pollination accompanied by recording of pedigree information should be practiced in the interbreeding phase of the cycle.

Commercial seed would be produced by crosses between the A and B selection material groups; however, the amount of inbreeding done prior to making the seed production cross could vary. At the one extreme relatively pure lines could be developed for use in double crosses of the type $(A_1 \times A_2) \times (B_1 \times B_2)$ where $A_1$ and $A_2$ are lines developed from source A material and $B_1$ and $B_2$ are lines from source B material. The other extreme would involve crossing noninbred descendants of random matings among the selected plants within each of the two groups.

## Comparison of the Potentialities of Reciprocal Selection with Those of Other Procedures

The two other breeding methods considered were the following:

*Method 1: Selection based on general combining ability with a common tester series.*—The tester series assumed was one consisting of at least two single crosses between pure lines. Use of double crosses or varieties would involve loss of efficiency due to genetic variation among tester parent plants involved in the top crosses. The idea of recurrent selection cycles is as applicable in this method as in that of reciprocal selection. It was therefore assumed that plants from two or more genetically divergent sources would be tested and that, as in the reciprocal selection method, selected plants of each source would be interbred and a new selection cycle initiated. Thus the only difference between the two methods is in the tester. In this method all selection material is tested against the same tester. In the reciprocal method source A plants are tested against source B plants and source B plants against source A plants.

*Method 2: Recurrent selection for specific combining ability as proposed by Hull.*—This involves one segregating pool of material from which plants are topcrossed to a tester. Selection is based on performance of the topcrossed progenies, selected plants are interbred using their selfed seed, and new topcrosses made for the next selection cycle using plants of the resulting population. The tester used is a pure line or single cross of pure lines and is one of the parents of the commercial hybrid, of which the improved selection material is the other parent. While there are practical objections to use of a pure line as the tester, it was chosen as the basis of this comparison since in the genetic situation for which this method was originated maximum potentialities for improvement in commercial hybrid performance would not be realized using a single cross tester.

Recurrent reciprocal selection, the procedure proposed in this manuscript, will be referred to as method 3.

## BASIS FOR THE COMPARISONS

The comparisons of the three methods were made subject to the following assumptions:
(a) No interactions of non-allelic genes.
(b) No more than two alleles per locus.
(c) Equilibrium state relative to the joint distribution of genotypes at linked loci.

It is highly improbable that any one of these assumptions is strictly valid for the general situation. However, the problem becomes more tractable when comparisons are made first for the simplified situation and the effects of non-validity of the assumptions are considered later. Some comments on these effects will be made later in the manuscript. It will suffice at this point to state that the general conclusions to be drawn concerning the methods are apparently not seriously affected as a consequence of basing the comparisons on a simplified genetic model.

## COMPARISON OF IMPROVEMENT LIMITS

The methods were compared first with respect to the potential limit of improvement by each. It was sufficient to consider the response at a single locus since the findings for one will cover possibilities with respect to any other locus.

Three specific situations were considered with respect to dominance. If the value of the genotypes possible at the locus of a gene pair (B, b) are symbolized as follows:

| Genotype | Value |
|---|---|
| BB | $2u$ |
| Bb | $u+au$ |
| bb | 0 |

$a$ is obviously a measure of the dominance involved in the action of the alleles, B and b.

If dominance is complete, $a = 1.0$; if dominance is partial, $a < 1.0$, and for over-dominance, $a > 1.0$. Values of $a$ considered were .5, 1.0, and 2.0.

Let B be a more favorable gene than its allele, b, $p$ be the frequency[4] of B in the material under selection and $q$ be its frequency in the material used as the tester parent.

The algebraic sign of the regression of $p$ on performance of test cross progenies indicates the direction of the selection effect on gene frequency. This regression is

$$\frac{\text{Cov. pY}}{V_Y}$$

where Y symbolizes the performance of test cross progenies; $V_Y$, the variance of progeny means; and

[4]The frequency of a gene refers to the number of that gene in a population as a fraction of the total number of loci occupied by it or an allele. For example, in a pure line all gene frequencies will be either 1.0 or zero depending on whether the line is homozygous for the gene in question or for an allele of that gene. In an $F_1$ between pure lines, gene frequencies will be 1.0, 0.5, or zero; in a tester composed of two such $F_1$'s they will be 1.0, .75, 0.5, .25, or zero, and in a variety they can vary from zero to 1.0.

Cov.$_{pY}$, the covariance of $p$ and Y. Since variances are always positive the sign of the regression is determined by its numerator. It can be shown that

$$\text{Cov. pY} = \frac{p(1-p)}{2}(1+f)[1+(1-2q)a]u \quad (1)$$

where $f$ is inbreeding of the material under selection in terms of Wright's coefficient. Since neither $p(1-p)$, $1+f$, nor $u$ can take a negative value, the sign of Cov. pY is determined by the factor $[1+(1-2q)a]$. It will be positive when $q < (1+a)/2a$ and negative when $q > (1+a)/2a$. When $q = (1+a)/2a$ it will equal zero and selection will have no effect on gene frequency. Values of $(1+a)/2a$ are listed below for the a-values to be considered.

| $a$ | $(1+a)/2a$ |
|---|---|
| .5 | 1.5 |
| 1.0 | 1.0 |
| 2.0 | .75 |

It follows that with two exceptions the average effect of selection in the cases to be considered will be to increase the frequency of B. When $a = 1.0$ and $q$, the frequency of B in the tester, $= 1.0$, selection will have no effect. When $a = 2.0$, selection will decrease $p$, the frequency of B in the selection material, whenever $q > .75$.

The comparison of limits of improvement for the three selection methods is summarized in Table 1. The tabled values can be verified by inspection keeping in mind the facts brought out above relative to the direction of the selection effect. For example, with method 1, $a = 2.0$, tester genotype BB, and both alleles present in both groups of selection material; selection would carry the frequency of B toward zero in both groups. When this limit has been reached all lines developed from the material would be of the bb genotype, hybrids between them would be the same, and the limiting genotypic value approached through selection would be zero.

The upper portion of the table relates only to methods 1 and 3 and to those situations in which the material of both source groups is completely homozygous for one of the alleles. Obviously selection will be ineffective in these situations regardless of method. The lower portion of the table covers all other situations relative to methods 1 and 3 and all situations relative to method 2.

The important points brought out by Table 1 are:
1. When $a < 1.0$ the improvement limit is the same for methods 1 and 3 but lower for method 2. The value $2u$ will be reached by the latter method only when B is present in both the tester and the selection material whereas it will be attainable by the other two methods whenever B is present in both selection material groups, an occurrence to be expected more frequently.

TABLE I.—*Limit of improvement in genotype at a single locus of hybrids.*

| Original genotype in | | a = 0.5 | | | a = 1.0 | | | a = 2.0 | | |
|---|---|---|---|---|---|---|---|---|---|---|
| Tester | Selection material* | | | | Method | | | | | |
| | | 1 | 2 | 3 | 1 | 2 | 3 | 1 | 2 | 3 |
| | BB  BB | 2u | | 2u | 2u | | 2u | 2u | | 2u |
| | BB  bb | 1.5u | | 1.5u | 2u | | 2u | 3u | | 3u |
| | bb  BB | 1.5u | | 1.5u | 2u | | 2u | 3u | | 3u |
| | bb  bb | 0 | | 0 | 0 | | 0 | 0 | | 0 |
| BB | B, b†  BB | 2u | 2u | 2u | 2u | 2u | 2u | 3u | 2u | 3u |
| | B, b  B, b | 2u | 2u | 2u | ‡ | 2u | 2u | 0 | 3u | 3u |
| | B, b  bb | 1.5u | 1.5u | 1.5u | ‡ | 2u | 2u | 0 | 3u | 3u |
| B, b (q<.75) | B, b  BB | 2u | | 2u | 2u | | 2u | 2u | | 3u |
| | B, b  B, b | 2u | | 2u | 2u | | 2u | 2u | | 3u |
| | B, b  bb | 1.5u | | 1.5u | 2u | | 2u | 2u | | 3u |
| B, b (q>.75) | B, b  BB | 2u | | 2u | 2u | | 2u | 3u | | 3u |
| | B, b  B, b | 2u | | 2u | 2u | | 2u | 0 | | 3u |
| | B, b  bb | 1.5u | | 1.5u | 2u | | 2u | 0 | | 3u |
| bb | B, b  BB | 2u | 1.5u | 2u | 2u | 2u | 2u | 2u | 2u | 3u |
| | B, b  B, b | 2u | 1.5u | 2u | 2u | 2u | 2u | 2u | 3u | 3u |
| | B, b  bb | 1.5u | 0 | 1.5u | 2u | 0 | 2u | 2u | 0 | 3u |

*With respect to methods 1 and 3, the two columns represent material from two genetically divergent sources; with respect to method 2, consider only the second column.
†B,b indicates that both alleles are present in the material, not that all individuals of the material are heterozygous.
‡Selection ineffective since all top-cross genotypes would have the same value.

2. When $a > 1.0$ the improvement limit will be essentially the same for methods 2 and 3 but much lower for method 1.

3. When $a = 1.0$ there would under no circumstances be much difference among the limits for the three methods. Method 1 would be somewhat less efficient if the tester involved were homozygous for the dominant allele at any considerable number of loci. Method 2 would suffer by comparison if there were insufficient genetic diversity represented in the selection material used.

### COMPARISON OF RATES OF IMPROVEMENT

In view of what has been learned from consideration of improvement limits, interest in expected improvement rates centers on comparison of reciprocal selection with selection based on general combining ability in the case of complete or partial dominance, and on comparison of the reciprocal method with Hull's method in the over-dominance case. Good estimates of absolute rates of improvement would require experimental data that are not now available. However, comparisons are possible on a relative basis.

Genetic change resulting from selection is a consequence of changes in gene frequencies. When selection is based on the performance of test cross progenies, as in all the methods under consideration, the regression of gene frequency in plants tested upon performance of their test cross progenies can, as indicated in the foregoing section, be shown to be

$$b_{pY} = \frac{p(1-p)(1+f)[1+a-2aq]u}{2V_Y} \quad (2)$$

The change in $p$ resulting from selection will be the product of this regression coefficient and the selection differential, the difference between the mean of all progenies and that of the progenies of selected plants. The selection differential will be symbolized as $s\sqrt{V_Y}$ where $s$ is the differential in units of the standard deviation of progeny means, $\sqrt{V_Y}$. Then the expected change in gene frequency as a consequence of selection is

$$\triangle p = p(1-p)[1+a-2aq](1+f)\frac{su}{2\sqrt{V_Y}}$$
$$= p(1-p)[1+a-2aq](1+f)c \quad (3)$$

where

$$c = \frac{su}{2\sqrt{V_Y}}$$

It will be convenient for our purpose to consider $c$ equal for all three methods. The only objection to so doing is the fact that $\sqrt{V_Y}$ will be a trifle larger for method 3 because of sampling variance among genotypes of tester plants. However, if each test cross progeny involves as many as three or four tester parent plants $\sqrt{V_Y}$ for method 3 would only under the most extreme conditions be more than 10% larger than for the other methods.[5]

---

[5]The basis for this conclusion was the following. The genotypic variance among progeny means was derived in terms of gene frequencies in the material under selection and in the tester. It was then evaluated relative to non-genetic variance on the basis of data available from the study reported by Robinson *et al.* and a liberal margin of safety allowed in arriving at the 10% figure indicated in the text. Space does not permit reporting on this in detail.

The mean genotypic value for a given locus in hybrids resulting from random crosses between two populations will be

$$X = [p_1(1+a) + p_2(1+a) - 2ap_1p_2] u \quad (4)$$

where $p_1$ and $p_2$ are frequencies of the favorable allele in the two populations. After selection has increased $p_1$ to $p_1 + \Delta_1$ and $p_2$ to $p_2 + \Delta_2$ this mean performance will be

$$X' = [(p_1+\Delta_1)(1+a) + (p_2+\Delta_2)(1+a) - 2a(p_1+\Delta_1)(p_2+\Delta_2)] u$$

The difference, $X'-X$, is the increase in hybrid performance resulting from the selection.

$$X'-X = [\Delta_1(1+a) + \Delta_2(1+a) - 2a\Delta_1 p_2 - 2a\Delta_2 p_1 - 2a\Delta_1\Delta_2] u \quad (5)$$

Since $\Delta_1$ and $\Delta_2$ resulting from one cycle of selection will be small fractions, the effect of ignoring the term involving their product will be inconsequential. Substituting in this expression on the basis of equation (3), and reducing, expressions for improvement in hybrid genotype resulting from one selection cycle are obtained for each of the selection methods. They are as follows:

Method $\quad \Delta_X = X' - X$

1 $[p_1(1-p_1)(1+a-2ap_2) + p_2(1-p_2)(1+a-2ap_1)]$ $(1+a-2aq)(1+f)cu$ (6)

2 $p(1-p)(1+a-2aq)^2 (1+f)cu$ (7)

3 $[p_1(1-p_1)(1+a-2ap_2)^2 + p_2(1-p_2)(1+a-2ap_1)^2]$ $(1+f)cu$ (8)

It is clear from the way $f$ enters these expressions that comparisons based on them are valid regardless of the amount of inbreeding of the selection material that precedes the making of test crosses, so long as we assume it constant for all methods.

If the value of $q$ which makes (6) and (8) equal is symbolized by $q'$, we have

$$q' = \frac{p_1 p_2 (1-p_1)(1+a-2ap_2) + p_1 p_2 (1-p_2)(1+a-2ap_1)}{p_1(1-p_1)(1+a-2ap_2) + p_2(1-p_2)(1+a-2ap_1)} \quad (9)$$

This result is obtained by equating (6) and (8) and solving for $q$. When $q = q'$ the expected improvement from one cycle of selection will be essentially[6] the same for methods 1 and 3. When $q > q'$, improvement will be more rapid by method 3; when $q < q'$ more rapid improvement will be expected with method 1. Thus the ratio of expected improvement rates for the two methods is not a constant but varies as $p_1$, $p_2$, and $q$ vary. Since (6) and (8) are equal when $q = q'$, (8) is equal to

$$[p_1(1-p_1)(1+a-2ap_2) + p_2(1-p_2)(1+a-2ap_1)](1+a-2aq')(1+f)cu \quad (10)$$

and the ratio of improvement rate by method 1 to that by method 3 is

$$(1+a-2aq) / (1+a-2aq') \quad (11)$$

---
[6] Not exactly since the term $2a\Delta_1\Delta_2$ was dropped in going from (5) to (6) and (8).

When $a = 1.0$, equation (9) reduces to

$$q' = \frac{2p_1 p_2}{p_1 + p_2} \quad (12)$$

which is less than $(p_1+p_2)/2$, the mean of $p_1$ and $p_2$, unless $p_1 = p_2$, in which case $q' = p_1 = p_2$. Thus when dominance is complete $q$ must be less than $(p_1+p_2)/2$ if method 1 is to be as effective as method 3. When there is only partial dominance ($0 < a < 1.0$), $q'$ will be less than $(p_1+p_2)/2$ when

$$p_1 + p_2 < 1 + \frac{1+a}{2a} p_1 p_2 \quad (13)$$

and for those loci at which this condition does not hold, improvement by method 1 will be greater than by method 3 if $q = (p_1+p_2)/2$.

If the lines used in the method 1 tester are chosen at random, it can reasonably be assumed that $q$ will, on the average for all loci, be about equal to the mean of the initial values of $p_1$ and $p_2$. Further, since the source material groups are to be as genetically divergent as possible, $p_1$ and $p_2$ may be assumed unequal for almost all loci. With these two points in mind consider a numerical example. Suppose that dominance were complete and that at a given locus, $p_1 = .6$ and $p_2 = .9$. Then from (12), $q' = .72$ and if $q = (p_1+p_2)/2 = .75$, as expected on the average if lines for the method 1 tester are chosen at random, the progress expected with method 1 will, from (11), be 89% of that expected with method 3. Computations of this sort for a range of values of $p_1$ and $p_2$ and for both partial and complete dominance have led to the following general conclusions relative to methods 1 and 3:

(1) In the case of complete or almost complete dominance, method 3 should at the outset yield a little more rapid improvement. However, as $p_1$ and $p_2$ are increased by the selection, a point is reached where on the average for all loci, they are sufficiently large relative to gene frequency in the method 1 tester so that improvement will be more rapid by method 1. It is unlikely that the initial advantage of method 3 would ever exceed 20 to 25%. The advantage possible in genotype improvement at a single locus approaches 100% as a limit, but for loci contributing most to total improvement it would be much less.

(2) In case of partial dominance the initial advantage of method 3 would be less than in the case of complete dominance and conceivably there might be a slight advantage for method 1.

(3) Choice of lines with low general combining ability, and hence presumably lower than average frequencies of favorable genes, for use in the method 1 tester might make method 1 equal to 2 even at the outset and would increase its superiority in later cycles.

The important point to be noted is that in the situation, *absence of over dominance*, to which selection for general combining ability is adapted, this method is at best not greatly superior to reciprocal selection. However, in the event that over-dominance prevails at any considerable number of loci selection for general combining ability will be unsatisfactory because of ineffectiveness with respect to improvement at the over-dominance loci. Reciprocal selection will then be the more effective of the two methods.

It should perhaps be pointed out here that equation (6) relates only to improvement in average performance of hybrids between lines of diverse origin expected to result from selection for general combining ability and not to that to be obtained by selection for specific combining ability among such hybrids. However, this does not invalidate the comparison since that source of final improvement is available no matter which of the two methods is used.

Comparison of methods 2 and 3, as noted earlier, need be made only under the assumption of over-dominance (a>1.0). As a basis for the comparison it seems reasonable to assume that the same genetic material would be used regardless of the method employed; that if method 2 were to be used all this material would be pooled to establish the segregating population to which selection was to be applied; and that if method 3 were to be used two such populations would be established dividing the genetic material in the manner believed most likely to result in the greatest genetic divergence between the two populations. We can then consider that the frequency (p) of a specific gene in the method 2 selection material would be the average of its frequencies ($p_1$ and $p_2$) in the two groups of selection material that would be used in method 3. Symbolically, this amounts to the following:

$$p = (p_1+p_2) / 2, \quad p_1 \neq p_2$$

That in general $p_1 \neq p_2$ is assured if the two groups of selection material used in method 3 are genetically divergent. Now consider a group of loci for all of which $p$ has the same value, and let $r$ be the proportion of these loci at which $q$ for the method 2 tester is 1.0. Then using equations (4), (7), and (8) it is possible to determine for any values of $p$ and ($p_1-p_2$),

a. The value of $r$ necessary if initial hybrid performance is to be as good for the method 2 material as for the method 3 material.

b. The value of $r$ necessary if improvement per selection cycle is to be as rapid by method 2 as by method 3.

It turns out that

a. When $p < \frac{1+a}{2a}$, $r$ must be larger than $p$ if initial hybrid performance is to be as large for method 2 as for method 3. The required value of $r$ increases as the difference ($p_1-p_2$), i.e., the genetic divergence between the selection material groups of method 3, increases. However, as $p_1-p_2$ is increased a point is reached where if $r$ is large enough so that initial hybrid performance will be equal for the two methods, expected improvement will be less by method 2.

b. When $p = \frac{1+a}{2a}$ initial hybrid performance will always be less for method 2, but expected improvement rate will be greater by method 2.

c. When $p > \frac{1+a}{2a}$, $r$ must be less than $p$ if initial hybrid performance is to be as great for method 2. However, the expected improvement rate will always be larger for method 2.

The most reasonable value for $r$ is the same as for $p$. On that basis, it is to be expected that initial hybrid performance would ordinarily be lower in method 2 than in method 3. If the method 2 tester line is chosen on the basis of its combining ability with the segregating selection material, the initial hybrid performance could be made better but improvement rate from selection would suffer if in general $p < \frac{1+a}{2a}$. The difference to be expected in improvement rate depends on the size of $p$ relative to $\frac{1+a}{2a}$ and the magnitude of ($p_1-p_2$). If ($p_1+p_2$) / 2 > $\frac{1+a}{2a}$, improvement will almost certainly be more rapid by method 2; if $p < \frac{1+a}{2a}$, improvement can be more rapid by method 3 if ($p_1-p_2$) is large enough i.e., if the genetic divergence between the two groups of selection material is sufficient. Moreover, the above is based only on results expected in the first cycle of selection. Since rate of improvement by method 3 increases as ($p_1-p_2$) is increased and since this difference is in general increased by the selection, the comparison of the methods would become more favorable to method 3 in subsequent cycles of selection.

The important over-all conclusions relative to methods 2 and 3 for selection when a >1.0 are as follows: If the two groups of selection material for the reciprocal selection method are well chosen (so that there is maximum genetic divergence between them) initial hybrid performance will tend to be better than for method 2. Which method will yield the most rapid improvement at the outset depends on gene frequencies, but in any event the ratio will become more favorable to the reciprocal method in later cycles of selection. Thus even assuming no partial dominance loci, it appears unlikely that the method proposed by Hull could have, at the most, more than a slight advantage over reciprocal selection.

## Nonvalidity of Assumptions

The assumptions (see Basis for Comparisons) under which the foregoing comparisons were made are in all probability not generally valid. The consequences of their being invalid upon the results of the comparisons must therefore be considered.

### THE EFFECT OF INTERACTIONS OF NON-ALLELIC GENES

When there are interactions of non-allelic genes, the effects of some genes are conditioned by genes present at other loci. As a consequence, the importance of a gene to performance in a test cross, and therefore the intensity of selection for it, depends on genes present at certain other loci in the tester. Clearly, when such interactions are important in the genetic mechanism, the advantage of testing against the same material to be used as the opposite parent of the commercial hybrid is increased. Hence, if the assumption of no interactions of nonallelic genes is invalid, the method of selection for general combining ability suffers by comparison with the others.

### THE EFFECT OF MULTIPLE ALLELES

The situation with multiple alleles is not greatly different from that with only two at each locus. It differs mainly in that at the outset of selection the frequencies of more than one allele of a series may be increased; though the ultimate effect of continued selection would be to increase the most favorable gene of a series and decrease the frequencies of all the others. The specific allele favored in all stages of continued selection will be the one having the best effect in the test crosses. Thus, if the genetic system were that proposed by East (3) it would again be advantageous to use the same material for tester that will later be used as the opposite parent of the commercial hybrid.

It would appear that multiple allelism would not seriously modify the comparison of the methods made subject to all three assumptions; that methods 1 and 3 would be superior to method 2 if the homozygote of one of the alleles were the best genotype possible at the locus; but that methods 2 and 3 would be superior to method 1 if the best genotype were (as postulated by East) the heterozygote involving two alleles of the series.

### THE EFFECT OF DISEQUILIBRIUM IN THE RELATIVE FREQUENCIES OF GENOTYPES FOR LINKAGE GROUPS

If the chromosome segment involved in a linkage group is considered as a single "locus", the various possible gametic combinations of the individual genes involved become analagous to the genes of a multiple allelic series. If the frequency distribution of these combinations, or "alleles", is that expected on the basis of random distribution of the individual genes the effect of selection is no different because of the linkage. However, if the genes for certain loci are associated predominantly in the repulsion phase, the situation is changed. Then among the "alleles" available in the population certain heterozygotes would very likely be superior to any homozygote (even though individual genes were no more than partially dominant to their alleles) and the situation is once more a less favorable one for method 1.

On the other hand, assuming that there is no over-dominance in the action of the individual genes and that some exhibit only partial dominance, method 2 would be deficient in not providing the means for taking advantage of recombinations in the improvement of both parents.

## Discussion

The most important aspect of the comparisons made above is the fact that reciprocal selection appears about as effective as either of the other two methods in all of the genetic situations considered. Further, it is clearly superior to selection for general combining ability for loci at which there is over-dominance and to the method proposed by Hull for loci at which there is partial dominance. Hence, in the event of partial dominance at some loci and over-dominance at others reciprocal selection would be more effective than either of the other methods. In addition, assuming no over-dominance, there appears a strong possibility that reciprocal selection would be superior to selection for general combining ability as a consequence of either interactions of nonallelic genes or repulsion phase linkages between certain loci, or both.

There remain a few points relative to the application of reciprocal selection which deserve some discussion.

The authors feel that it offers potentialities for further improvement of good double crosses. It is known that advanced generations of the single crosses of a double cross combine as well as the singles themselves. It appears entirely possible that reciprocal selection applied to segregating generations of the single crosses would affect improvement in the hybrid between them. This would be an important application of the method since it would offer the possibility of working improvement from a starting point of already superior performance.

The importance of basing selection among test cross progenies on all agronomic characters in proportion to their importance and not on yield alone deserves reemphasis. Since the test-crosses are between the same parent stocks as will be used in the commercial hybrid this would appear particularly pertinent and logical.

A problem that may well arise in application of the method is that of avoiding inbreeding within the two selection stocks. This was mentioned earlier, but the importance of controlled pollination and pedigree recording as an aid to avoiding a serious amount of inbreeding must definitely be recognized. If such inbreeding were allowed it would clearly result in decreased opportunity for effective selection. However, this same problem also arises with other methods.

## Summary

A breeding and selection technique for improvement of commercial hybrids in diploid organisms has been outlined, and designated as recurrent reciprocal selection. Theoretical comparisons have been made of the limits of improvement and improvement rates to be expected of this method, selection for general combining ability, and the method proposed by Hull (4). They indicate that under no circumstances would reciprocal selection be more than slightly inferior to the better of the other two. However, it would be definitely superior to selection for general combining ability for loci at which there is over-dominance or if a situation analogous to that with over-dominance exists due to linkage; and it would be definitely superior to the method proposed by Hull for loci at which there is partial dominance. In view of the reality of linkage and the improbability of partially dominant genes being of minor importance in real genetic situations, there is good reason to believe reciprocal selection will in practice be more effective than the other methods discussed in this paper.

## Literature Cited

1. COMSTOCK, R. E., and ROBINSON, H. F. The components of genetic variance in populations of biparental progenies and their use in estimating the average degree of dominance. Biometrics, 4:254–266. 1948.
2. CROW, JAMES F. A consequence of the dominance hypothesis of hybrid vigor. Records Genet. Soc. Amer., No. 16, p. 28. 1947.
3. EAST, E. M. Heterosis. Genetics, 21:375–397. 1936.
4. HULL, FRED H. Recurrent selection for specific combining ability in corn. Jour. Amer. Soc. Agron., 37:134–145. 1945.
5. JENKINS, MERLE T. The effect of inbreeding and of selection within selfed lines of maize upon the hybrids made after successive generations of selfing. Iowa State Col. Jour. Sci., 9:429–450. 1935.
6. ROBINSON, H. F., COMSTOCK, R. E., and HARVEY, P. H. Estimates of heritability and the degree of dominance in corn. Agron. Jour., 41:353–359. 1948.
7. SPRAGUE, G. F., and TATUM, LLOYD A. General vs. specific combining ability in single crosses of corn. Jour. Amer. Soc. Agron., 34:923–932. 1942.

# 13

Copyright © 1960 by The Royal Society
Reprinted from R. Soc. (London) Proc. **B153**:234-249 (1960)

## A theory of limits in artificial selection

By A. Robertson

(1) The paper presents a theory of selection limits in artificial selection. It is, however, developed primarily in terms of single genes.

(2) For a single gene with selective advantage $s$, the chance of fixation (the expected gene frequency at the limit) is a function only of $Ns$, where $N$ is the effective population size. In artificial selection based on individual measurements, where the selection differential is $\bar{i}$ standard deviations, the expected limit of individual selection in any population is a function only of $N\bar{i}$.

(3) For low values of $N\bar{i}$, the total advance by selection is, for additive genes, $2N$ times the gain in the first generation but may be much greater than this for recessives, particularly if their initial frequency is low.

(4) The half-life of any selection process will, for additive genes, not be greater than $1 \cdot 4\,N$ generations but may for rare recessives equal $2N$.

(5) The effect of an initial period of selection or inbreeding or of both together on the limits in further selection is discussed. It appears that the effects of restrictions in population size on the selection limit may be a useful diagnostic tool in the laboratory.

(6) The treatment can be extended to deal with the limits of further selection after the crossing of replicate lines from the same population when the initial response has ceased.

(7) In a selection programme of individual selection of equal intensity in both sexes, the furthest limit should be attained when half the population is selected from each generation.

(8) The treatment can also be extended to include selection based on progeny or family records. Consideration of the optimum structure, as far as the limit is concerned, shows that the use of the information on relatives is always a sacrifice on the eventual limit for the sake of immediate gain in the early generations. The loss may, however, be small in large populations.

Current theories of artificial selection have been concerned almost entirely with the prediction of the rate of response of the population mean to selection pressures of different kinds. The selection may be expected to increase the frequency of favourable alleles until, in a large population, they eventually reach fixation. But if the population size is finite there is a possibility that one allele may be fixed by chance even though there is a more desirable one in the population. The smaller the population, the greater will this possibility be. In this paper, a theory of selection limits is developed taking into account at the same time the forces of selection and those of chance fixation. Some of the conclusions are not very precise, but, in so far as they indicate qualitatively what we might expect, they may be of value. They bear on many practical problems, for instance the attainment of the maximum possible advance from the existing variation with a population and also the retention of the potential for change in a population while keeping it in 'cold storage'.

It may of course happen that a population will reach a selection limit while still retaining genetic variation, due to continued selection for heterozygotes. It is hoped to deal with the effects of restricted population size in such a situation in a separate paper.

We have first to develop the theory with respect to individual genes whose selective advantage we know and then to modify the conclusions to deal with the situation when we are merely selecting the animals extreme for some quantitative measurement. Two kinds of agencies affecting gene frequency are at work in any selection programme. The first is genetic sampling, or genetic drift, causing a random change from generation to generation in the frequency, $q$, of any gene. The mean change is zero and its variance is $[q(1-q)]/2N$, where $N$ is the effective population size. The second is the directed change in gene frequency due to selection which we can write in the case of two alternative alleles as $sf(q)$, where $s$ depends on the relative selective advantages of the three genotypes at that locus and $f(q)$ is a function of the gene frequency, depending on the type of gene action involved. The approach to such problems is essentially due to Wright (1931). He introduced the concept of the distribution of gene frequencies, which may be considered alternatively to refer to loci of the same kind and magnitude of effect in one population or to an individual locus in many such populations. The central concept in this present approach is that of the chance of fixation of the gene in question, to which Kimura (1957) has given the symbol $u(q)$ where $q$ is the frequency of the gene in the initial population. $u$ is then alternatively the proportion of equivalent loci which would be expected to be fixed in any line, perhaps the easiest model in considering artificial selection, or the proportion of replicate selected lines in which an individual gene would be expected to be fixed. $u$ may then have a value intermediate between 0 and 1 even though homozygosis is complete. We may refer to $u(q)$ as the 'expected limit' where the extreme possible limit will be a value of 1. $u(q) - q$ is then the expected total change in gene frequency and, corresponding to it in artificial selection, we have the change in the character under selection when fixation is reached, which we shall describe as the 'total advance'. At the extreme, we have a population in which all desirable genes in the initial population have been fixed and we shall describe by the phrase 'possible advance' the difference in the selected character between this and the initial population.

Genetic sampling causes a broadening of the gene-frequency distribution leading to eventual fixation at the extreme values and selection causes a general shift of the distribution in the direction of selection. If we write $\phi(q,t)$ for the distribution of gene frequencies at time $t$, we can, using a continuous model, write down the process of change with time as

$$\frac{\partial \phi}{\partial t} = \frac{\partial^2}{\partial q^2} \frac{(q(1-q)\phi)}{4N} - \frac{\partial}{\partial q}(\phi sf(q)) \qquad (1)$$

the first term on the right-hand side representing drift and the second selection. This may be rearranged as

$$\frac{\partial \phi}{\partial(t/N)} = \frac{\partial^2}{\partial q^2} \frac{(q(1-q)\phi)}{4} - \frac{\partial}{\partial q}(\phi Nsf(q)). \qquad (2)$$

Thus the change in $\phi$ at a particular value of $q$ in an amount of time $t/N$ is dependent only on $Ns$ and on the initial function $\phi(q, o)$. It then follows that the pattern of the change is determined by $Ns$ and its time scale is directly proportional to $N$.

For a given value of $q$, the value of $\phi$ is a function of $Ns$ and $t/N$, the exact form of the function being determined by the initial gene-frequency distribution. Now the pattern when $t = \infty$ is merely the proportion of loci (or replicate populations) in which the allele in question has been fixed. The chance of fixation of the allele is then determined by the initial distribution and $Ns$. If we start with a single locus in a single population, we can say that the chance of fixation is a function solely of $Ns$ and the initial gene frequency.

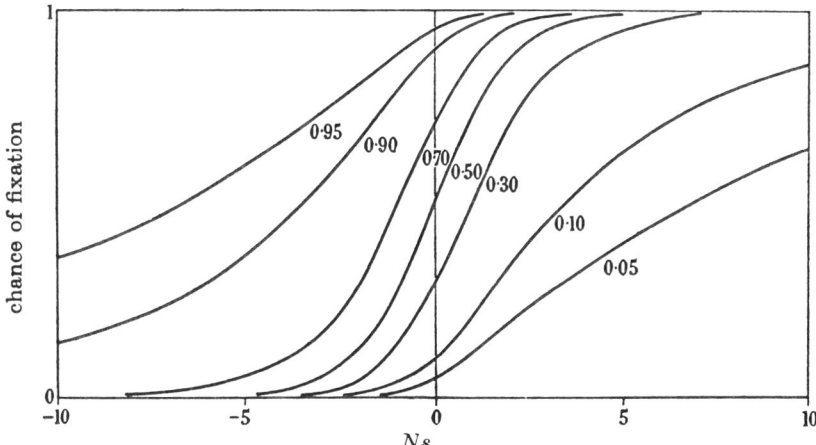

FIGURE 1. The chance of fixation of a gene acting additively. The curves are drawn for different initial gene frequencies.

The treatment that follows is an extension of some results of Kimura (1957) who developed explicit expressions for the chance of fixation. In the case of a pair of alleles with additive effects on selective advantage, i.e. the three genotypes have selective advantage $1-\tfrac{1}{2}s$, $1$ and $1+\tfrac{1}{2}s$, respectively, the chance of fixation of a gene whose initial frequency is $q$ is given by

$$u(q) = \frac{1-e^{-2Nsq}}{1-e^{-2Ns}}, \qquad (3)$$

which is shown in figure 1. When $Ns = 0$, $u(q)$ equals $q$. This merely says that if there is no selection, the mean gene frequency is not changed. An expansion in terms of $Ns$ gives.

$$u(q) = q+q(1-q)Ns+\dots \qquad (4)$$

Thus the slope of the curves, when $Ns = 0$, is $q(1-q)$. As the total possible advance is $1-q$, it follows that the greater part of this will be achieved if $Ns$ is greater than $1/q$. In fact, if $Nsq > 1$, it can be shown that more than 70% of the possible advance will be achieved and if $Nsq > 2$, more than 93%.

The value of $u(q)$ when $Ns$ is small can be derived by a completely different approach. We then assume that the mean frequency will change little during the fixation process and that the mean heterozygosity declines by the usual proportion $1/2N$ each generation. In the first generation of selection, it can be shown that the change in gene frequency is $(\frac{1}{2}s)q(1-q)$. But we know that the average value of $q(1-q)$ will decline by a fraction $1/2N$ each generation. Thus

$$u(q) - q = \sum_{t=0}^{\infty} \frac{s}{2} q(1-q) \left(1 - \frac{1}{2N}\right)^t$$
$$= Nsq(1-q). \quad (5)$$

The total advance is thus $2N$ times the change in the first generation. For larger values of $Ns$, the total advance may be larger than this because at low initial gene frequencies the mean value of $q(1-q)$ may well increase during selection, in spite of the inbreeding, as the mean gene frequency increases. For low values of $q$ and large values of $Ns$ we have, by expansion of (3), $u(q) = 2Nsq$ so that the total advance approaches $4N$ times the change in the first generation.

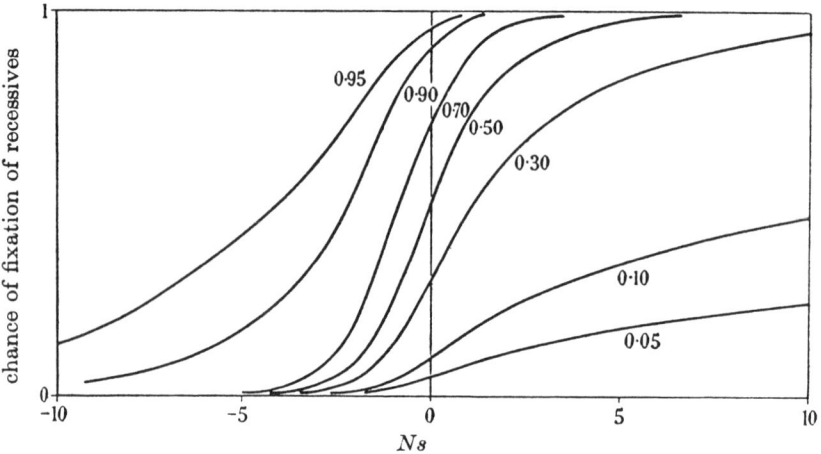

FIGURE 2. The chance of fixation of a recessive gene. The curves are drawn for different initial recessive frequencies.

If there is no additive gene action, the algebra is more complex. With selection for a recessive gene with frequency $q$ where the homozygous recessive has a selective advantage $s$, we have $f(q) = q^2(1-q)$ and Kimura (1957) has shown that the chance of fixation is given by

$$u(q) = \frac{\int_0^q e^{-2Nsq^2} dq}{\int_0^1 e^{-2Nsq^2} dq}. \quad (6)$$

This function is shown in figure 2. An expansion is possible, giving

$$u(q) = q + \tfrac{2}{3} Nsq(1-q^2) + \dots. \quad (7)$$

The alternative approach, assuming that $u(q) - q$ is small, necessitates the use of the moment generating matrix (Robertson 1952). This gives

$$u(q) - q = s \sum_{t=0}^{\infty} \left[ \tfrac{1}{2} q(1-q) \left(1 - \frac{1}{2N}\right)^t - \tfrac{1}{2} q(1-q)(1-2q) \left(1 - \frac{3}{2N}\right)^t \right]$$
$$= \tfrac{2}{3} Nsq(1-q^2). \quad (8)$$

Formulae (7) and (8) have obvious relevance to the chance that an inbred line will rid itself of deleterious recessives during inbreeding as $u(q)$ is the proportion of such recessives which will be fixed. We are then dealing with negative values of $s$. Thus if $\frac{2}{3}Ns$ is more negative than $-1$, $u(q)$ will be close to zero, the gene will have been selected out and will not contribute to the depression of fitness on inbreeding. As would be expected, the lower the rate of inbreeding and the greater the gene effect, the greater the chance that a harmful recessive will be selected out. Then the lower the rate of inbreeding the smaller the effects of the individual harmful genes fixed by chance and the less the inbreeding depression when complete homozygosis is reached.

The ratio of expected total response to initial change in the first generation when $Ns$ is small is $\frac{2}{3}Nsq(1-q^2)/sq^2(1-q) = 2N \cdot (1+q)/3q$, rather than $2N$ as in the additive case. This ratio can then be very much greater for recessives at low frequencies than for additive genes. This is a consequence of the increase of the genetic variance within lines due to low-frequency recessives up to inbreeding coefficients of 0·5 noted by Robertson (1952).

The case of equilibrium at an intermediate gene frequency resulting from selection for heterozygotes, $(f(q) = q(1-q)(q-q_0)$, where $q_0$ is the equilibrium frequency), is more complex as the selection may effectively prevent fixation. It is hoped to deal in detail with this and some related problems in another paper.

### THE TIME SCALE OF THE SELECTION PROCESS

We have been discussing so far the limits of selection. How long will it take to get there? As the approach to the limit will be asymptotic, this question has not much meaning but it is useful to ask how long it will take the mean gene frequency to get half-way to the limit. In the language of physics, what is the 'half-life' of the selection process?

It was shown earlier (2) that the time scale is proportional to $N$, the effective population size, and the pattern of change is dependent only on $Ns$. The expected half-life in generations will then be a multiple of $N$, determined by $Ns$ and the initial gene frequency. In the case of low values of $Ns$, we can obtain a very simple expression. Writing $q_t$ as the mean gene frequency after $t$ generations, we have from (5)

$$q_t - q = \sum_0^t \tfrac{1}{2}sq(1-q)\left(1-\frac{1}{2N}\right)^t$$
$$= Nsq(1-q)(1-e^{-t/2N}) \text{ approx.} \qquad (9)$$

The half-life is then given by $e^{-t/2N} = \tfrac{1}{2}$, or $t = 1\cdot 4N$ generations. In the case of a recessive gene, we have

$$q_t - q = s\sum_0^t \left[\tfrac{1}{2}q(1-q)\left(1-\frac{1}{2N}\right)^t - \tfrac{1}{2}q(1-q)(1-2q)\left(1-\frac{3}{2N}\right)^t\right] \qquad (10)$$
$$= Nsq(1-q)[(1-e^{-t/2N}) - \tfrac{1}{3}(1-2q)(1-e^{-3t/2N})] \text{ approx.}$$

When $q = \tfrac{1}{2}$, the second term vanishes and the half-life is the same as in the additive case. As $q$ approaches zero (i.e. the recessive is at low frequency) it can be shown

161

that the half-life is $2 \cdot 12N$ generations and as $q$ approaches unity, the factor approaches $1 \cdot 03N$. We may then expect that, if $Ns$ is low, the half-life will probably be between $N$ and $2N$ generations.

If $Ns$ is not small, the problem becomes extremely difficult to solve explicitly. It can, however, be explored by evaluating empirically the change in the gene-frequency distribution as the generations proceed. We then make use of the fact that the pattern of change is determined only by $Ns$. For a given value of $N$, there are $2N+1$ possible gene frequencies. If a population has a certain frequency at a particular generation, the probability distribution of its frequency in the next generation can be calculated as a binomial distribution with index $2N$ and mean correspondingly modified by selection. We can then write down the transformation matrix to convert the gene-frequency distribution in one generation into that in the next. It only proved possible to deal with values of $N$ up to 5 on a desk calculator because the total increases as $N^3$.

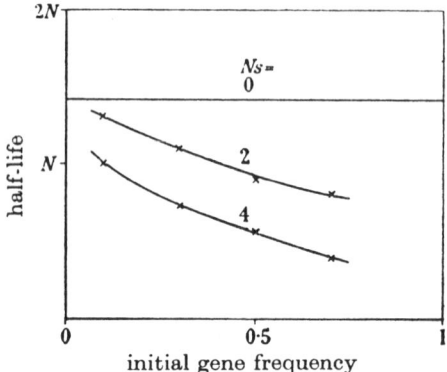

FIGURE 3. The average 'half-life' in generations of the selection process for a gene acting additively.

Some of the half-lives so calculated with $N = 5$ are given in figure 3. It appears that the half-life decreases continuously as $Ns$ increases. In a sense this might have been expected. The higher $Ns$, the greater is the chance that the favourable allele becomes fixed by selection before it is lost from the line by sampling. It seems then that the values calculated theoretically for low $Ns$ are probably upper limits. If the half-life of a selection programme is reached well before the range of $N$ to $2N$, expected when the chance of fixation is not high, we may perhaps conclude that we have fixed all the desirable alleles.

THE EFFECT OF SELECTION AND INBREEDING ON SUBSEQUENT SELECTION LIMITS

We have been discussing so far the chance of fixation of a gene with a known initial frequency and have shown that it is dependent only on $Ns$. We now proceed to ask how this dependence on $Ns$ is affected by an initial period of inbreeding or selection or of both together. It may be of advantage here to think of the gene-frequency distribution as referring to genes at different loci, all with the same selective advantage. In the initial population we shall assume that they all have

240                       A. Robertson

the same gene frequency. This frequency will be altered by the initial period of selection or inbreeding and we wish to know how the chance of fixation (or the proportion of the genes likely to be fixed) considered as a function of $Ns$, has been altered in this initial period. The detailed discussion will be devoted to the case of additive action.

There are three alternative treatments in the initial period that we shall consider.

### (a) *Selection in a very large population*

The frequency of all genes will have been altered by the same amount and the curve of chance of fixation against $Ns$ is that corresponding to the new frequency.

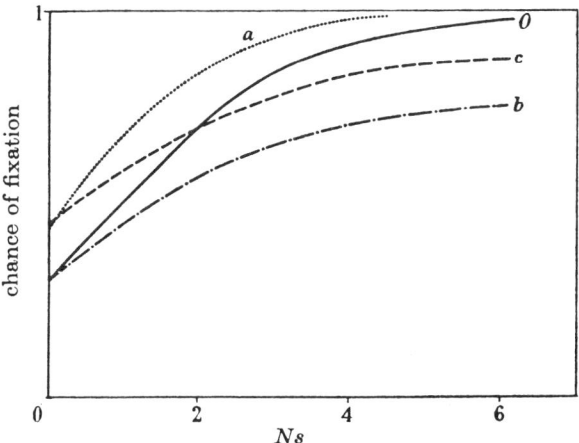

FIGURE 4. The effect of various treatments on the curve of chance of fixation against $Ns$ for a gene with initial frequency of 0·3. The treatments are three generations of (a) selection with $s = 0·4$ in a large population, (b) restriction of effective population size to 5, (c) selection with $s = 0·4$ and effective population size 5. $O$ = original.

### (b) *Restriction of population size without selection*

The mean gene frequency will stay the same but different genes will have different initial frequencies. Some may be lost altogether from the population so that the ultimate limits of selection at high values of $Ns$ will be reduced. This effect will be most marked at low initial gene frequencies. For additive genes, the slope of $u(q)$ plotted against $Ns$ at low values of $Ns$ will be reduced due to the decline in heterozygosity. The slope will also be less for recessives.

### (c) *Restricted population size with selection*

We have now a mixture of the two effects. The mean gene frequency will be higher when $Ns = 0$ because of the selection but the ultimate limit will be reduced because of chance fixation in the initial period. But the new curve of $u(q)$ against $Ns$ must intersect the old one at the point corresponding to the value of $Ns$ used in the initial period. The further selection process is then merely a continuation of the old and will have the same expected limit.

Figure 4 shows the curves of chance of fixation against $Ns$ for an initial gene frequency of 0·3 for the initial population and after three generations of (a) selection with $s = 0.4$ in a large population (b) restriction of population size to $N = 5$ and (c) selection and restriction together. The curves for the last two were obtained by calculating the gene-frequency distribution after the three generations using the transformation matrix as in the calculation of the half-life. The curves illustrate well the points made earlier.

The effect of restriction or 'bottleneck' in populations for some generations may be enlarged upon in relation to initial gene frequency. The lower the initial frequency the greater the chance a gene will be lost from the population by a sudden reduction in population size. Figure 5 shows the curve of limit against $Ns$ for the 'bottleneck' of restriction of parents to a single pair, and of such a restriction for three consecutive generations, in each case followed by expansion. The effect of

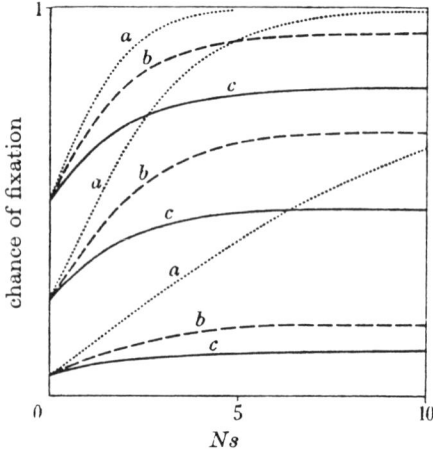

FIGURE 5. The effects of 'bottlenecks' in population size on the curve of chance of fixation against calculated for initial gene frequencies of 0·1, 0·3 and 0·5. (a) initial population, (b) restriction to a single mating for one generation only, (c) restriction to a single mating for three consecutive generations.

the bottleneck on the maximum advance possible in further selection is very marked for the genes with low initial frequency. If the initial frequency is 0·5, a single restriction of parents to one pair reduces the possible further advance to 0·80, so that the possible advance is only 60% of that in the initial population. If the initial gene frequency is 0·05, such a restriction reduces the expected ultimate advance to 0·185 so that the possible advance is only 14% of that in the initial population. But a further two generations of restriction only reduce the possible advance by a further factor of two. In the first generation, many such genes would be lost altogether, but those that were retained would have their mean frequency increased to at least 0·25 so that further restriction would not have such a large effect. In selecting from a previously unselected large population, it is the first generation that is critical in losing low-frequency genes. Thereafter the population has its segregating genes at higher frequencies.

As a consequence, for genes with an initial low frequency in populations that have been through a bottleneck, the values of $Ns$ necessary to attain a major part of the possible advance will be less than before the restriction. In this they are similar to selected lines. In both cases, the genes that are still segregating are at a higher frequency than they were before. The figure shows that to get 70 % of the distance to the possible limit, the $Ns$ value should be 12·2 if the initial gene frequencies are 0·05 but declines to 3·0 after one restriction to a single pair mating.

Highly selected populations or those which have passed through a severe 'bottleneck' in population size will be tolerant of any further size restrictions in the sense that the desirable alleles will be harder to lose because, if they are present at all, they will have a reasonable frequency. This has a paradoxical practical consequence to the storage of populations for possible use in a selection programme, a problem now facing many poultry breeders. The more highly selected a strain the smaller the numbers needed for keeping it. Any desirable genes that are still segregating are probably at a high frequency. Much more care should be given to a completely unselected strain because there the desirable alleles are more likely to be at low frequency and therefore to be lost by accident. The extreme type of population of this kind is one in which the desirable alleles are all at very low frequency, as would happen in attempting to produce new useful variation by irradiation. As Dempster (1958) has pointed out, it is then very important to keep the population size high in storage and in the early generations of selection.

### THE CROSSING OF SELECTED LINES

If we cross together two selected lines from the same population after they have reached fixation, we may expect to make further progress in the cross if either line contained a desirable gene which the other did not. Let us assume that we take a series of lines which reached fixation with a given value of $Ns$ and cross them together in pairs. What can we predict about the limits to further selection as a function of $Ns$? If the chance of fixation in the initial selection is $u$, then this will be the expected gene frequency when $Ns$ is zero. We may expect further that a proportion $u^2$ of the pairs will both have the gene in question, in $2u(1-u)$ one will have it and one will not and in $(1-u)^2$ neither will have it. The possible limit for high values of $Ns$ is $1-(1-u)^2$. Any further gain will come from the $2u(1-u)$ pairs in which the gene frequency is one-half.

We therefore know the values for $Ns = 0$ and $Ns = \infty$. A surprising result holds for additive genes if the further selection from the cross has twice the value of $Ns$ as had the original selection. The expected limit is then exactly the same as the limit for selection from the original population with twice the original value of $Ns$. The result in fact holds more generally. If we make populations by crossing any number of lines together and then use a proportionally higher value of $Ns$, we have the same expected limit as if we had used the higher value all the time.

These results only hold approximately for recessive genes. The discrepancies are not large but, in single crosses, the expected limit on selecting the cross with double the original value is for high-frequency recessives rather higher than if all the selection had been at the higher value and rather lower for low-frequency recessives.

## SELECTION FOR A QUANTITATIVE CHARACTER

We have been dealing so far with genes whose selective advantage we know. But in many of our laboratory selection experiments, we select animals on the basis of their measurement for some 'metric character'. We presume that we increase the frequency of the desirable genes, but we know neither their frequency nor the selective advantage that we confer on them by our artificial selection. We merely observe that the mean of the population changes.

We can take the first step in applying the earlier results by using a formula originally derived by Haldane (1931). He showed that under artificial selection the selective advantage associated with any small difference on the metric scale, on which selection is based, is equal to that difference multiplied by $I/\sigma^2$, where $I$ is the superiority of chosen parents above the mean of the population and $\sigma^2$ is the phenotypic variance in the population. If we express the intensity of selection as a dimensionless character, by putting $\bar{i} = I/\sigma$, the factor becomes $\bar{i}/\sigma$. For a gene acting additively with a difference of '$a$' units on the metric scale between the mean of the two homozygotes, we then know that it will act additively as far as concerns selective advantage under artificial selection and that

$$s = \frac{\bar{i}a}{\sigma}.$$

We saw that, for all individual genes, the chance of fixation is a function of $Ns$ and the initial frequency $q$. In artificial selection, we may write

$$u(q) = f(N\bar{i}a, q)$$

and since the mean phenotype at fixation can, in the absence of interaction between genes at different loci, be written by summing over loci $\Sigma au(q) = \Sigma af(N\bar{i}a, q)$ it follows that in any population the expected limit of selection is a function only of $N\bar{i}$. The exact form of this function will depend on the distribution of gene frequencies and effects and on the type of gene action involved.

We saw earlier that, for an additive gene, at least 70% of the possible gain in gene frequency at the limit would be obtained provided $Nsq > 1$. In artificial selection, the condition may be written $N\bar{i}aq > \sigma$. At low values of $N\bar{i}$ we can only be sure to fix genes either with large effect or with high frequency. As $N\bar{i}$ increases we begin to fix the rarer genes with smaller effects.

But in an artificial selection programme we do not observe changes in gene frequency, we only see a change in the mean of the measurement in the population. The total change in the mean, $X - \bar{X}$, will for additive genes be given by

$$\begin{aligned}
X - \bar{X} &= \Sigma a(u(q) - q) \\
&= \Sigma a(q(1-q)Ns + \ldots) \\
&= 2N\bar{i} \Sigma \frac{a^2 q(1-q)}{2\sigma} \quad \text{if } N\bar{i} \text{ is low} \\
&= N\bar{i} \frac{2\sigma_y^2}{\sigma},
\end{aligned}$$

where $\sigma_g^2$ is the additive genetic variance. We can therefore predict the slope of the curve of $X - \bar{X}$ plotted against $N\bar{\imath}$ for low values. As in the case of single genes, the total advance is $2N$ times the response in the first generation ($\bar{\imath}\sigma_g^2/\sigma$) as Dempster (1955) has pointed out.

Again, as with single genes, this does not hold for recessives. Then

$$X - \bar{X} = \Sigma a(u(q) - q^2)$$
$$= \Sigma \left( aq(1-q) + \frac{2N\bar{\imath}}{3} \frac{a^2 q(1-q^2)}{\sigma} + \ldots \right)$$

of which the first term is the inbreeding depression. The coefficient of $N\bar{\imath}$, $[2\Sigma a^2 q(1-q^2)]/3\sigma$, may be much greater than $2\sigma_g^2/\sigma$, which equals $[4\Sigma a^2 q^3(1-q)]/\sigma$, especially for recessives at low frequency, so that the total advance due to recessives may be much greater than $2N$ times the response in the first generation.

The possible advance in the case of additive genes is given by the expression $\Sigma a(1-q)$ and this we cannot predict merely by measuring the genetic variance in the initial population. The smaller the number of genes contributing to any given additive genetic variance (and in consequence the greater their individual average effect) the lower will be the possible advance by selection and the quicker will it be reached. Though we can in some measure predict the initial slope of the curve of limit against $N\bar{\imath}$ there is no way in which we can predict the possible limit at high values of $N\bar{\imath}$.

Some of these points are illustrated in figures 6 and 7 which show the curves of advance plotted against $N\bar{\imath}$ for two imaginary populations. In figure 6, all genes have the same initial frequency of 0·5, but it is assumed that the advance at high values of $N\bar{\imath}$ is contributed equally by three classes of genes with $a/\sigma$ values of 0·5, 0·3 and 0·1, respectively. In figure 7, it is assumed that $a/\sigma = 0·3$ for all genes but that the ultimate advance is contributed equally by genes with frequencies of 0·5, 0·3 and 0·1, respectively. It will be seen that the further increase in limit when $N\bar{\imath}$ is high is due almost entirely to the genes with small effects or low initial frequencies.

We are now in a position to take over into the context of artificial selection the results that were obtained for single genes. We cannot now observe the chance of fixation, as we cannot observe gene frequency. If $u(q)$ is very different from unity for many genes, we will notice that replicate lines from the same initial population will be very different in the limit they reach. The variation will be greatest when the average value of $u$ for the different genes controlling the character is in the neighbourhood of 0·5. The variation between replicates will again be a function of $N\bar{\imath}$, but the type of function will depend on the gene frequencies in the initial population. If they are low, the variation in limit between replicates may pass through a maximum as $N\bar{\imath}$ increases but will decline to zero at high values. This is important in any experimental work on this problem as many replicates may be needed at low values of $N\bar{\imath}$. The possibility of utilizing the variation between replicate programmes may too have important consequences in practical animal breeding.

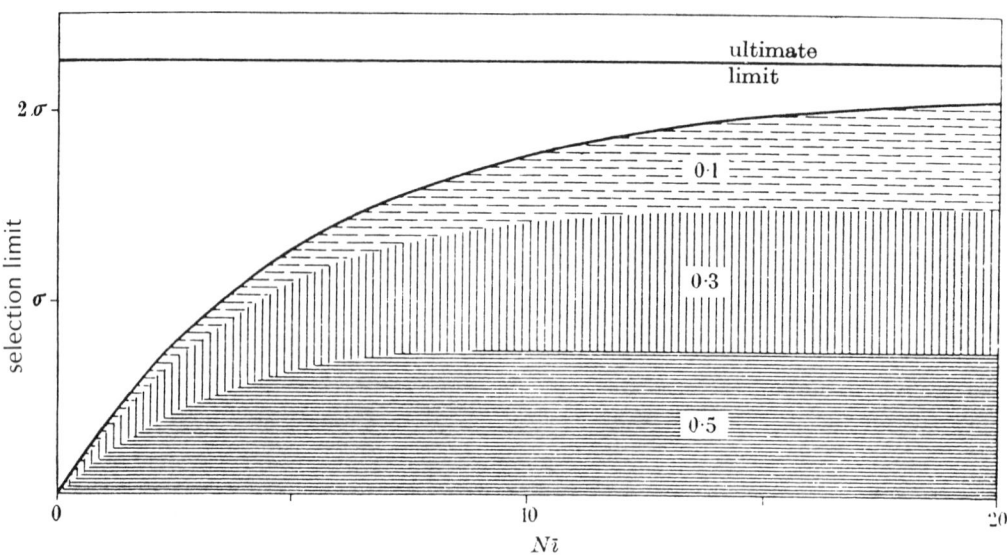

FIGURE 6. The expected limits to artificial selection in a population in which all genes have initial frequency 0·5 and in which the possible advance is contributed equally by genes with $a/\sigma = 0\cdot1$, $0\cdot3$ and $0\cdot5$, respectively.

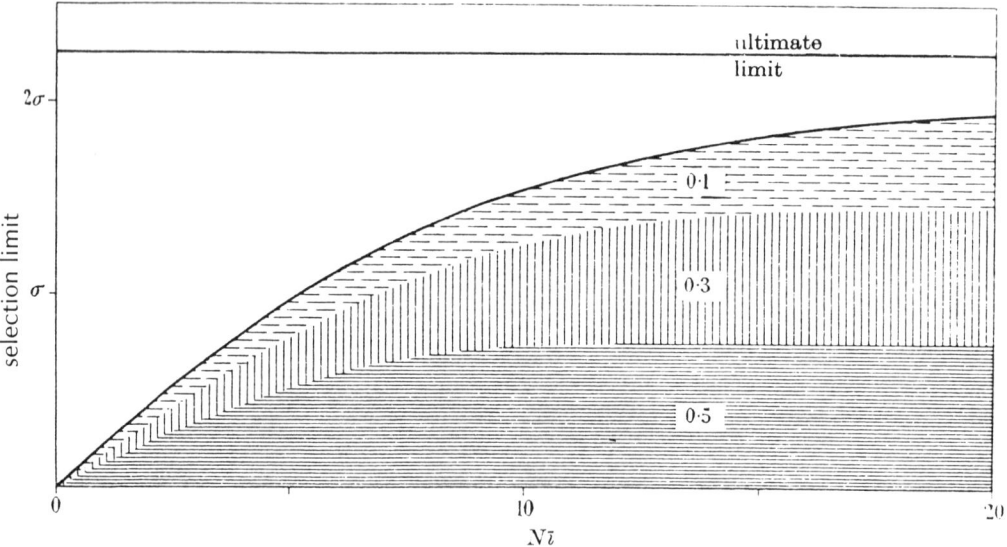

FIGURE 7. Similar to figure 6 except all genes now have $a/\sigma = 0\cdot3$ and the possible advance has equal contributions from genes with frequency $0\cdot1$, $0\cdot3$ and $0\cdot5$.

The discussion of the time-scale of the selection process is immediately relevant to the quantitative case. We may then expect that selection programmes should have an upper limit to their half-life in the region of $1\cdot4N$ generations if the genetic variation is mainly additive. If the half-life is reached earlier than this, we may presume that the majority of desirable alleles have been fixed and that there is little point in crossing two such lines derived from the same base population.

The effect of the initial treatment of the population on the curves of advance as a function of $Ns$ is directly referable from the single gene to the quantitative situation. The section on the effects of a restriction in population size on the possible limits at high $N\bar{\imath}$ values may be important here. In dealing with continuous variation, it is usual to describe the situation in terms of variance components. But it is very difficult to penetrate these to discover anything about the effects and frequencies of the genes which give rise to them. In the effect of a 'bottleneck' in population size on the ultimate limit of selection, we have a phenomenon which depends almost entirely on the frequencies of the desirable genes in the initial population and hardly at all on the magnitude of their individual effects. The investigation of the effects of bottlenecks on selection limits may thus be valuable in the analysis of laboratory populations.

These results are also relevant to the problem of how to get the maximum possible advance from a given initial population. To stand a high chance of fixing all the rare but desirable genes we shall start obviously with as high a value of $N\bar{\imath}$ as we could. How large this should be will depend on the previous history of the population. If it has been previously selected or restricted in population size, the need would not be so great. Then as selection proceeded, the size of the programme could be reduced because the frequency of the desirable genes would continually be increasing. Unfortunately we cannot give a recipe of how rapid this reduction could be.

The control we have over $N\bar{\imath}$ lies mostly in $N$, the effective population size, rather than in $\bar{\imath}$, the selection intensity. In most selection programmes that are at all efficient, $\bar{\imath}$ lies between 1 and 2, these figures corresponding to a proportion selected, $p$, of 40 % and 6 %, respectively.

Occasionally, a selection programme is arranged so that the genetic contribution of all parents to the next generation is equal in order to minimize the increase in inbreeding each generation. It is rather doubtful whether this would affect the total advance by selection. It prevents the loss of variation by chance fixation but at the price of exposing only a part of the genetic variation to selection. In the simplest case of each mating contributing one male and one female as parents of the next generation, the effective breeding size of the population is twice the actual size. But the selection acts only on the variation within full-sib groups, i.e. half the total additive variance. The limit would not then be changed as the increase in $N$ is exactly balanced by the decrease in the effective value of $\sigma_g^2$. This argument would not hold if there were considerable non-genetic differences between full-sib groups when the accuracy of the selection might thereby be increased.

### THE OPTIMUM INTENSITY OF INDIVIDUAL SELECTION

Suppose we have a selection programme in which we measure $T$ animals and select the proportion $p$ that are highest for some metric character. What is the optimum value of $p$?

If the character is normally distributed, we may put $\bar{\imath} = z/p$, where $z$ is the ordinate of the unit normal curve at the point where the area cut off is $p$. But $N$, the number of parents selected, is equal to $Tp$ so that $N\bar{\imath} = Tz$. $z$ has a maximum when

# A theory of limits in artificial selection

$p = \frac{1}{2}$ so that the greatest advance will be attained when on half the population is selected each generation, as Dempster (1955) has shown. But the maximum of the curve may be very flat, especially when $T$ is large, because of the asymptotic nature of the curve of limit against $N\bar{\imath}$. Figure 8, which is based on the results for the synthetic population given in figure 7, shows the expected limit plotted against $p$ for two different values of $T$. When $T = 50$, the curve has become extremely flat-topped. In practice, the problem is to find the value which will combine a high rate of initial response with a reasonable approach to the ultimate possible limit.

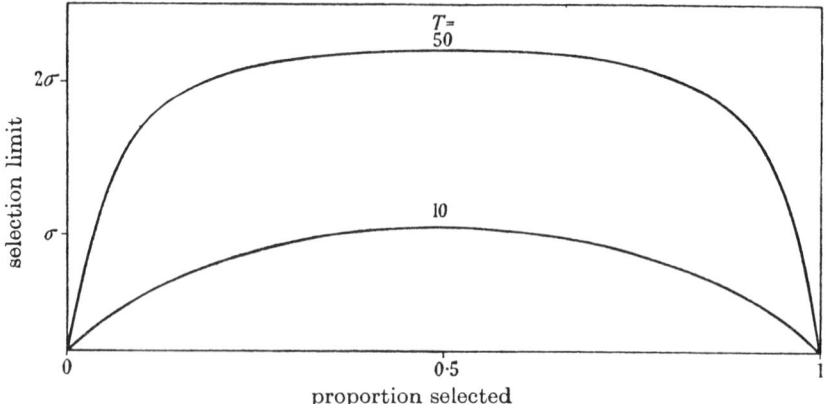

FIGURE 8. The expected limits to individual selection in the population in figure 6 when the number of animals measured is $T$.

## THE EXTENSION TO FAMILY SELECTION

It is a frequent practice in a selection programme to use the records of the relatives of the animals under selection. It can be shown that, if the information on which selection is based has a correlation of $r_{IG}$ with the animal's breeding value, the selective advantage of an additive gene is $\bar{\imath} a r_{IG}/h$, where $h^2$ is the heritability of the measurement under selection. Thus the use of measurements on relatives in a selective programme changes by the same factor the selection pressure on all the genes affecting the character. If we have the same response in two large selection programmes of different kinds from the same initial population, we may expect the gene frequencies in the two lines to be very similar. We can now generalize our earlier results to state that the selection limit is a function of $N\bar{\imath}r_{IG}/h$. This will not apply if other characters are taken into account in selection. The relative selection pressures on the different genes may then be altered, and the effect on selection limits cannot be predicted. Finally, if the value of $\bar{\imath}$ is different in the two sexes the value to be inserted in the formula is the arithmetic mean of the two.

The effect of different kinds of selection programme may be illustrated by the following table which gives the value of $N\bar{\imath}r_{IG}/h$ for some of the different kinds of selection programmes for bristles in *Drosophila* which have been carried out in this laboratory. Following Crow (1954), we shall assume that in mass-mating populations of *Drosophila* the effective population size is 0·55 of the actual size.

The character selected for will be assumed to have a heritability of 0·50. The symbol 20/100 means that the extreme 20 of the 100 flies measured were chosen as parents.

| selection method | intensity | | $N\bar{\imath}r_{IG}/h$ |
|---|---|---|---|
| individual | 20/100 | | 31 |
| | 20/50 | | 21 |
| | 20/25 | (each sex) | 8 |
| | 10/25 | | 10 |
| | 1/5 | | 2·4 |
| half-sib family (size 20) | 2/10 | | 9 |
| full-sib family (size 10) | 4/20 | — | 10 |

This leads naturally to the question of the optimum values of selection intensity and family size, if selection is based on family average, in order to achieve the greatest advance. The not very useful answer turns out to be that, for a given number of animals measured, the highest limit (determined by the maximum value of $N\bar{\imath}r_{IG}$) is reached when one-half of the families are selected and the family size is one, which would mean that selection was based only on the individual's own measurement. In other words, a family selection or progeny testing programme always involves some sacrifice of ultimate response for the sake of greater immediate gain.

## Defects of the model

From the point of view of the frequencies of single genes, the model is clearly of direct applicability. But, in transferring the conclusions to selection for quantitative characters, several hidden assumptions have been made which should now be mentioned.

The first one is that $a/\sigma$ will remain constant throughout the selection. On simple theory, we would expect the genetic part of $\sigma$ to decline gradually as the genes at other loci become fixed. On the other hand, as the level of homozygosity rises, the environmental part of $\sigma$ might be expected to increase in some characters. And in many selection experiments there is evidence that even the genetic part does not decline greatly as the limit is reached because of selection for heterozygotes. Rather of the same kind are problems of scale. These are almost impossible to cope with in a completely logical and satisfactory manner and each case has to be treated on its merits. But scaling difficulties should be more of a nuisance in discussing the extent of the selection limit rather than how long it takes to get there.

Linkage presents the second problem. One might expect *a priori* that the greater the number of generations over which a given selection differential was spread, the greater the response because of the high probability of recombination within chromosomes. If this is true, one might expect that the optimum proportion selected in individual selection would be rather less than the value of one-half which holds for the independent segregation of genes. But the theoretical treatment of linkage and selection in a population of limited size is a complex one. The use of Monte Carlo methods with digital computers, now being tried by Fraser (1957) and by Cockerham & Martin (1960), is probably the most satisfactory way of handling the problem.

We must also consider natural selection, probably for heterozygotes, in the sense that any population may have many gene frequencies held in equilibrium. The effect of this tendency to return to the original set of gene frequencies will be to reduce the effective value of $\bar{i}$ by a constant proportion (see Robertson 1956, p. 246). The optimum proportion selected in individual selection would then remain in the neighbourhood of 0·5.

The problem has been discussed entirely in terms of two alternative alleles. The existence of many alleles at each locus tremendously complicates the detailed treatment but some useful statements can still be made. It will still be possible to recast the basic differential equations, similar to (1), into a form similar to (2) so that again the pattern of change will be determined in artificial selection by $N\bar{i}$ and the time scale will be proportional to $N$. The limits of selection in any population will then still be a function of $N\bar{i}$. When $N\bar{i}$ is small, we can still say that the slope of the curve of advance against $N\bar{i}$ will have slope $2\sigma_g^2/\sigma$.

Finally, the effective population size in any artificial selection programme may be dependent on the intensity of selection. All parents chosen in one generation will not have an equal chance of contributing progeny to the group chosen in the next. The parents with a higher breeding value will have a higher chance of contributing than those with a low breeding value. A theoretical examination has been made of this effect in individual selection which will be published elsewhere in detail. It has been found that the ratio of the actual number of parents to the effective number is very roughly equal to $1 + K\bar{i}^2 h^2$ where $K$ depends on the relationship within families. This factor would have to be borne in mind in a more detailed discussion of the problem.

It must be emphasized that this theoretical investigation has probably its real value not in predicting exactly what is going to happen in reality but in enabling one to design experiments on selection limits and to interpret them when we have done this.

I am much indebted to my colleagues for their critical reading of the manuscript, in particular to Dr D. S. Falconer.

## REFERENCES

Cockerham, C. C. & Martin, F. G. 1960 *The IBM 650 and its uses in Selective Studies: in Biometrical Genetics.* Pergamon Press (In the Press).

Crow, J. F. 1954 Breeding structure of populations; in *Statistics and Mathematics in Biology*, ed. Kempthorne, Bancroft, Gower and Lush, Iowa State College Press, Ames, Ia.

Dempster, E. R. 1955 Genetic models in relation to animal breeding problems. *Biometrics*, **11**, 535–6.

Dempster, E. R. 1958 The fate of mutations in closed populations: in *Proc. 11th Pacific Poultry Breeders Roundtable.* University of California, California: Davis.

Fraser, A. S. 1957 The simulation of genetic systems by automatic digital computors. II. The effects of linkage on the rates of advance under selection. *Aust. J. Biol. Sci.* **10**, 492–9.

Haldane, J. B. S. 1931 A mathematical theory of natural and artificial selection. VII. Selection intensity as a function of mortality rate. *Proc. Camb. Phil. Soc.* **27**, 131–6.

Kimura, M. 1957 Some problems of stochastic processes in genetics. *Ann. Math. Stat.* **28**, 882–901.

Robertson, A. 1952 The effect of inbreeding on the variance due to recessive genes. *Genetics*, **37**, 189–207.

Robertson, A. 1956 The effect of selection against extreme deviants based on deviation or on homozygosis. *J. Genet.* **54**, 236–48.

Wright, S. 1931 Evolution in mendelian populations. *Genetics*, **16**, 97–159.

# RESULTS FROM SELECTION EXPERIMENTS

# Editor's Comments
# on Papers 14 Through 20

**14  DUDLEY**
76 Generations of Selection for Oil and Protein Percentage in Maize

**15  MATHER**
Excerpts from Variation and Selection of Polygenic Characters

**16  LERNER and DEMPSTER**
Excerpts from Attenuation of Genetic Progress Under Continued Selection in Poultry

**17  ROBERTSON**
Selection Response and the Properties of Genetic Variation

**18  FALCONER**
Excerpts from The Genetics of Litter Size in Mice

**19  CLAYTON, MORRIS, and ROBERTSON**
An Experimental Check on Quantitative Genetical Theory. I. Short-Term Responses to Selection

**20  BELL, MOORE, and WARREN**
Excerpt from The Evaluation of New Methods for the Improvement of Quantitative Characteristics

When the first selection experiments were conducted in the early 1900s (reviewed in Paper 3), there was no theory for predicting responses to selection for quantitative traits. All these experiments could establish was that changes did or did not take place. We now turn to experiments that were, with the exception of the Illinois corn experiment, established and in all cases analyzed after a solid body of theory was available. All the papers in this section are concerned with directional selection; consideration of stabilizing selection is deferred to the last section.

The Illinois corn experiment is probably the best known study of the effects of selection on quantitative traits. It was started in 1896 and continues to the present; a number of reports have been published over the years. Paper 14 by J. W. Dudley is recent and puts the results into a quantitative genetic context because it was given at the International Conference on Quantitative Genetics in Ames in 1976. Dudley's paper covers seventy six generations of selection; an earlier report by Winter (1929) on the first twenty eight generations was used by "Student" (Paper 27) to estimate minimum numbers of genes determining oil content in corn.

Selection started from an outbred population, but the lines have subsequently remained closed, there being four main ones, high and low for percent oil and for percent protein in the kernel. Details are given by Dudley (p. 460), but in general terms, mass selection has been practised throughout, with selection of twelve ears from sixty analyzed over the last fifty generations when realized heritabilities have averaged a little over 10 percent for oil content, and the same, but more erratically, for protein content. While there are obvious plateaux in the low-selected lines, they can be largely accounted for by the lack of variation and possibilities of progress, the oil content now being down to 0.3 percent from an initial mean of nearly 5 percent. In the high lines there is no indication of any plateaux, although reverse selection in the high-protein lines produced very rapid downward responses (RHP, Figure 2). The Illinois corn experiment has been an important demonstration to plant breeders and others of the power of selection to change a population steadily and continuously.

An analysis of the results is undertaken both in terms of Robertson's theory of limits (see Paper 13) and using the analysis previously adopted by "Student" (Paper 27). Dudley's calculations, particularly of initial gene frequencies, are rendered doubtful by the problem of scale effects—that is, the fact that the low lines were much less variable as they approached zero percent oil or protein. They also assume that all variability utilized by the selection was present at the start; in view of the continued linear response, it seems likely that new mutations were contributing substantially in later generations (Hill, 1982).

In quantitative genetics, as in the analysis of individual genes and their relation on the chromosome, *Drosophila melanogaster* has been the most widely used experimental species. Of the traits of the fruit fly, bristle or chaeta number on one part of the body or another has been most studied, primarily because it is easy to score, but also because its inheritance is "simple," tending to exhibit mostly additive variance with high heritability and almost no inbreeding depression. Bristle number was used by Sturtevant (Paper 3), subsequently by Payne (1920), and widely thereafter by Mather and colleagues at the

*Editor's Comments on Papers 14 Through 20*

John Innes Institute and Birmingham University (Papers 15 and 26) and A. Robertson and colleagues at Edinburgh University (Papers 19 and 25). It seems appropriate, therefore, to start the collection of more modern papers on selection experiments with an important study using bristle number in *Drosophila*.

Paper 15 by K. (later Sir Kenneth) Mather, reports a number of selection experiments from a series of crosses among stocks and from within the inbred parents. Of note are the large responses obtained from the cross-bred bases, despite using rather small lines, and apparent (although not very certain) plateaux after very few generations of selection followed by substantial responses later (Figures 2 and 4). The inbred base populations, on the other hand, yielded little or no responses over twelve generations (Figure 6). Because of shortage of space, some results have been omitted, including a table giving the results shown graphically in Figure 6; a description of the selection of the **BB** inbreds shown in Figure 6, where fertility problems caused loss of the high line and the low line to be represented in small numbers; a report on a second **BB** × + selection experiment, which gave largely similar answers to the first; and a report on selection in which selected flies each generation were back-crossed to inbreds.

Mather's study is, perhaps, as well known for its discussion and proposed models as for the experiments *per se*. He argued that the short periods of plateau subsequent to crossing the lines followed by subsequent responses were due to "balanced polygenic combinations" released by recombination (see Figure 5), certainly a feasible explanation for crosses of lines of *Drosophila* of similar mean level. He then went on to argue that the presence of such balanced combinations would be the norm in natural populations because they would allow the genetic variability expressed in the population to be reduced in the presence of stabilizing selection, but at the same time retain genic variation lest a major environmental change require new adaptations. This requires some foresight on the part of the individual: "Thus the organism is faced with a paradoxical situation in which variation results in an immediate loss of fitness while lack of variation means deferred but none the less serious disadvantage. In general then long term selection will effect a compromise between the two extremes. The mechanism of this compromise is linkage leading to balanced polygenic combinations" (p. 185). Mather goes on to show how balanced combinations could build up, but he does not invoke any mathematical analysis.

In subsequent papers Mather expanded these hypotheses (particularly Mather, 1943, and for a recent review see Mather and Jinks, 1982). The hypotheses have been the subject of some controversy, and were criticized by Wright (1952, see Part I, Paper 9), for example, who calculated that the selection pressures exerted against extremes was insufficient to maintain any degree of tight internal balance. Nevertheless, subsequent analyses, for example those by Bulmer and by Lande (Papers 23 and 24 of this volume), have shown that stabilizing selection can lead to a substantial reduction in the expressed variance through generating negative disequilibrium. But to put the magnitude of this discussion into perspective, with additive genes of equal effect and frequency at $n$ loci, an average correlation of gene frequencies at different loci of $-1/(n-1)$ is sufficient to give a zero genotypic variance. This correlation is tiny if many genes influence the trait.

The selection from the parental inbred lines of which the first twelve generations were reported by Mather (Paper 15, pp. 174-175) was continued for a total of fifty-three generations (Mather and Wigan, 1942). Eventually large responses were obtained, presumably as a consequence of accumulated mutations (see Paper 25 for a discussion of these results).

In the Illinois corn experiment steady responses have been made without apparent plateaux in the high-selected lines (Paper 14), but in most other experiments continued for large numbers of generations some plateaux or at least substantial reductions in response have been obtained. Paper 16 discusses such cases and describes methods of analysis and interpretation.

I. M. Lerner of the University of California, Berkeley, was one of the pioneers in promoting quantitative genetic methods in poultry breeding. Although Lush, following Wright, derived the first direct applications, his text, *Animal Breeding Plans* (Lush, 1937, and later editions), lacked the analytical detail of some of his papers (e.g., Paper 5). However Lerner's text, *Population Genetics and Animal Improvement* (1950) is a classic exposition of the principles and practices of applied quantitative genetics. He followed this text by others (Lerner, 1954, 1957), of which the latter includes further details of the experiments reported in Paper 16. Unfortunately, space did not permit inclusion of more of Lerner's work, either under single authorship or jointly with his colleague, E. R. Dempster (e.g., Dempster and Lerner, 1947), and an appendix to Paper 16 on "Calculated Reduction in Gain Due to Differential Fecundity of Dams" had to be omitted.

In the experiments on poultry reported in Paper 16, Lerner and Dempster find a plateau for increased shank length after about seven generations of selection, and in a separate line a reduced rate of response for egg production after about ten generations of selection. In neither case, however, were unselected controls maintained, although they used the production line as a control for the shank length line.

Lerner and Dempster show that genetic variability remained in the shank length line (Table 2), but that an unfavorable correlation between shank length and fitness, as evidenced by the difference between expected and realized selection differentials, was at least partly responsible for the small responses. Further evidence was provided (Lerner, 1957, p. 134), for shank length declined substantially in a line, started in 1950, in which selection was suspended. There were also presumably some effects of inbreeding on mean, if not on genetic variance, if Lerner and Dempster's estimates are to be accepted, because the shank length line was of very small size. Overall, however, the results are not easily interpreted. As Lerner (1957, p. 144) concludes on the shank length experiment: "The case history . . . although useful in assessing how much is known about selection, may be even more valuable in showing how incomplete our understanding of the underlying processes is."

The apparent slowing of rates of response in the production flock is of more direct practical importance, for selection was being practised for the major economic trait of poultry for table egg production. In this case, Lerner and Dempster find no evidence of counteracting natural selection. The issue of achieved rates of response in egg production has generated much debate. Dickerson (1955) reported that little response was being made in a commercial flock; he called this phenomenon "genetic slippage" and attributed it first to unfavorable genetic correlations, but subsequently to epistatic effects (Dickerson, 1965). Some thirty years after Lerner and Dempster's report, there remains evidence, however, that commercial populations of poultry are continuing to improve, as Flock (1980) has indicated.

Forbes W. Robertson and E. C. R. Reeve conducted extensive selection experiments in Edinburgh on body size in *Drosophila melanogaster*, taking measurements on wing length or thorax length. Their results contrast somewhat with those undertaken on bristle number in that body size is clearly more strongly influenced by nonadditive genes and natural selection (see, for example, Paper 22). Paper 17 summarizes many of their results; regrettably for the breadth of authorship in this collection of reprints, it is by Robertson alone, but many of the major papers in the series with Reeve are rather long (e.g., Reeve and Robertson, 1953; Robertson and Reeve, 1952, 1953, 1955). Contributions to the open discussion of Paper 17 at the Cold Spring Harbor symposium have been omitted.

A striking feature of Robertson and Reeve's results (see Figure 1) is that though responses in the direction of selection were made quite rapidly, reverse selection brought the populations back to their original level more quickly in the large lines even though plateaux had been reached. Further evidence that plateaux were not associated with fixation of all loci was obtained from the regression to the origin on relaxation, but that some fixation was involved was indicated by the additional responses on crossing the lines. Other results of Robertson and Reeve's series of papers are summarized in Paper 17. Among them was the observation that "exchange of chromosomes between lines reveals widespread interaction between non-homologous chromosomes and a non-additive relation between homologues" (p. 173).

Robertson and Reeve's results are complex and difficult to review briefly—indeed, one of the major conclusions from their work is that an apparently simple trait like body size, on which straightforward directional selection experiments are practised, does not yield nice additive results. In correspondence, Robertson has also emphasized some further features of the results given by Robertson and Reeve (1955) on crosses among inbred lines, referred to in Paper 17. Among these are the evidence for epistatic effects in contributing to heterosis, the evidence for more marked interactions the closer the trait is to fitness, and the inverse relation between the mean level of the trait and environmental variance. Also, Robertson (Paper 17, p. 176) notes that "Recent experiments suggest that such environmental variation cannot be regarded as independent of the expression of genetic differences even in the unselected stock, and probably less so in strains undergoing selection or inbreeding," and in correspondence states that these sentiments were the starting point for his series of experiments on "Ecological Genetics of Growth in *Drosophila*" (F. W. Robertson, 1960 et seq.).

In using laboratory animals as models for domestic species, it is obviously desirable that their biology should be as close as possible. Growth or egg production in *Drosophila* seems far removed from growth or reproduction in mammals and birds. Reproductive rate (per litter!) is, however, very similar in the mouse and the pig, so it seems reasonable to assume that studies of the genetics of litter size in the mouse may give useful information about that in the pig, in addition to whatever general information about quantitative inheritance is obtained. Paper 18, by D. S. Falconer, is such a classic model experiment. It includes studies on selection and other analytical techniques such as inbreeding and gives results that have, at least in part, been helpful in interpreting data on farm animals. The text of Paper 18 is reproduced in full, but contributed discussion from a conference has been omitted. Paper 7, also by Falconer, refers to another selection experiment, that on growth of mice on different nutritional regimes (Falconer and Latyzsewski, 1952), which also used the mouse as a model to answer animal breeding questions.

Response to selection was obtained both for high and for low litter size for at least twenty generations (Paper 18, Figure 2). Although the responses appeared fairly symmetric to high and low selection, the cumulative selection differential was much less in absolute terms in the low lines (Figure 3), so the heritability for selection for low litter size was much greater. In view of the lack of replication it is not certain that this asymmetry was significant, but when Falconer investigated the changes in the components of litter size at birth—namely, ovulation rate and embryonic mortality—there was striking asymmetry (Table 5). Selection for high litter size led primarily to increased ovulation rate, while selection for low litter size led primarily to increased embryonic mortality. These results show the value of the detailed analysis of these responses, for the complex relationship between ovulation rate, embryonic mortality, and litter size has featured in a number of subsequent experiments. For example, Land and Falconer (1969) and Bradford (1968) with mice and Cunningham et al. (1979) with pigs have shown that successful selection for increased ovulation rate has not led to an increase in litter size, and the work with mice has helped in the interpretation of data on pigs and other large animals.

In Paper 18, Falconer computes "realized heritabilities," regressions of response on selection differential, as have many other authors (e.g., in Papers 17 and 19) following his pioneering use of the statistic in an earlier paper (Falconer, 1954).

Papers 16 through 18 have revealed that selection responses for a single trait frequently involve complex interactions with other traits or with fitness. The reader may be left wondering whether the simple prediction formulae obtained in Section II or summarized in Chapter 11 of Falconer (1981) have any value. A thorough investigation of the basic quantitative genetic theory was undertaken by G. A. Clayton, J. A. Morris, and A. Robertson in Edinburgh and is included as Paper 19. This paper covers results on direct responses up to, usually, five generations, and generally the analysis is straightforward. In subsequent papers in the series further results on long-term responses (Clayton and Robertson, 1957) and on correlated responses were given (Clayton, Knight, Morris, and Robertson, 1957), and there the outcome was rather more involved. Here we need deal only with the short-term responses in abdominal bristle number reported in the first paper.

The checks on theory in Paper 19 were of several kinds. Clayton et al. confirmed that, as predicted from the Mendelian model (see papers in Part I), similar heritability estimates were obtained by offspring-parent regression and half-sib correlation, (and also by full-sib correlation, because the trait of abdominal bristle number shows little or no dominance or maternal environment effects). They found that the variance among inbred lines accorded well with expectation for an additive trait. Several checks on selection theory were carried out by Clayton et al., as outlined on p. 137. In general terms, responses to mass selection were at the rates predicted from the estimates of parameters in the base population, except that the low lines responded rather less rapidly that expected. Similarly, with family selection, the responses in the high direction agreed well with theory, those in the low direction were rather less than predicted, with the ranking of alternative schemes as expected. The authors conclude: "In the main, selection for bristles in this population has given results in fair agreement with theory, though this is in some measure a subjective judgment" (p. 146). It should be noted that in their predictions of response they were not able to take account of the effects of selection on variance, as discussed subsequently by Bulmer (Paper 23). With initial heritabilities around 0.5, a reduction to about 0.4 would be expected after two generations of selection. The interpretation of the results is also, as the authors mention, complicated by scale effects, the variance dropping more rapidly in the low lines. A reanalysis of the first few generations of response might be justified to take account of these two effects on variance.

The results of Paper 19 also illustrate (Figure 1) a feature of selection experiments that we have not discussed in the previous papers in this volume, namely, the variability among replicates. This was one of the few experiments conducted with sufficient replicates to enable this variation to be detected and estimated (p. 139), where it was found to accord roughly with predictions for unselected lines. Subsequently, the sampling properties of selection experiments have been investigated in more detail both theoretically (e.g., Hill, 1971) and experimentally (Falconer, 1973). It suffices to say that not too many conclusions should be drawn from small, unreplicated selection experiments.

The final paper on directional selection experiments is another from the Twentieth Cold Spring Harbor Symposium, that by A. E. Bell, C. H. Moore, and D. C. Warren of Purdue University. Their report is reproduced as Paper 20; the contributed discussion has been omitted. Their study is a model experiment different from those included previously, in that they set out to use an experimental animal *(Drosophila melanogaster)* to compare alternative breeding programs. The work was stimulated in part by Comstock et al.'s proposal (Paper 12) of the method of reciprocal recurrent selection (RRS) for improving crossbreds, and it was a thorough attempt to evaluate this proposal against alternative procedures using both egg number, a trait of low heritability exhibiting much heterosis, and egg weight, a highly heritable and not heterotic trait. Although this experiment was conducted with *Drosophila*, in later work on RRS and related problems the Purdue group under Bell turned to *Tribolium* as an experimental animal, partly for reasons given on p. 210 of their paper.

Paper 20 is so extensive as to prohibit a brief summary, and since the main questions it tackled related to practical application rather than general principle, only a few points will be made. First, Bell et al. showed that for egg weight, direct selection was most effective and that inbred performance was a good predictor of inbred cross performance. For fecundity, cross performance was only poorly predicted by inbred performance; and both RRS and recurrent selection to a tester (Hull, 1945) were superior to selection on pure-line record, but both methods were generally inferior to crosses among inbred lines. This study has been succeeded by others in experimental and farm species that have tested the relative efficiency of alternative breeding programs in improving cross performance. Their result, that simple mass or family selection is best for traits of high heritability, has been substantiated; their somewhat less clear conclusions on more lowly heritable reproductive traits have been reflected by subsequent conflicting reports. For a summary and discussion of some more current views on the role of RRS, see Eberhart (1977) or Hallauer and Miranda (1981) for work in corn, and Flock (1980) for poultry.

Finally, let us consider other important selection experiments. One major area omitted in this section is any test of the theory of limits proposed by Robertson in Paper 13, and extensive study was conducted by Jones, Frankham, and Barker (1968), earlier generations of this study also providing checks on short-term theory (Frankham, Jones, and Barker, 1968). Apart from the Illinois corn experiment, there have been others in which selection has been continued for a long time without obvious plateaux, perhaps due to new mutations, as will be discussed in the last section. Despite long, continued responses, some experiments have produced evidence of substantial epistatic interaction, the best example being that of Enfield (1980 and earlier reports). No study directly on the selection of correlated traits in separate lines or as an index has been included; notable among these are experiments by Eisen (e.g., 1977).

For commercial species, important studies of plants other than the Illinois corn oil experiment, include those of Gardner and colleagues in Nebraska (Gardner, 1977). On farm animals, a model selection study was carried out by Hetzer in pigs (Hetzer and Harvey, 1967).

Most of the papers in this section have reported experiments conducted in the 1940s and 1950s. Although a number of important studies were conducted earlier, they largely lacked a theoretical basis. Nevertheless, it is hoped that those that have been reprinted show some of the breadth of analysis possible with selection experiments.

## REFERENCES

Bradford, G. E., 1968, Selection for Litter Size in Mice in the Presence and Absence of Gonadotropin Treatment, *Genetics* **58:**283-295.

Clayton, G. A., G. R. Knight, J. A. Morris and A. Robertson, 1957, An Experimental Check on Quantitative Genetical Theory. III. Correlated Responses. *J. Genet.* **55:**171-180.

Clayton, G. A., and A. Robertson, 1957, An Experimental Check on Quantitative Genetical Theory. II. The Long-term Effects of Selection, *J. Genet.* **55:**152-170.

Cunningham, P. J., M. E. England, L. D. Young, and D. R. Zimmerman, 1979, Selection for Ovulation Rate in Swine: Correlated Response in Litter Size and Weight, *J. Anim. Sci.* **48:**509-516.

Dempster, E. R., and I. M. Lerner, 1947, The Optimum Structure of Breeding Flocks, I. Rate of Genetic Improvement Under Different Breeding Plans, *Genetics* **32:**555-566.

Dickerson, G. E., 1955, Genetic Slippage in Response to Selection for Multiple Objectives, *Cold Spring Harbor Symp. Quant. Biol.* **20:**213-224.

Dickerson, G. E., 1965, Experiment Evaluation of Selection Theory in Poultry, in *Genetics Today, Proceedings of the Eleventh International Congress of Genetics,* Vol. 3, S. J. Geerts, ed., Pergamon Press, Oxford, pp. 747-761.

Eberhart, S. A., 1977, Quantitative Genetics and Practical Corn Breeding, in *Proceedings of the International Conference on Quantitative Genetics,* E. Pollak, O. Kempthorne, and T. B. Bailey, Jr., eds., Iowa State University Press, Ames, pp. 491-502.

Eisen, E. J., 1977, Antagonistic Selection Index Results with Mice, in *Proceedings of the International Conference on Quantitative Genetics,* E. Pollak, O. Kempthorne, and T. B. Bailey, Jr., eds., Iowa State University Press, Ames, pp. 117-139.

Enfield, F. D., 1980, Long Term Effects of Selection; The Limits to Response, in *Selection Experiments in Laboratory and Domestic Animals,* A. Robertson, ed., Commonwealth Agricultural Bureaux, Slough, pp. 69-86.

Falconer, D. S., 1954, Asymmetrical Responses in Selection Experiments, in *Symposium on Genetics of Population Structure,* Int. Union Biol. Sci. (Naples) series B, no. 15, pp. 16-41.

Falconer, D. S., 1973, Replicated Selection for Body Weight in Mice, *Genet. Res. (Cambridge)* **22:**291-321.

Falconer, D. S., 1981, *Introduction to Quantitative Genetics,* 2nd ed., Longmans, London.

Falconer, D. S., and M. Latyszewski, 1952, The Environment in Relation to Selection for Size in Mice, *J. Genet.* **51:**67-80.

Flock, D. K., 1980, Genetic Improvement of Egg Production in Laying Type Chickens, in *Selection Experiments in Laboratory and Domestic*

*Animals,* A. Robertson, ed., Commonwealth Agricultural Bureaux, Slough, pp. 214-224.

Frankham, R., L. P. Jones, and J. S. F. Barker, 1968, The Effects of Population Size and Selection Intensity in Selection for a Quantitative Character in *Drosophila.* I. Short-term Response to Selection, *Genet. Res. (Cambridge)* **12**:237-248.

Gardner, C. O., 1977, Quantitative Genetic Studies and Population Improvement in Maize and Sorghum, in *Proceedings of the International Conference on Quantitative Genetics,* E. Pollak, O. Kempthorne, and T. B. Bailey, Jr., eds., Iowa State University Press, Ames, pp. 475-486.

Hallauer, A. R., and J. B. Miranda, 1981, *Quantitative Genetics in Maize Breeding,* Iowa State University Press, Ames.

Hetzer, H. O., and W. R. Harvey, 1967, Selection for High and Low Fatness in Swine, *J. Anim. Sci.* **26**:1244-1251.

Hill, W. G., 1971, Design and Efficiency of Selection Experiments for Estimating Genetic Parameters, *Biometrics* **27**:293-311.

Hill, W. G., 1982, Rate of Change in Quantitative Traits from Fixation of New Mutations, *Natl. Acad. Sci. (U.S.A.) Proc.* **79**:142-145.

Hull, F. H., 1945, Recurrent Selection for Specific Combining Ability in Corn, *J. Am. Soc. Agron.* **37**:134-145.

Jones, L. P., R. Frankham, and J. S. F. Barker, 1968, The Effects of Population Size and Selection Intensity in Selection for a Quantitative Character in *Drosophila.* II. Long-term Response to Selection, *Genet. Res. (Cambridge)* **12**:249-266.

Land, R. B., and D. S. Falconer, 1969, Genetic Studies of Ovulation Rate in the Mouse, *Genet. Res. (Cambridge)* **13**:25-46.

Lerner, I. M., 1950, *Population Genetics and Animal Improvement,* Cambridge University Press, Cambridge.

Lerner, I. M., 1954, *Genetic Homeostasis,* Oliver and Boyd, Edinburgh.

Lerner, I. M., 1957, *The Genetic Basis of Selection,* Wiley, New York.

Lush, J. L., 1937, *Animal Breeding Plans,* Iowa State College Press, Ames.

Mather, K., 1943, Polygenic Inheritance and Natural Selection, *Biol. Rev.* **18**:32-64.

Mather, K., and J. L. Jinks, 1982, *Biometrical Genetics,* 3rd ed., Chapman and Hall, London.

Mather, K., and L. G. Wigan, 1942, The Selection of Invisible Mutations, *R. Soc. (London) Proc.* **B131**:50-64.

Payne, F., 1920, Selection for High and Low Bristle Number in the Mutant Strain "Reduced," *Genetics* **5**:501-542.

Reeve, E. C. R., and F. W. Robertson, 1953, Studies in Quantitative Inheritance. II. Analysis of a Strain of *Drosophila melanogaster* Selected for Long Wings, *J. Genet.* **51**:276-316.

Robertson, F. W., 1960, The Ecological Genetics of Growth in *Drosophila* I. Body Size and Developmental Time on Different Diets, *Genet. Res. (Cambridge)* **1**:288-304.

Robertson, F. W., and E. C. R. Reeve, 1952, Studies in Quantitative Inheritance, I. The Effects of Selection on Wing and Thorax Length in *Drosophila melanogaster, J. Genet.* **50**:414-448.

Robertson, F. W. and E. C. R. Reeve, 1953, Studies in Quantitative Inheritance. IV. The Effects of Substituting Chromosomes from Selected Strains in Different Genetic Backgrounds in *Drosophila melanogaster, J. Genet.* **51**:586-610.

Robertson, F. W., and E. C. R. Reeve, 1955, Studies in Quantitative Inheritance. VII. Crosses Between Strains of Different Body Size in *Drosophila melanogaster, Z. Indukt. Abstamm. Vererbungsl.* **86**:424–443.

Winter, F. L., 1929, The Mean and Variability as Affected by Continuous Selection for Composition in Corn, *J. Agric. Res.* **39**:451–475.

Wright, S., 1952, The Genetics of Quantitative Variability, in *Quantitative Inheritance,* E. C. R. Reeve, and C. H. Waddington, eds., Her Majesty's Stationery Office, London, pp. 5–41.

# 14

Copyright © 1977 by Iowa State University Press
Reprinted from pages 459-473 of Proceedings of the International Conference on Quantitative Genetics, E. Pollak, O. Kempthorne, and T. B. Bailey, Jr., eds., Iowa State University Press, Ames, 1977, 872 p.

# 76 generations of selection for oil and protein percentage in maize

## J. W. Dudley
AGRONOMY DEPARTMENT
UNIVERSITY OF ILLINOIS, URBANA, IL 61801

### INTRODUCTION

Seventy-six generations of selection for high oil, low oil, high protein, and low protein in corn (Zea mays L.) have been completed. A recent paper (Dudley, Lambert, and Alexander, 1974) summarized a detailed analysis of the data from 70 generations of selection. The purpose of this paper is to present the results of 76 generations of selection and their implications to the questions of:

1) limits of selection,
2) population gene frequencies,
3) number of loci,
4) effects of linkage on selection progress,
5) types of gene action, and
6) genetic diversity.

### MATERIALS AND METHODS

Details of the selection procedure, chemical and statistical analyses have been presented (Dudley, Lambert, and Alexander, 1974) and only pertinent parts are included here. Selection was initiated in 1896 by analyzing 163 open-pollinated ears of the variety 'Burr's White' for percent oil and percent protein. The 24 ears highest in protein, the 12 ears lowest in protein, the 24 ears highest in oil, and the 12 ears lowest in oil were selected to initiate the Illinois high protein (IHP), low protein (ILP), high oil (IHO), and low oil (ILO) strains respectively. The forward selection phase of the experiment was divided into 4 segments as follows:

Segment 1. Generations 0-9, mass selection based on chemical composition. Number of ears analyzed and selected varied but approximately 20% of the ears analyzed were selected. Each strain was grown in a separate isolated field.

Segment 2. Generations 10-25. 120 ears per strain were analyzed and 24 were saved. Seed from each ear was planted ear-to-row. Alternate rows were detasseled and 20 ears were analyzed from each of the six highest yielding rows. Four ears were saved per row.

Segment 3. Generations 26-52 in IHP and ILP; generations 26-58 in IHO and ILO. Twelve selected ears were arbitrarily divided into two lots (A and B) of six ears. Seed within each lot was bulked and planted in the nursery. Silks in lot A were pollinated by a bulk sample of pollen from 15-20 plants in B while silks in B were pollinated with pollen from A. Thirty ears from each lot were analyzed and the 12 most extreme of the 60 ears analyzed were saved.

Segment 4. Generations 53-76 in IHP and ILP; 59-76 in IHO and ILO. Procedure was the same as in segment 3 but 80 lbs of N per acre were added to the soil.

In generation 48, reverse selection was initiated in each strain (Leng, 1962b) to form four new strains: reverse high protein (RHP), reverse low protein (RLP), reverse high oil (RHO), and reverse low oil (RLO). The procedure was the same as in the forward selection experiment except that selection was for low protein in the high protein strain, high oil in the low oil strain, etc. After seven generations of selection in RHO, a new strain, designated switchback high oil (SHO) was initiated by selecting the 12 ears in RHO that were highest in oil. The reverse phase of the experiment was divided into two segments based on the date of application of N as follows: (1) generations 0-4 in RHP and RLP and generations 0-10 in RHO and RLO; (2) generations 5-28 in RHP and RLP and generations 11-28 in RHO and RLO.

## STATISTICAL ANALYSIS

Each generation a bulk sample of IHO and ILO was analyzed for percent protein. Because protein was affected by year-to-year weather fluctuations and the correlated response between protein and oil was negligible, the protein means for each generation were adjusted by subtracting the mean protein percentage of IHO and ILO. The protein data are presented as adjusted percent protein. The oil data were not adjusted because of the small year-to-year variation in percent oil.

Realized heritabilities for each strain were calculated from regression of cumulative selection differential on generation means for each segment using the procedure of dummy variables in multiple regression described by Draper and Smith (1966, p. 140, ex. 2).

Additive genetic variance ($\sigma_A^2$) for each segment was estimated from the realized heritabilities and the pooled estimates of phenotypic variance ($\sigma_P^2$) from the generations included in the segment. Estimates of $\sigma_A^2$ for the original population were calculated as the average of the estimates from segment 1 of the high and low strains. Because percent oil in a kernel is determined by the genotype of the kernel, selection among ears is equivalent to selection among half-sib family means. Thus, $\sigma_A^2 = 8\sigma_P^2 h^2$ for segments 1, 3 and 4 for percent oil instead of $2\sigma_P^2 h^2$ as given by Dudley, Lambert, and Alexander, 1974. Because percent protein in the kernel is determined by the genotype of the mother plant, selection among ears is equivalent to mass selection and $\sigma_A^2 = 2\sigma_P^2 h^2$ as previously reported.

RESULTS

Limits to Selection

The limit to selection has not been reached in any of the 9 strains as evidenced by:

1) effectiveness of reverse selection and switchback selection (Figs. 1, 2, 3, 4);
2) significant estimates of realized heritability in the last segment of all 9 strains (Tables I and II);
3) significant estimates of genetic variation among half-sib family means in IHP, ILP, IHO, and ILO after 65 generations of selection (Dudley and Lambert, 1969).

FIG. 1

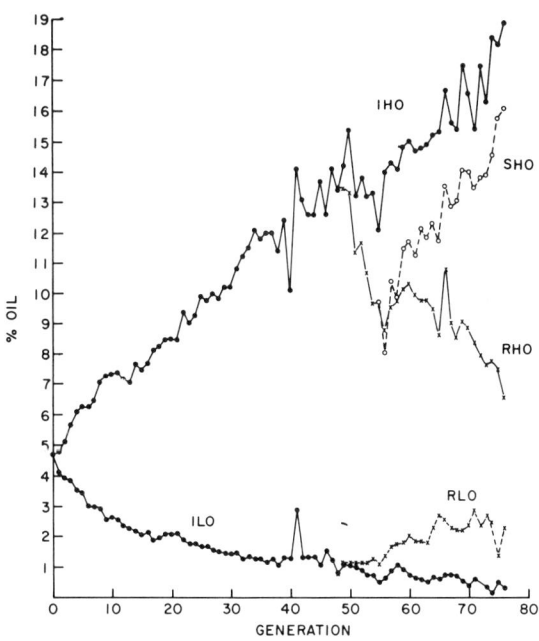

Mean percent oil for IHO, ILO, RHO, RLO, and SHO plotted against generations.

FIG. 2

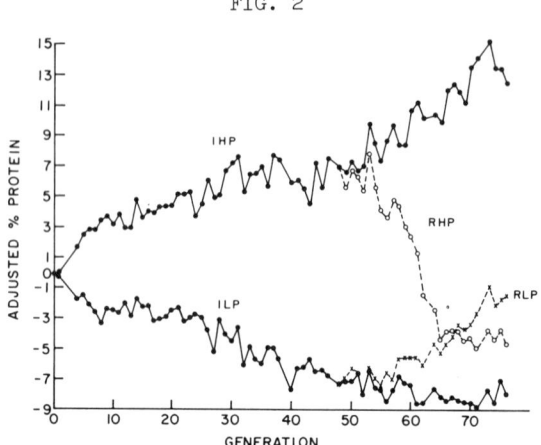

Mean adjusted percent protein for IHP, ILP, RHP, and RLP plotted against generations.

FIG. 3

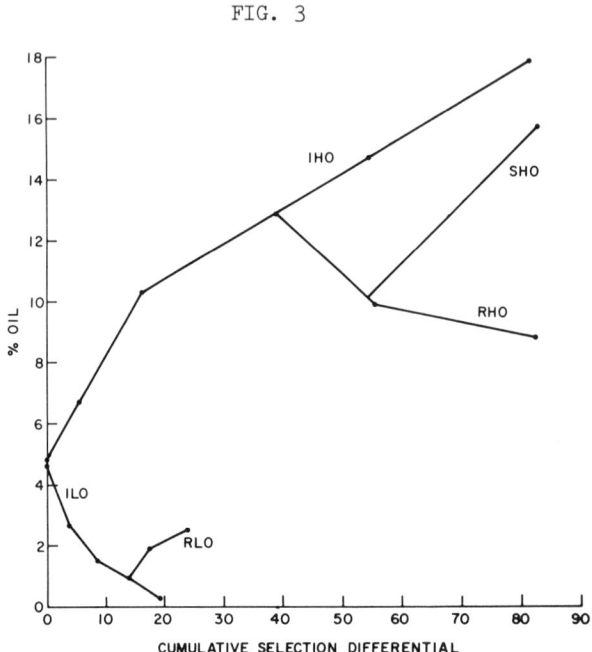

Plot of realized heritabilities for the oil strains.

FIG. 4

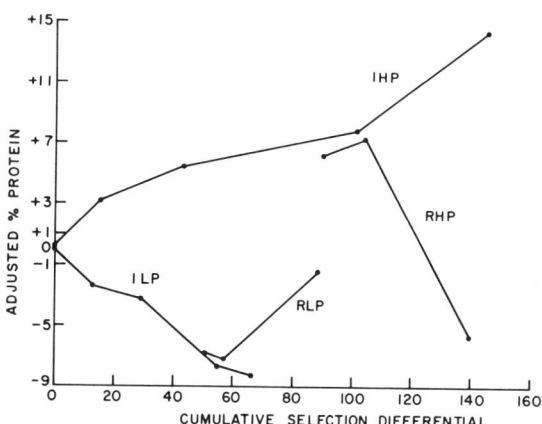

Plot of realized heritabilities for the protein strains.

TABLE I

Realized Heritabilities for the Oil Strain

| Segment | Strain | | | | |
| --- | --- | --- | --- | --- | --- |
| | IHO | RHO | SHO | ILO | RLO |
| 1 | .32±.10 | | | .50±.05 | |
| 2 | .34±.04 | | | .23±.03 | |
| 3 | .11±.01 | .18±.04 | | .10±.01 | .37±.08 |
| 4 | .12±.02 | .07±.02 | .19±.01 | .15±.04 | .09±.03 |

TABLE II

Realized Heritabilities for the Protein Strains

| Segment | Strain | | | |
|---|---|---|---|---|
| | IHP | RHP | ILP | RLP |
| 1 | .20±.05 | | .18±.04 | |
| 2 | .08±.02 | | .05±.02 | |
| 3 | .04±.01 | | .17±.01 | |
| 4 | .15±.01 | .37±.02 | .07±.02 | .17±.01 |

Although there is still genetic variation for percent oil in ILO, it is apparent that, with a mean of .3% oil, progress is nearly exhausted since the lower limit cannot be less than zero.

At intermediate gene frequencies, realized heritability for reverse selection should be similar to realized heritability for forward selection in the segment preceding initiation of reverse selection assuming environmental variance remains constant. This was true for RLP. However, for RHP and RLO realized heritability for reverse selection was significantly higher than for the preceding segment of forward selection. For RHO realized heritability was significantly lower than in the preceding segment of IHO. The reasons for this are not clear. In the case of IHP and RHP, progress in segment 3 of IHP was low, presumably because of limiting soil N levels, since realized heritability in segment 4 was significantly higher than in segment 3.

Gains at the end of 76 generations were 20, 6, 20 and 12 x $\sigma_A$ for IHO, ILO, IHP, and ILP respectively. These gains, expressed as percent of the original mean are 279, 92, 133, and 78 percent. Although gains of this magnitude at first seem surprising, calculation of expected ultimate limits to selection with free recombination using the equation:

$$G/\sigma_A = [2n(1-q)/q]^{1/2} \quad \text{(Robertson, 1970)}$$

where G is the ultimate gain, n is the number of loci, and q is the frequency of the favorable allele, for varying values of n and q (Table III) shows that to gain $20\sigma_A$ requires 200 loci if q = .5 or between 50 and 100 if q is .25. Thus, the gains achieved are explainable when reasonable gene frequencies and numbers of loci are assumed.

TABLE III

Ultimate Limits to Selection in Terms of $\sigma_A$ with Varying Values of n (Number of Loci) and q (Frequency of the Favorable Allele).

| n | q | | | | |
|---|---|---|---|---|---|
|   | .1 | .25 | .5 | .75 | .9 |
| 10 | 13 | 8 | 4 | 3 | 2 |
| 50 | 30 | 17 | 10 | 6 | 3 |
| 100 | 42 | 24 | 14 | 8 | 5 |
| 200 | 60 | 35 | 20 | 12 | 7 |

Estimates of Gene Frequency

No good method of estimating gene frequency for quantitative traits is available. However, if divergent selection experiments are carried to the limit, it should be possible to obtain an estimate (of unknown reliability) of average gene frequency using the equation for ultimate gain with no linkage given by Robertson (1970) as follows:

Let ultimate gain in the upward direction be $G_H$ and in the downward direction be $G_L$. Since $G_H = [2n(1-q)/q]^{\frac{1}{2}}\sigma_A$ and $G_L = [2nq/(1-q)]^{\frac{1}{2}}\sigma_A$ then $G_H/G_L = [2n(1-q)/q]^{\frac{1}{2}}/[2nq/(1-q)]^{\frac{1}{2}} = (1-q)/q$ and $q = 1/[G_H/G_L) +1]$.

Thus, q can be estimated knowing only the gain at the limit for selection in both directions. Assuming that the 9 strains in this experiment had reached the limit, estimates of q were obtained for the three possible comparisons among the protein strains and the four possible comparisons among the oil strains (Table IV).

## TABLE IV

Estimates of q at Different Stages in the Selection Experiment Assuming the Ultimate Limit was Reached.

| % Oil | | % Protein | |
|---|---|---|---|
| Source | q | Source | q |
| IHO/RHO | .51 | IHP/RHP | .62 |
| SHO/RHO | .32 | IHP/ILP | .37 |
| IHO/ILO | .25 | RLP/ILP | .19 |
| RLO/ILO | .28 | | |

Estimates of q = .25 for % oil and q = .37 for % protein in the original population were obtained. After 48 generations of selection in IHP (see IHP/RHP comparison) q = .62 and in ILP q = .19. The results for percent oil were inconsistent in that the estimated q after 48 generations of downward selection (RLO/ILO) was similar to the estimate for the original population. These estimates are tentative since the limits to selection have not been reached. However, the results do suggest that initial average gene frequencies were less than .5 for both traits and for percent oil considerably less than .5 since the present estimate is .25, and appreciable progress in the upward direction can likely still be obtained while little additional progress downward is possible. Thus, the ratio of $G_H/G_L$ may increase which will reduce the estimate of q.

### Number of Effective Factors

Student (1934) estimated that 33 loci controlled percent oil based on data through 28 generations. However, his estimate of $\sigma_A^2$ was crude. Assuming that the high selected strains are homozygous for the (+) alleles for the selected traits, estimates of the number of effective factors controlling percent oil and percent protein were obtained using the expression

$$K_1 = [\tfrac{1}{2}(\overline{P}_1 - \overline{P}_2)]^2 / [1/2q(1-q)](\sigma_A^2)$$

which is a modified form of the equation given by Mather and Jinks

(1971) where $K_1$ is the number of effective factors, $\bar{P}_1$ is the mean of the high strain, $\bar{P}_2$ is the mean of the low strain, $\sigma_A^2$ is estimated as outlined in Materials and Methods, and values of q are those given in Table IV. This procedure assumes no dominance but removes the restriction that q = .5.

Estimates obtained suggest that the number of effective factors controlling percent protein (122) is larger than the number controlling percent oil (54) (Table V).

TABLE V

Estimates of Number of Effective Factors Controlling Percent Oil and Percent Protein.

| % Oil | | % Protein | |
|---|---|---|---|
| Source | $K_1$ | Source | $K_1$ |
| IHO vs ILO | 54 | IHP vs ILP | 122 |
| IHO vs RHO | 9 | IHP vs RHP | 47 |
| SHO vs RHO | 3 | ILP vs RLP | 18 |
| RLO vs ILO | 6 | | |

The estimates of numbers of factors differentiating the reverse strains from the strain from which they were selected were made using estimates of $\sigma_A^2$ from the segment of the experiment following initiation of reverse selection. The sum of the number of effective factors differentiating IHP from RHP and ILP from RLP is less than the total estimated from IHP vs. ILP even though RHP has lower percent protein than RLP and all the genetic difference between IHP and ILP is accounted for by the comparisons IHP vs. RHP plus ILP vs. RLP.

Additional estimates of the number of factors controlling percent oil were obtained from a Design III experiment (Comstock and Robinson, 1952) using estimates of $\sigma_A^2$ and the dominance variance $\sigma_D^2$ (Moreno-Gonzalez, Dudley, and Lambert, 1975) from the random-mated $F_6$ generation of the cross IHO x ILO. From the $F_6$

data, estimates of $\sigma_A^2$ and $\sigma_D^2$ were .812 and .189 respectively. Estimates of $K_1$, $K_2$, and k using the procedure described by Mather and Jinks (1971) (p. 323) were 35, 14, and 71 respectively, where $K_1$ is as previously described, $K_2 = (\overline{F}_1 - MP)^2/4\sigma_D^2$ and $k = 3K_1K_2/(4K_2 - K_1)$. The $K_1$ estimates from the Design III and from the IHO/ILO comparison are in reasonable agreement when one considers that linkage disequilibrium in the IHO x ILO cross may not be completely dissipated by the $F_6$.

Because of the simplifying assumptions necessary (see Mather and Jinks, 1971) these are minimum estimates of the number of loci involved. However, the data suggest that the number of effective factors controlling percent protein may be higher than the number controlling percent oil.

Effects of Linkage

Although there is no apparent way of measuring the effect linkage has had on progress in this experiment, a few observations can be made. Following Hanson (1959) the average unbroken segment length of chromosome after 76 generations was calculated as .013 crossover units. Thus, there has been ample opportunity for essentially free recombination to occur. Using Robertson's (1970) result that linkage has little effect before $2/ih^*$ generations where i = approximately 1.4 for this experiment and $h^* = h/H^{\frac{1}{2}}$, where H is the haploid number of the species, the number of generations before linkage should show an effect was calculated as 7 for IHO and ILO and 10 for IHP and ILP. Since the breeding system was changed at generation 9, it is not clear whether the reduced realized heritability in segment 2 of both IHP and ILP (Table III) resulted from the change in breeding system or effects of linkage. For IHO it is apparent that if linkage had an effect it was much later than 7 generations since realized heritability remained the same through 25 generations.

Evidence from the design III study (Moreno-Gonzalez, Dudley, and Lambert, 1975) clearly demonstrated the presence of coupling phase linkages in the cross of IHO x ILO since the additive genetic variance in $F_2$ was nearly twice the $F_6$ estimate. Thus, linkage for genes controlling percent oil exists but its effect on progress is unknown.

Types of Gene Action

Analysis of a diallel cross (including parents) among all 9 strains in the experiment using design II of Gardner and Eberhart (1966) showed 5% and 7% of the entry sum of squares were accounted for by heterosis for percent oil and percent protein, respectively (Dudley, de la Roche, and Lambert, 1976). $F_1$-MP values for percent oil indicated that dominance for high oil was present in crosses of RHO with IHO and SHO while dominance for low oil was present in crosses of ILO and RLO with IHO and SHO (Table VI).

TABLE VI

Estimates of Heterosis ($F_1$-MP) for Percent Oil from Crosses among Oil Strains.

|  | RHO | SHO | ILO | RLO |
|---|---|---|---|---|
| IHO | 1.9* | -.2 | -3.3* | -2.0* |
| RHO |  | 1.6* | -.8* | .1 |
| SHO |  |  | -2.4* | -2.2* |
| ILO |  |  |  | .2 |

* = Significant at the .05 probability level.

Although the low percentage of the entry sum of squares accounted for by heterosis in the diallel analysis suggests that dominance is relatively unimportant for percent oil, evaluation of the estimates of $\sigma_A^2$ and $\sigma_D^2$ from the Design III suggests a relatively high proportion of the genetic variance results from loci showing dominance. In the $F_6$ of the cross IHO x ILO, gene frequency is approximately .5. Only loci showing dominance contribute to $\sigma_D^2$ (ignoring epistasis). For loci with complete dominance, $\sigma_D^2 = 1/3$ of the total genetic variance and $\sigma_A^2 = 2/3$ when q = .5. Thus the additive genetic variance contributed by loci showing dominance equals twice the dominance variance. In the $F_6$ of IHO x ILO $\sigma_D^2$ = .189 and $\sigma_A^2$ = .812. Therefore, the additive genetic variance contributed by loci showing dominance is .378 or 46.5% of the total additive genetic variance. If all loci contribute equally to the genetic variance, then nearly half the loci by which IHO and ILO differ show dominance. If some loci show only partial dominance, then the proportion of loci showing dominance is greater than indicated.

Genetic Diversity

Much has been written about the lack of genetic diversity in major cultivated plant species as a barrier to continuing genetic improvement. The implication is that future progress is being hampered by a lack of genetic variability in existing germplasm which could be remedied by collection and use of exotic germplasm or other means. What do the results of this experiment imply for the future improvement of economic species? If the mean of an open-pollinated corn variety could be increased to a level of 3-4 times its original mean as was done for oil in IHO and protein in IHP, selected populations could, conservatively, yield in the 300 bushel per acre range with possible additional improvement from use of appropriate hybrids between populations.

There appears to be little need for concern about exhausting genetic variability for grain yield since with gene frequencies between .25 and .5 only from 50-200 loci are required to allow a gain of $20\sigma_A$. One would expect that there are more loci segregating for yield in most random-mating populations than for percent oil or protein. Thus, so far as quantitative traits are concerned, the most pressing problems may not be those of obtaining more genetic variability, but those of learning how to most efficiently concentrate the favorable alleles we now have and then proceeding to do it.

The preceding argument, of course, does not deny the importance of having available diverse genetic materials as potential sources of major genes for traits such as resistance to disease and insect pests. In some cases, resistance to such pests may not be present in adapted germplasm and availability of divergent germplasm may be extremely valuable.

## SUMMARY AND CONCLUSIONS

1. 76 generations of high and low selection for percent oil and protein in corn has not exhausted genetic variation for either trait in either direction even though progress of 20 $\sigma_A$ was made in the high direction for both oil and protein. Evaluation of theoretical limits to selection suggest that these results are well within the range of response expected with reasonable gene frequencies and numbers of loci.

2. A method of estimating gene frequency from divergent selection experiments which have reached their upper limits is presented. Application of this method to the data from this experiment indicate gene frequencies for favorable alleles in the original population were below .37 for percent protein and below .25 for percent oil.

3. Estimates of number of effective factors suggest a minimum of 54 loci differentiating IHO and ILO and 122 loci differentiating IHP and ILP.

4. Evidence suggests dominance for both high and low percent oil and that nearly half the loci differentiating IHO and ILO show some degree of dominance.

5. The results of this experiment suggest that the potential for improvement of random mating corn populations may be well beyond what is commonly considered possible.

## BIBLIOGRAPHY

(1) Comstock, R.E. & Robinson, H.F. (1948). Estimates of average dominance of genes. p. 494-516. In J. W. Gowen (ed.) *Heterosis*. Iowa State College Press, Ames.

(2) Draper, N.R. & Smith, H. (1966). *Applied Regression Analysis*. John Wiley & Sons, Inc., New York.

(3) Dudley, J.W. & Lambert, R.J. (1969). Genetic variability after 65 generations of selection in Illinois high oil, low oil, high protein, and low protein strains of Zea mays L. Crop Sci. 9, 179-181.

(4) Dudley, J.W., Lambert, R.J. & Alexander, D.E. (1974). Seventy generations of selection for oil and protein concentration in the maize kernel. p. 181-212. In J. W. Dudley (ed.) Seventy Generations of Selection for Oil and Protein in Maize. Crop Sci. Soc. of Am., Madison, Wis.

(5) Dudley, J.W., Lambert, R.J. & de la Roche, I. (197-). Genetic analysis of crosses among corn strains divergently selected for percent oil and protein. (Accepted for publication by Crop Sci.)

(6) Gardner, C.O. & Eberhart, S.A. (1966). Analysis and interpretation of the variety cross diallel and related populations. Biometrics 22, 439-452.

(7) Hanson, W.D. (1959). Theoretical distribution of the initial linkage block lengths intact in the gametes of a population intermated for n generations. Genetics 44, 839-846.

(8) Mather, K. & Jinks, J.L. (1971). Biometrical Genetics. Cornell University Press, Ithaca, New York. 382 p.

(9) Moreno-Gonzalez, J., Dudley, J.W. & Lambert, R.J. (1975). A design III study of linkage disequilibrium for percent oil in maize. Crop Sci. 15, 840-843.

(10) Robertson, A. (1970). A theory of limits in artificial selection with many linked loci. In Ken-ichi Kojima (ed.) Mathematical Topics in Population Genetics. Biomathematics 1, 246-288. Springer-Verlag, New York-Heidelberg-Berlin.

(11) 'Student'. (1934). A calculation of the minimum number of genes in Winter's selection experiment. Ann. Eugenics 6, 77-82.

**15**

Copyright © 1941 by Naomi Mitchison
Reprinted from pages 159–175 and 183–193 of *J. Genet.* **41**:159–193 (1941)

# VARIATION AND SELECTION OF POLYGENIC CHARACTERS

BY K. MATHER

*John Innes Horticultural Institution, Merton, London*

(With Plate 6 and Ten Text-figures)

## CONTENTS

|  |  | PAGE |
|---|---|---|
| 1. | Introduction | 159 |
| 2. | Preliminary investigations | 161 |
| 3. | The cross $\underline{y} \times f\,B^iB^i$ | 164 |
|  | (a) Technical | 164 |
|  | (b) Results | 165 |
| 4. | The cross $BB \times +$ | 168 |
|  | (a) Technical | 168 |
|  | (b) Results | 172 |
| 5. | Selection in the parental lines | 174 |
|  | (a) Oregon + : Technical | 174 |
|  | (b) Oregon + : Results | 175 |
|  | (c) BB | 177 |
| 6. | The second $BB \times +$ experiment | 177 |
|  | (a) Technical | 177 |
|  | (b) Results | 178 |
| 7. | The backcross lines | 178 |
|  | (a) Technical | 178 |
|  | (b) Results | 180 |
| 8. | The development of polygenic combinations | 183 |
| 9. | The interaction of polygenic combinations | 187 |
| 10. | Polygenic combinations and polymorphism | 189 |
| 11. | Conclusions | 190 |
| 12. | Summary | 191 |
|  | References | 192 |
|  | Explanation of Plate 6 | 193 |

## 1. INTRODUCTION

GENETICS has been mainly concerned with the inheritance and behaviour of genes by means of which individuals can be classified into categories showing sharp phenotypical differences. These are the so-called "qualitative" genes. It has, however, long been recognized that there exists another type of heritable variation, termed "quantitative" or "metrical".

Such variation does not allow of individuals being separated into distinct types; all gradations between certain limits are to be observed. The only classifications possible are arbitrary and dependent on the accuracy of the available means of measurement. Stature in man is typical of this kind of character. Human stature is, indeed, of special interest, as it is one of the characters whose study led to the formulation of Galton's law of ancestral inheritance and to the dispute on the nature of inheritance between Pearson and Bateson.

Early researches on these quantitative or metrical characters were devoted mainly to discovering the mechanism of their inheritance, whether blending, as the biometricians supposed, or particulate, as in the case of the sharply distinguished variants. It is now generally accepted that these characters are (a) controlled in inheritance by an indefinitely large number of genes, many of which have approximately equal effects, and (b) markedly subject to environmental variation. Thus it is clear that characters of this kind will be difficult to study by the common techniques of genetics. Some method of distinguishing between environmental and genetical variation must be used, and the analysis of the genetical variation itself will be complicated by the impossibility of separating the large number of genes involved. Complex statistical methods are needed. For these reasons the genetical analysis of such characters has progressed very slowly, in spite of their vital importance for the theory of evolution and the practice of selection.

The distinction between qualitative and quantitative variants is itself not a final one. It is possible that, if some organism could be grown in a constant environment and rendered homozygous for all but one of the genes affecting a quantitative character, this one gene might be observed to segregate and give sharply distinct classes just as a qualitative gene does. Nor do qualitative and quantitative genes affect different characters. Stature, for example, is usually a quantitative character, but in many organisms dwarf forms are known to segregate sharply from the normal type, so falling into the qualitative class. In any case most variations are in some sense quantitative, and so the terms qualitative and quantitative are far from ideal. A better classification can be made on the basis of the mode of inheritance. Qualitative variation is usually monogenic or digenic in inheritance. Cases of trigenic and tetragenic inheritance are known, but are relatively rare. In contrast with these, quantitative variation may be said to be polygenic, and this term will be adopted.

The importance of a better knowledge of polygenic inheritance can hardly be overestimated. Geneticists have found it convenient to use monogenic characters in the study of chromosome behaviour, of mutation, and of gene action; but to those who are concerned with applied genetics, in plant and animal improvement as well as in the study of evolutionary changes, such characters are of minor importance, most of the variation being polygenic. It is true that wild populations of a number of organisms are known to be heterogeneous for monogenic characters, but these are usually rare variants. It is, on the other hand, highly probable that all populations are heterogeneous for polygenic characters. Furthermore, specific differences are always polygenic if the species are biologically isolated. Some cases of monogenic specific differences have been described, notably in wheat (Watkins, 1930), but in most cases, as for example *Antirrhinum* (Baur, 1924) and *Nicotiana* (East, 1935), polygenic differences have been observed. This would also appear to apply to the distinguishing features of the two races of *Drosophila pseudoobscura*, especially to the mechanism determining inter-racial sterility (Dobzhansky, 1936). This view derives further support from the observation of homologous monogenic variants, both naturally occurring and mutants from culture, in the *Drosophila* species (Gottschewski & Tan, 1938; Sturtevant & Tan, 1937). If such monogenic changes were the materials for specific differentiation, as their occurrence in the wild has suggested to some authors, at least some of the species should differ by these genes. Their occurrence as homologous mutations in several of the forms shows that the latter do not, however, differ in this way. Thus a study of polygenic rather than monogenic inheritance is the prime need of applied genetics.

There are a number of ways in which the problem can be approached, but perhaps the most appropriate is by an analysis of the effects of selection. Clearly the kind of result of selective breeding is a compound of the nature of the selection applied and the nature of the genetic variation available for the action of selection. The type of selection can be controlled by the experimenter, within reasonable limits, and so definite inferences about the nature of the available variation can be drawn from the results of such a programme. A series of such experiments is described and discussed below.

## 2. Preliminary investigations

*Drosophila melanogaster* was chosen as the material for this investigation, for two reasons. In the first place it has the technical advantages of being easy of culture and a rapid breeder. Secondly, the existence of tester stocks makes possible the further genetic analysis of any selected lines, should such analysis be desirable.

Of the various characters which are most probably polygenic in inheritance the one eventually chosen was the combined number of chaetae on the ventral surfaces of the fourth and fifth abdominal segments.[1] The numbers of hairs on these segments are easily determined under a high-power binocular microscope, as will be seen from the photographs of Pl. 6, and as there are usually from thirty-five to forty-five hairs on the segments taken together, the discontinuity of the distribution leads to no serious difficulty. The joint number of hairs of the two segments was used as the metric throughout all the experiments. In the original stocks there was no evidence of differences between the means of the segments taken separately, but this point was not tested in the selected material. Hence it is not possible to say whether the segments made equal contributions to the joint effect of selection.

The number of hairs per fly varies continuously from about thirty to about fifty in the original stocks, in the sense that flies with all numbers of hairs between these limits can be found. (These limits were later widely transgressed in selected material.) Some examples of the frequency distribution of hair number are shown in Text-fig. 1. It will be

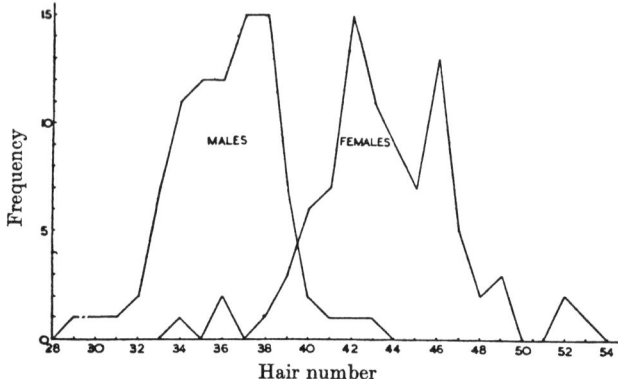

Text-fig. 1. Frequency distributions for hair number in males and females of the BB stock.

seen that the curves approximate to normality, being in fact typical of the frequency distributions observed for polygenic characters. In particular there is no consistent evidence of skewness, as tested by the $k_3$ statistic (Fisher et al. 1932). Nor is bimodality regularly encountered.

[1] According to Bridges' illustrations, the fifth is the most posterior hair-bearing segment in the male; but in the female the sixth also carries some hair.

Hence the distributions are reasonably well described by calculations of their means and variances. These are tabulated for a number of the original stocks in Table 1. It will be seen that the mean number of hairs is consistently lower in males than in sister females. Furthermore, the stocks show different means, thus providing *prima facie* evidence of hair number being a polygenic character. It is, however, also clear that some non-heritable variation occurs, since the Oregon+ line, inbred for at least seventy generations, still showed considerable variation.

Two questions now arise, viz. as to whether heritable variation occurs and, if so, whether it is polygenic. That the variation within other stocks was not wholly non-heritable was shown by a series of single generation selection experiments. Pairs of flies with high, medium and low hair number were chosen from certain stocks and mated. It was observed that, in all but the Oregon+ line, there was a positive regression of the progeny mean on the parental values, thus showing that the variation in hair number is partly heritable. It was also observed that the progeny means did not fall into clearly delimited groups, so showing that the character is polygenic. The progeny means showed no relation to the parental values in the Oregon+ line, thus indicating that the inbreeding this line had undergone had in fact made it homozygous for the hair-number polygenes. Thus the character is quite suitable for a study of the action of selection in polygenic inheritance.

It may be added that the hair number also constitutes a specific difference. Table 1 shows the mean numbers of various species other than *D. melanogaster*. These means show marked differences. The sex

Table 1

| Stock | Males | | | Females | | |
|---|---|---|---|---|---|---|
| | No. counted | Mean | Variance | No. counted | Mean | Variance |
| f B$^i$B$^i$ | 20 | 31·85 | 7·5026 | 20 | 36·95 | 10·8921 |
| B | 20 | 36·85 | 8·0289 | 20 | 42·25 | 8·6184 |
| BB | 20 | 35·45 | 6·0500 | 20 | 42·60 | 5·9368 |
| Oregon+ | 20 | 39·55 | 4·1553 | 20 | 44·40 | 4·8842 |
| y × w m f | 20 | 37·90 | 5·8842 | 20 | 42·40 | 12·9895 |
| *D. simulans* | 10 | 31·50 | 5·1667 | 13 | 40·7692 | 9·1923 |
| *D. subobscura* | 20 | 30·75 | 7·8816 | 20 | 33·25 | 5·7763 |
| *D. virilis virilis*\* | 20 | 59·90 | 12·7263 | 20 | 34·90 | 3·9895 |

All figures refer to the sum of the hair numbers of the fourth and fifth segments.

\* In *D. virilis virilis* only one segment (no. 5) was counted. Hence the means in the table should be multiplied by two to render them comparable.

difference in hair number is also variable between species. In *D. virilis* it is the reverse of that found in *D. melanogaster*. So species differentiation in *Drosophila* has involved the production of hair number differences, presumably by the action of selection, as such a difference as that between *D. virilis* and *D. subobscura* cannot reasonably be ascribed to chance fluctuations. Thus selection experiments on hair number are in some sense a repetition of the action of natural selection in species formation.

### 3. The cross $\underline{y} \times f\ B^lB^l$

#### (a) *Technical*

Two crosses were used in the main experiments, viz. **y** females from the $\underline{y} \times$ **w m f** stock by **f $B^lB^l$** males, and Oregon+, inbred, reciprocally crossed with **BB**. Though actually the second in time, the former cross will be described first as it proved less complex than the latter.

The initial cross was of **y** females by **f $B^lB^l$** from stock. The mating was made in a vial. The flies were transferred to a half-pint milk bottle after a day, and were allowed to lay in this bottle for several days. Their progeny were counted and two $F_2$ cultures set up, each comprising a mating of two females by two males taken at random from the $F_1$. These $F_2$ matings were made in vials and after 1 day transferred to bottles, where they were allowed to lay for 2 days. This procedure was followed in all the subsequent matings, with the exception that single pair matings were used to give the $F_3$ and later generations.

The hair number of the $F_2$ flies was counted and the two highest females mated individually to the two highest males to give a high selected $F_3$. Similarly, the two lowest females were mated individually to the two lowest males to start the low selection line. These two lines, the high and low selections, were maintained separately in later generations by selecting the two highest, or lowest, females and the two highest, or lowest, males in each generation in order to make the two single-pair matings from which the next generation was bred. Thus the two progenies in the fourth generation of the high line were from the two highest males and females of the third generation of that same line, and so on. As far as possible each mating was made between a female and a male from different bottles of the previous generation, in order to avoid inbreeding. In the later stages of the experiment a considerable degree of sterility was encountered. The progenies were small and many matings failed completely. To mitigate this difficulty, three matings per generation were made in each line, though more than two progenies were seldom counted or used to provide parents of the next generation. Where all three matings were fertile, the two with the most extreme parents, viz. the highest in the high line, and the lowest in the low line, were counted and selected for the next generation parents. Finally, the high and low lines were kept in step throughout the experiment, i.e. the corresponding generations of each line were always raised side by side in the incubator, which was maintained at a temperature of $25 \pm 1°$ C. Never more than twenty flies of each sex were counted from a single culture.

The means of the cultures in the various generations of each line, together with those of the parental stocks, are given in Table 2. The males and females are given separately. Text-fig. 2 shows the means

Table 2. *The cross* y × f $B^lB^l$

| Generation | Males No. counted | Males Mean | Females No. counted | Females Mean | Generation | Males No. counted | Males Mean | Females No. counted | Females Mean |
|---|---|---|---|---|---|---|---|---|---|
| Parent f $B^lB^l$ | 20 | 31·85 | 20 | 36·95 | H 8—1 | 14 | 38·08 | 12 | 46·50 |
| Parent y × w m f | 20 | 37·90 | 20 | 42·40 | L 1—1 | 9 | 34·67 | 9 | 42·56 |
| $F_1$ | 14 | 34·36 | 18 | 38·72 | —2 | 5 | 34·80 | 7 | 40·00 |
| $F_2$—1 | 20 | 35·00 | 20 | 41·10 | L 2—1 | 20 | 34·95 | 20 | 41·60 |
| —2 | 20 | 35·40 | 20 | 42·95 | L 3—1 | 20 | 36·05 | 18 | 44·33 |
| H 1—1 | 9 | 35·67 | 13 | 44·31 | —2 | 20 | 34·70 | 20 | 39·95 |
| —2 | 9 | 38·56 | 11 | 47·45 | —3 | 20 | 34·00 | 20 | 40·35 |
| H 2—1 | 12 | 38·83 | 9 | 46·22 | L 4—1 | 13 | 34·23 | 10 | 39·90 |
| —2 | 13 | 37·69 | 10 | 46·30 | —2 | 20 | 34·35 | 20 | 39·55 |
| H 3—1 | 10 | 40·50 | 6 | 45·40 | —3 | 20 | 35·00 | 19 | 40·42 |
| —2 | 15 | 38·47 | 11 | 44·81 | L 5—1 | 20 | 33·90 | 20 | 39·35 |
| H 4—1 | 5 | 37·20 | 11 | 46·73 | —2 | 13 | 33·62 | 15 | 38·67 |
| —2 | 13 | 38·39 | 7 | 45·43 | L 6—1 | 8 | 34·13 | 8 | 40·88 |
| H 5—1 | 13 | 38·54 | 20 | 45·05 | —2 | 9 | 34·32 | 20 | 40·35 |
| —2 | 10 | 38·50 | 9 | 45·33 | L 7—1 | 9 | 34·56 | 13 | 39·92 |
| H 6—1 | 13 | 39·15 | 14 | 45·00 | L 8—1 | 10 | 34·30 | 14 | 40·29 |
| H 7—1 | 13 | 38·08 | 12 | 46·50 | —2 | 13 | 32·23 | 15 | 40·47 |

H = high line. L = low line. The number of the generation is the number of occasions in its ancestry at which selection has been exercised.

The various cultures in each generation are shown separately.

plotted against generation. Each dot represents a culture mean, except in a few cases where the two cultures each gave so few flies that they had to be added together to give a reasonable number of individuals. The continuous lines in Text-fig. 2 join the means of the means. These were not obtained by weighting according to the number of individuals involved, i.e. they are *not* the grand means of each generation in each line. They were calculated by taking the unweighted culture means, adding, and dividing by the number of cultures involved, irrespective of how many flies each culture had yielded.

(*b*) *Results*

The $F_1$ flies were intermediate in hair number between the parental lines, but the $F_2$'s showed an increase over the $F_1$'s. This is most probably

## 166  *Variation and Selection of Polygenic Characters*

due to the fact that the $F_1$ culture was overcrowded, as the parents were allowed to lay in the bottle for several days, whereas egg-laying in the $F_2$ and later generations was limited to 2 days. The hair numbers of the

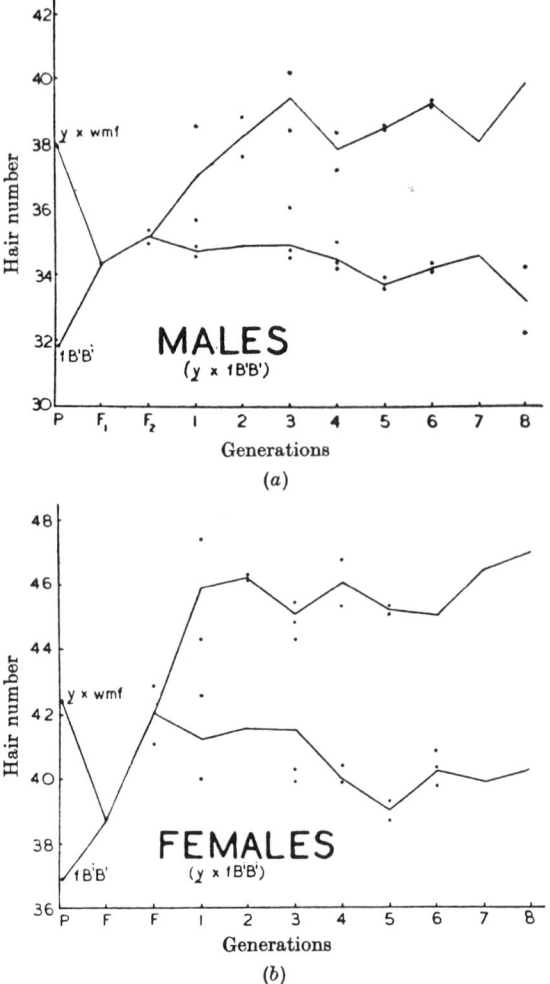

Text-fig. 2. The effect of selection for high and low hair number in $\underline{y} \times f\,B^iB^i$. Each dot is a culture mean and the lines join means of means. (See also Text-fig. 3.)

parental stocks are comparable with that of the $F_1$, as they were somewhat crowded too.

The selection for hair number had an immediate effect, the high and low $F_3$'s differing markedly. The second selected generation ($F_4$) showed an even larger difference between high and low, but after that point little change is apparent in Text-fig. 2. It will be seen, however, that there is considerable fluctuation in each line from generation to generation, especially in the females. This is almost certainly due to environmental causes and can be largely eliminated by plotting the difference between the high and low lines instead of the actual means. This is done in

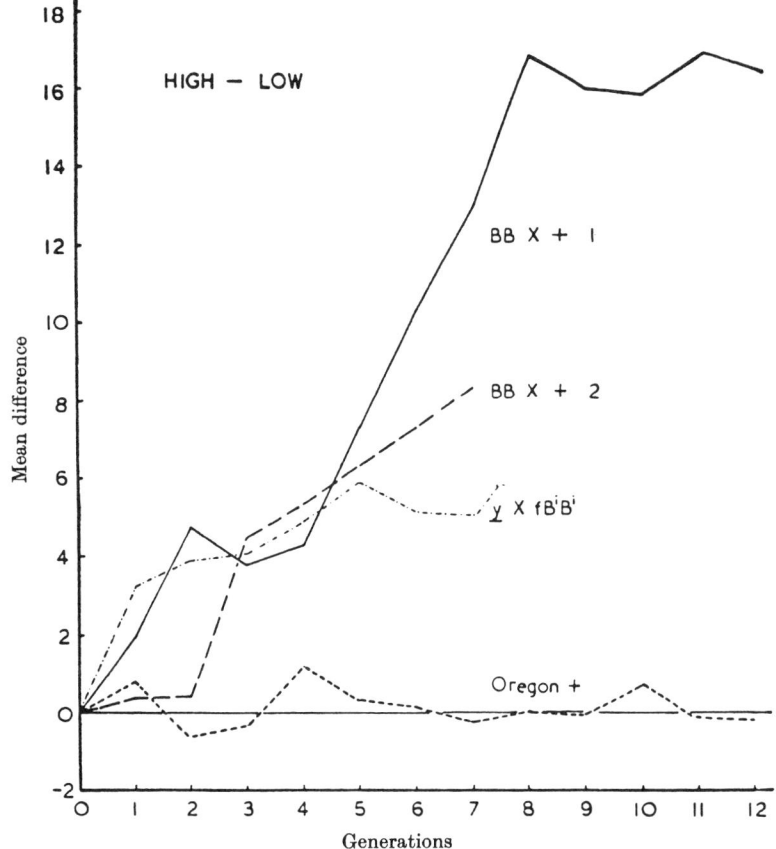

Text-fig. 3. The mean difference between high and low selection lines in various stocks plotted against generations of selection. (Generation $0 = F_2$ in the case of crosses.)

Text-fig. 3. The average difference for each generation was obtained by subtracting the mean of the low females of that generation from the corresponding female mean in the high line. The males were treated similarly and the two differences averaged. This smooths out most of the environmental fluctuation and makes interpretation of the results materially easier. The first generation plotted in Text-fig. 3 is the $F_2$ from which the high and low lines originated. The average difference in

168 *Variation and Selection of Polygenic Characters*

$F_2$ clearly must be zero. The difference clearly increases up to the second selected generation and afterwards is fairly stable, though perhaps showing a slight tendency to increase. The effect of selection was however almost complete after two generations, and was presumably confined to choosing new combinations of whole chromosomes.

Two points must be added, firstly that the main effect of selection was to increase the high line hair number, rather than to decrease that of the low line and secondly that the final difference between the high and the low lines was no greater than that shown by the parental stocks.

### 4. THE CROSS BB × +

#### (a) *Technical*

The technique used for this cross and the breeding lines selected from it was precisely the same as for y × f $B^iB^i$ with the exception that in the initial matings, as well as in all the subsequent ones, single pairs were allowed to remain only 2 days in the culture bottle. Thus only the hair numbers of the parental stocks, BB and inbred Oregon+, were obtained from relatively crowded cultures.

The two stocks used differed slightly in their mean hair numbers, BB being somewhat lower in both sexes than the inbred Oregon+ stock. The reciprocal $F_1$'s were both intermediate on the whole between the parental stocks (Table 3 and Text-fig. 4). The means of the twelve $F_2$ cultures, six from each of the reciprocal $F_1$'s, resembled those of the $F_1$. One point of special interest was observed in the $F_2$ cultures. The males were showing segregation for BB and +, and the former kind showed a higher hair mean number than the latter (Table 4). A suitable test of significance was applied to the results. If there is no real difference between the hair numbers of BB and + males, the mean difference between the values of the two classes in individual cultures will be zero. Twelve such differences are obtainable from the twelve $F_2$ families. The mean of the twelve values ($\bar{d}$) is 1·1350. The variance of this mean ($V_{\bar{d}}$) is 0·1182, and $t = \bar{d}/\sqrt{V_{\bar{d}}} = 3·301$ for eleven degrees of freedom. This has a probability of less than 0·01 and so it must be judged that BB males have a higher hair number than their + brothers.

In six of the $F_2$ cultures, viz. those from the cross BB/+ × +, the BB segregation was also followed in the female progeny. In this case too the BB/+ females had a higher mean hair number than their + sisters, as $\bar{d} = 0·8238$, $V_{\bar{d}} = 0·08619$ and $t = 2·806$ for five degrees of freedom, giving a probability of between 0·05 and 0·02; the difference between differences of the two sexes is not significant. Thus although the **BB** parental stock had a lower hair number than the Oregon+ to which

Table 3. *The cross* **BB** × +

| Generation | | No. counted (Males) | Mean (Males) | No. counted (Females) | Mean (Females) | Generation | | No. counted (Males) | Mean (Males) | No. counted (Females) | Mean (Females) |
|---|---|---|---|---|---|---|---|---|---|---|---|
| Parent **BB** | —1 | 20 | 35·45 | 20 | 42·60 | H 8 | —1 | 20 | 46·20 | 20 | 51·65 |
| | —2 | 20 | 37·60 | 20 | 46·45 | | —2 | 20 | 42·85 | 20 | 48·60 |
| | —3 | 19 | 35·05 | 16 | 41·75 | H 9 | —1 | 20 | 44·65 | 20 | 49·60 |
| | —4 | 7 | 36·00 | 16 | 42·75 | | —2 | 13 | 44·38 | 20 | 48·90 |
| | —5 | 20 | 36·20 | 13 | 43·85 | | —3 | 20 | 43·75 | 20 | 49·25 |
| Or+ | —1 | 20 | 40·65 | 20 | 45·20 | H 10 | —1 | 20 | 42·70 | 20 | 49·35 |
| | —2 | 20 | 39·55 | 20 | 44·40 | | —2 | 20 | 44·15 | 20 | 47·30 |
| | —3 | 20 | 38·45 | 20 | 43·75 | | —3 | 20 | 44·55 | 20 | 48·70 |
| | —4 | 20 | 40·90 | 20 | 45·05 | H 11 | —1 | 20 | 43·65 | 20 | 47·65 |
| | —5 | 20 | 39·85 | 20 | 44·55 | | —2 | 7 | 44·86 | 15 | 50·20 |
| $F_1$: + × **BB** | —1 | 20 | 37·75 | 20 | 42·30 | H 12 | —1 | 20 | 44·80 | 20 | 49·00 |
| | —2 | 20 | 37·50 | 20 | 44·70 | | —2 | 9 | 41·89 | 10 | 49·30 |
| | —3 | 20 | 38·60 | 20 | 42·95 | H 13 | —1 | 20 | 44·85 | 20 | 51·40 |
| **BB** × + | —1 | 20 | 38·05 | 20 | 43·00 | | | | | | |
| | —2 | 20 | 37·25 | 20 | 44·05 | | | | | | |
| $F_2$: *ex*+ × **BB** | —1 | 23 | 39·35 | 6 | 45·43 | L | —1 | 20 | 37·60 | 20 | 44·00 |
| | —2 | 40 | 38·65 | 29 | 44·76 | | —2 | 15 | 37·67 | 15 | 44·13 |
| | —3 | 38 | 38·87 | 40 | 44·30 | | —3 | 20 | 37·05 | 20 | 42·90 |
| | —4 | 21 | 39·00 | 13 | 45·15 | L 2 | —1 | 20 | 36·55 | 20 | 41·60 |
| | —5 | 35 | 38·77 | 21 | 43·90 | | —2 | 20 | 36·75 | 20 | 40·60 |
| | —6 | 32 | 38·78 | 26 | 44·88 | | —3 | 20 | 35·55 | 20 | 40·82 |
| *ex***BB** × + | —1 | 30 | 36·87 | 21 | 44·05 | L 3 | —1 | 11 | 36·30 | 9 | 41·44 |
| | —2 | 40 | 38·78 | 36 | 43·50 | | —2 | 20 | 35·75 | 18 | 41·24 |
| | —3 | 40 | 38·45 | 31 | 43·74 | | —3 | 20 | 35·00 | 20 | 43·20 |
| | —4 | 40 | 36·63 | 17 | 43·24 | L 4 | —1 | 20 | 36·85 | 20 | 42·15 |
| | —5 | 22 | 39·14 | 16 | 44·06 | | —2 | 16 | 35·25 | 20 | 40·60 |
| | —6 | 40 | 36·90 | 30 | 43·47 | | —3 | 16 | 34·25 | 20 | 40·50 |
| H 1 | —1 | 16 | 39·31 | 20 | 44·95 | L 5 | —1 | 13 | 35·31 | 20 | 40·50 |
| | —2 | 20 | 40·80 | 20 | 44·40 | | —2 | 20 | 34·90 | 20 | 40·30 |
| | —3 | 20 | 39·40 | 20 | 46·40 | | —3 | 20 | 32·20 | 20 | 38·05 |
| H 2 | —1 | 20 | 40·05 | 20 | 46·10 | L 6 | —1 | 20 | 32·85 | 15 | 38·53 |
| | —2 | 20 | 40·00 | 20 | 46·40 | | —2 | 20 | 32·65 | 15 | 38·93 |
| | —3 | 18 | 40·33 | 20 | 44·25 | L 7 | —1 | 14 | 30·64 | 14 | 34·93 |
| H 3 | —1 | 20 | 40·15 | 20 | 46·45 | | —2 | 15 | 30·33 | 20 | 36·05 |
| | —2 | 20 | 39·65 | 20 | 45·00 | L 8 | —1 | 20 | 26·30 | 20 | 31·60 |
| | —3 | 20 | 38·40 | 20 | 45·95 | | —2 | 20 | 28·65 | 20 | 34·50 |
| H 4 | —1 | 16 | 38·94 | 10 | 44·80 | | —3 | 20 | 28·40 | 20 | 34·25 |
| | —2 | 20 | 39·70 | 20 | 46·45 | L 9 | —1 | | Very few flies | | |
| H 5 | —1 | 20 | 39·60 | 20 | 45·65 | L 10 | —1 | 12 | 27·67 | 23 | 32·70 |
| | —2 | 20 | 41·25 | 20 | 50·20 | L 11 | —1 | 3 | 23·67 | 19 | 30·89 |
| H 6 | —1 | 14 | 43·21 | 13 | 48·85 | | —2 | 16 | 27·31 | 20 | 33·00 |
| | —2 | 7 | 42·29 | 7 | 48·86 | | —3 | 20 | 26·55 | 20 | 31·35 |
| | —3 | 13 | 42·69 | 9 | 48·78 | L 12 | —1 | 7 | 25·71 | 12 | 32·92 |
| H 7 | —1 | 26 | 43·47 | 38 | 49·58 | | | | | | |
| | —2 | 20 | 42·80 | 20 | 48·00 | | | | | | |

Nomenclature as in Table 2.

it was mated, the $X$ chromosome of **BB** had genes potentially able to give a higher hair number than that on the Oregon $X$-chromosome. The autosomes of **BB** must then have had sufficient low hair number genes

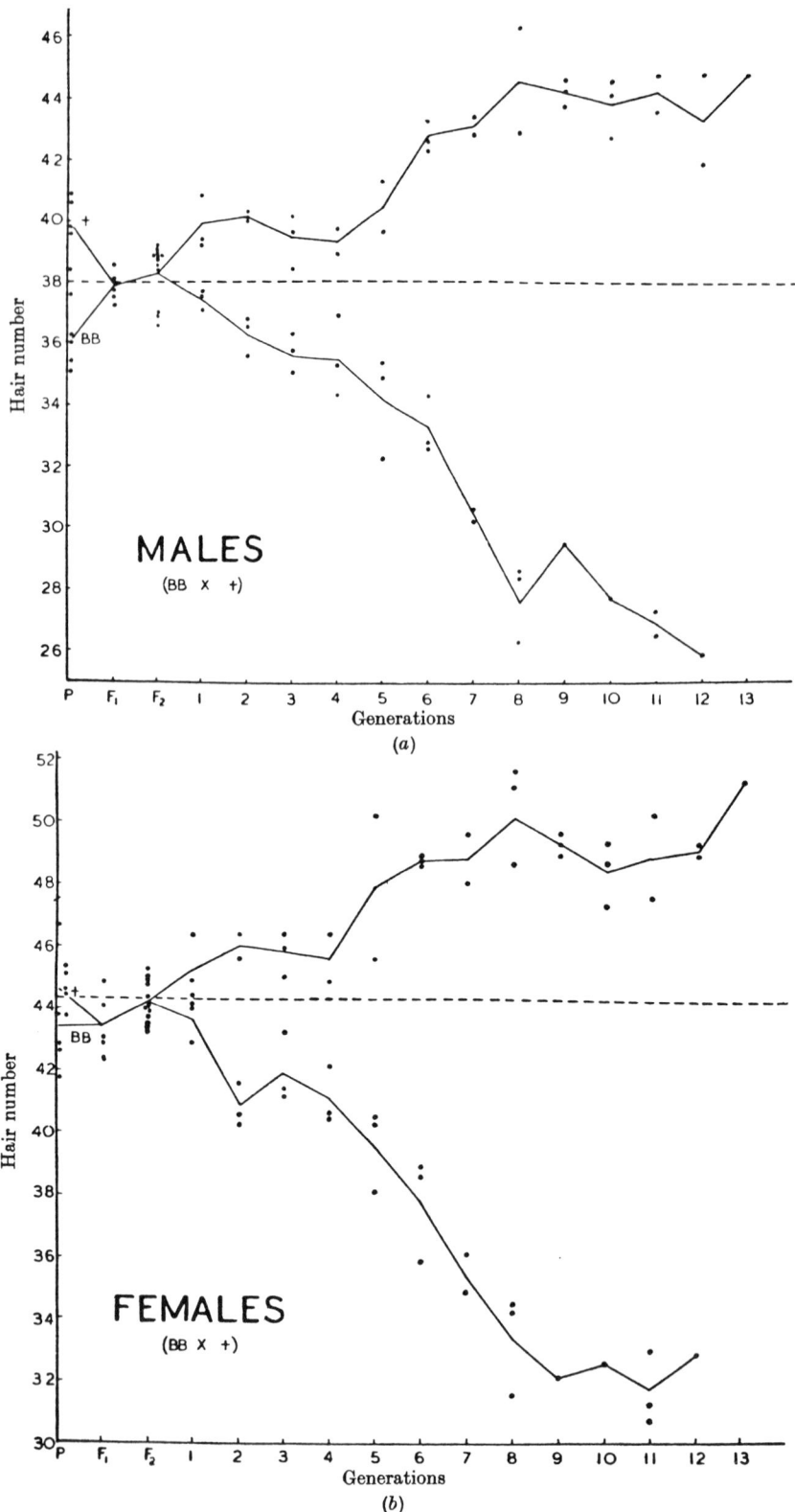

Text-fig. 4. The effect of selection for high and low hair number in **BB** × **+**. Each dot is a culture mean and the lines join means of means. (See also Text-fig. 3.)

to outweigh the effect of the $X$. The existence of such a difference between the gene contents of **BB** and **+** sex chromosomes was confirmed by the fact that, though eye shape was neglected in selection, the high line was soon homozygous **BB** and the low line **+**.

In order to raise the $F_3$ generation a large number of matings were made between $F_2$ males and females chosen at random. The mean hair numbers of the various $F_3$ progenies was determined and can be used to calculate the regression of hair number in offspring on that in parents. These calculations are, however, outside the scope of the present account

Table 4. *Hair numbers of* **BB** *and* **+** *flies in $F_2$ of* **BB** × Or **+** *and reciprocal*

|  | Culture | Mean of BB | Mean of + | (d) difference | |
|---|---|---|---|---|---|
| Males: | $F_2\, ex+ \times$ **BB**—1 | 41·3300 | 37·3636 | 3·9664 | $\bar{d} = 1·1350$ |
|  | —2 | 39·5500 | 37·7500 | 1·8000 | $V_d = 1·4187$ |
|  | —3 | 39·2000 | 38·5000 | 0·7000 | $V_{\bar{d}} = 0·11822$ |
|  | —4 | 39·0000 | 39·0000 | 0·0000 | $s_{\bar{d}} = 0·3438$ |
|  | —5 | 39·8000 | 38·0000 | 1·8000 | $t = 3·301$ |
|  | —6 | 39·2500 | 38·5000 | 0·7500 | $df = 11$ |
|  | $F_2\, ex$ **BB** × **+**—1 | 36·5400 | 37·1200 | −0·5800 | $P < 0·01$ |
|  | —2 | 39·0500 | 38·5000 | 0·5500 | |
|  | —3 | 38·5500 | 38·3500 | 0·2000 | |
|  | —4 | 37·6000 | 35·6500 | 1·9500 | |
|  | —5 | 40·0000 | 38·4167 | 1·5833 | |
|  | —6 | 37·3500 | 36·4500 | 0·9000 | |
| Females: | $F_2\, ex+ \times$ **BB**—1 | 46·0000 | 45·2000 | 0·8000 | $\bar{d} = 0·8238$ |
|  | —2 | 44·7690 | 44·7500 | 0·0190 | $V_d = 0·5171$ |
|  | —3 | 44·8000 | 43·8000 | 1·0000 | $V_{\bar{d}} = 0·08619$ |
|  | —4 | 46·4000 | 44·3750 | 2·0250 | $s_{\bar{d}} = 0·2936$ |
|  | —5 | 44·4000 | 43·4550 | 0·9450 | $t = 2·806$ |
|  | —6 | 44·5830 | 44·4290 | 0·1540 | $df = 6$ |
|  |  |  |  |  | $P = 0·05$–$0·02$ |

and must be reserved for later discussion. The actual $F_3$'s used in the selection experiments were chosen from this large number of random mated cultures. The three with the highest sum of parental hair numbers were taken to start the high line and the correspondingly lowest three to start the low line. Subsequently the method of selection and mating was exactly the same as that described for the cross $\underline{y} \times f\, B^l B^l$ except that three matings were set up in each selected line in every generation, all giving progeny being counted and used for the next selection. Table 3 and Text-fig. 4 show the results of selection in this cross. Text-fig. 3 shows the results of plotting the difference between high and low line against generations, exactly as for $\underline{y} \times f\, B^l B^l$. It is less subject to environmental fluctuation and so easier to read than is Text-fig. 4.

## (b) Results

The striking feature of the results from this cross is that after an initial response to selection for two generations, stability is attained exactly as in $\underline{y} \times f\ B^l B^l$, but this period of stability is itself succeeded, in both high and low lines, by an even more marked advance in the direction of selection than that which occurred in the first place. The second advance persisted for several generations before a second stable level was reached. It is to be regretted that such marked sterility had appeared in the selected lines that it was impossible to carry the experiment on after thirteen generations. So it is impossible to say whether a third advance would have occurred at a later point.

The second advance in the direction of selection had much greater results than did the first one. It led to the production of lines with more and less hairs, respectively, than have ever been seen in any stock culture, except for those which carry known major mutations affecting hair number, such as **sc**. The selected lines had gone outside the limits of the normal wild or cultured populations.

It was the second advance which produced this enormous selective effect. The first change with selection is most reasonably interpreted as due to recombinations of whole chromosomes present in the hybrid, as in the case of $\underline{y} \times f\ B^l B^l$. That such a change with selection is to be expected follows from the demonstration that the genes on the $X$ chromosome of **BB** are on the whole plus genes as compared with those of Oregon **+**, and its corollary that the **BB** autosomes are preponderantly minus in effect relative to the Oregon stock. Thus assortment of chromosomes in the hybrid should lead to an initial advance under the action of selection. The most favourable combinations will, however, soon be sorted out and stability will follow.

The second and larger advance cannot be accounted for in this way; it must depend on the presence of a different type of heritable variation. There seem to be two obvious ways in which such variations could arise, by mutation and by recombination of genes within chromosomes. Careful consideration must however lead to the conclusion that mutation is an unlikely origin. The advance occurred simultaneously in both lines and was of a magnitude greater than that ascribable to recombination of whole chromosomes. Thus the necessary mutation would be of frequent occurrence and having an effect of a greater magnitude than seems reasonable from what we know of the process. Recombination within the chromosomes is a much more plausible explanation. Intra-chromosome recombination is an inevitable occurrence, but to be effective in releasing variation it must lead to a marked redistribution of the plus and minus genes. The way in which such a reorganization could take place is shown in Text-fig. 5. In Text-fig. 5A a simple single cross-over would lead to the desired effect; but in the situations of 5B and 5C more complex recombination would be necessary to produce an equivalent reorganization. The chief point to note is, however, that for recombination to produce the necessary variation, the chromosome must originally have carried polygenic combinations which were more or less balanced, i.e. which had more or less equal numbers of plus and minus genes. Recombination thus leads to unbalancing of the combinations by producing an excessive

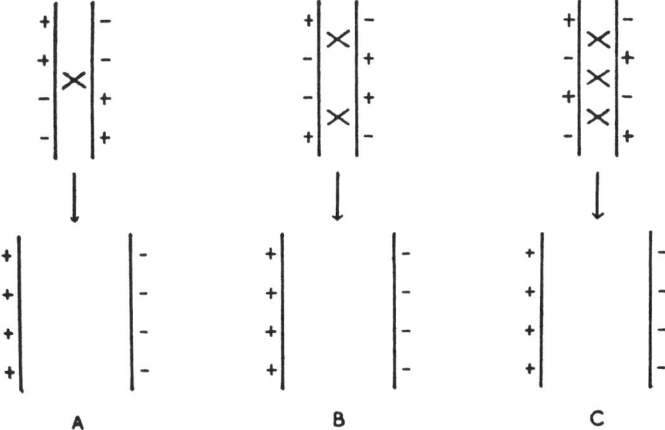

Text-fig. 5. Diagram to illustrate how balanced polygenic combinations may release variation by recombination. As the intermingling of the genes becomes more complex, higher order crossing-over becomes necessary for the full release of the potential variation; some of it may, however, be released by lower-order crossing-over.

number of plus modifiers on one and minus modifiers on the other derived chromosome. The origin of such balanced polygenic combinations will be discussed later.

The hypothesis that the second advance was due to the release of variation by recombination and not by mutation was tested in three ways. These were: (a) to select within the Oregon+ line, which being highly inbred could show variations only as a result of mutation and so provides a measure of the mutation rate; (b) by repeating the **BB** × **+** selection experiment, which would not be expected to give the same result if the relatively uncommon process of mutation were the cause but which should show the same behaviour if the relatively frequent occurrence of recombination released the variation; and (c) by following the effects of selection in lines in which heterozygous females, showing recombination, and males, which fail to show recombination, respectively were continually backcrossed to a homozygous stock. The results of these tests are described in the next three sections.

Before leaving the **BB** × **+** cross, it may be added that after the high and low lines had been selected for thirteen and twelve generations respectively, they were mass cultured without selection and later intercrossed reciprocally (Table 5). Two features of interest were observed in

Table 5

|  |  | Original cross | | Intercross of selected lines | |
|---|---|---|---|---|---|
|  |  | Mean | Variance | Mean | Variance |
| **BB** Parent: | Males | 36·06 | 6·0065 | 43·10 | 4·1667 |
|  | Females | 43·48 | 8·2708 | 49·13 | 7·1137 |
| **+** Parent: | Males | 39·88 | 6·5922 | 29·91 | 6·2909 |
|  | Females | 44·59 | 9·0700 | 35·79 | 12·7310 |
| $F_1$: | Males | 37·83 | 5·7710 | 35·86 | 6·6869 |
|  | Females | 43·40 | 4·9080 | 40·34 | 8·3389 |
| $F_2$: | Males | 38·35 | 6·8201 | 37·92 | 16·8555 |
|  | Females | 44·20 | 7·0436 | 42·80 | 18·0826 |

The means and variances given are the average values of all the appropriate cultures.

the $F_2$ of these crosses. In the first place the variance of these $F_2$'s was much greater than that of the original $F_2$ from **BB** × **+** (Table 5) as would be expected if released variation had been picked up and fixed in the course of selection. Secondly the segregation of the $X$-chromosome could be followed in this cross, as the high selection line was homozygous **BB** and the low line **+**. In the $F_2$ the difference between the **BB** and **+** flies was determined as described for the original cross. The average difference between **BB** and **+** males in the first five cultures was 0·8749, which is to be compared with a value of 1·1350 in the original $F_2$. Thus there is no evidence that the gene content of the sex chromosome has been affected by selection. Whatever produced the change which led to the second advance it must have occurred in one or more of the autosomes.

## 5. Selection in the parental lines

### (a) Oregon + : Technical

The method of selection practised in the Oregon **+** line was just the same as that used in the cross-bred material. At the time when selection commenced the line had been kept by brother-sister mating for at least seventy generations, and almost certainly had been inbred for a considerable time before the available records began. Hence it should be homozygous except for the effect of mutation.

A single culture was used to commence the experiment. The single pair matings were made in each of the two selection lines, high and low. In the former case the matings were between the highest three males and females, and in the latter case between the lowest three of each sex. In each generation the highest three males and females in the high line were chosen as parents of the three cultures of the next generation, and correspondingly in the low line. Occasional departures were made in that sometimes a slightly lower, in the high line, or higher, in the low line, individual would be taken in order to keep all the cultures represented in the next generation. The matings were made between flies from different cultures of the same line as far as possible, just as in the crossbred material of the previous sections.

*(b) Oregon + : Results*

The results of the selection are given in Table 6 and shown graphically in Text-fig. 6. It will be seen that selection had no effect in the twelve

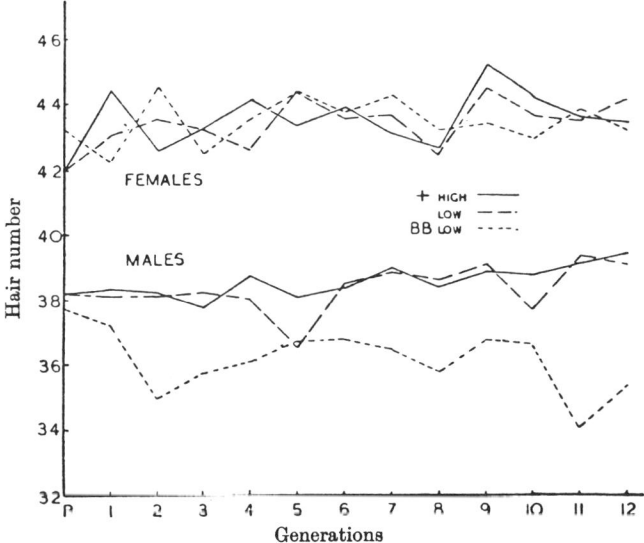

Text-fig. 6. Selection in the Oregon+ and **BB** lines. Only the means of each generation are shown. (See also Text-fig. 3.)

generations for which it was practised. This is shown even more strikingly in Text-fig. 3 where, as in the previous cases, the mean difference between high and low lines is plotted against generations. This difference is as often negative as positive, i.e. the low line frequently has a higher hair number than the high line.

[*Editor's Note:* Material has been omitted at this point.]

## 8. The development of polygenic combinations

It seems clear from the experimental evidence that strains of *D. melanogaster* can differ by balanced polygenic combinations within chromosomes. These combinations have approximately the same effect as each other on hair number, but recombination following crossing-over between them results in marked upset of the balance and in the release of considerable variation on which selection can act.

Having shown that balanced combinations occur, the next question which arises is that of how such balanced combinations come into being. Some opinion of their origin can be formed from a comparative consideration of the various selection experiments which have been conducted using *D. melanogaster* as material. These fall into two groups, those concerned with selection for naturally variable characters and those in which selection was exercised on the expression of a mutant character. The experiments described above fall into the first group, as hair number is now known to be genetically variable in wild flies. One other large experiment is in this class too, viz. Payne's (1918) selection for extra scutellar bristles. Such extra bristles are found as a rarity both in culture and in wild material. Selection for increased scutellars had little effect for several generations, but this stability was followed by a sharp advance, just as in the present hair number selections. Jumps and stability alternated for a considerable time in Payne's flies. His mating system is not clearly described, but it seems quite likely that even after thirty generations the degree of inbreeding in his material was quite small, so that recombinations could still be effective. Thus his results agree with my own in two important aspects, viz. (*a*) early stability was followed by a marked response to selection, and (*b*) the final result of selection was to give flies showing a degree of expression of the character well outside the range observable in the material from which the experiments started. Both these results would follow from the breaking-up of balanced polygenic combinations by crossing-over.

The second group of selection experiment, that concerned with the manifestation of mutant characters, shows results in marked contrast to those mentioned above. Three main pieces of research seem to fall in this class, viz. selection for (*a*) bristle number in Dichaete flies (Sturtevant, 1918), (*b*) bristle number in a stock homozygous for a recessive bristle increaser (MacDowell, 1917), (*c*) facet number in Bar (Zeleny, 1922; Zeleny & Mattoon, 1915). In each case a large advance in the direction of selection was achieved at the beginning of the experiment, and there

# 184 *Variation and Selection of Polygenic Characters*

was no evidence of marked delayed advances. Furthermore, in each case there was no marked transgression of the original limits of the character in the later generations. Thus these results would be expected if balanced polygenic combinations did not exist in the original material.

The position of Payne's (1920) later experiments on selection for the number of bristles in scute flies is obscure. One of his lines seems to show delayed and the other immediate response to selection. This experiment will require repetition before it can be classified.

So, on reviewing the evidence, it appears that balanced polygenic combinations are encountered when naturally variable characters are

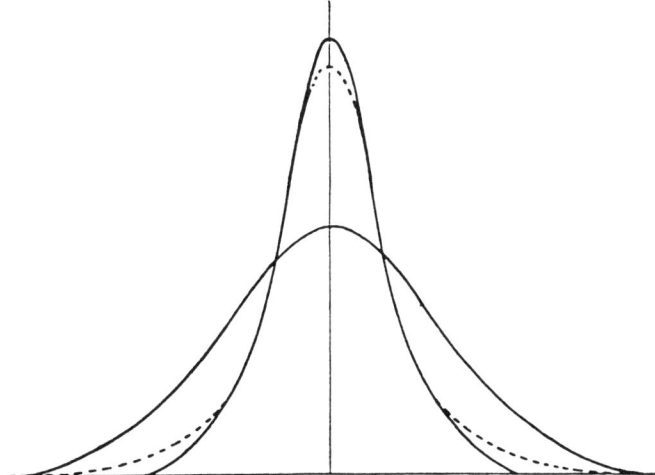

Text-fig. 10. Diagram to illustrate the relation between polygenic variation and linkage. If the outer full curve is that of variation when the genes are unlinked, and the inner full curve that obtained when variability is lowered by reduction in the number of segregating genes, the dotted curve is that obtained when variability is reduced by linkage. Marked variability reduction is obtained, but the two extremes of the curve are still as wide as they were originally.

used, but that the modifiers of expression of mutants are not so organized. This clearly suggests that the balanced combinations are themselves the product of natural selection, which would be relatively inoperative on the modifiers of very rare mutants.

In order to see how natural selection would bring about balanced combinations we must consider the effect of linkage on variation in the wild. Text-fig. 10 shows a series of frequency distribution curves of the kind characteristic of polygenic characters as found in natural material. Most of the individuals show expressions of the character near the mean, and as the degree of expression becomes more remote from the mean the frequencies of individuals showing it become less. In order to simplify our argument let us suppose that the curve is symmetrical and let us consider at present only the genetical variation for the character. The conclusions reached are not invalidated by skewness, and environmental fluctuations merely reduce the efficiency of selection. Hence our model is in no wise deceptive. Now such curves have very definite selective implications. Of these the most·important is that the mean of the observed distribution is near to the optimum value for the environment under consideration, the individuals with the more extreme high and low values being at a selective disadvantage. If this were not so and if the optimum were, say, markedly higher than the mean the individuals near the optimum would leave more descendants than those near the mean and this would lead to the piling up of numbers near the optimum to which, in a few generations, the mean would closely approximate. Thus the mean and the optimum will be near to one another, but they will never quite coincide, as the environment, and hence the optimum, is constantly changing slightly. The mean fluctuates with the optimum, though less widely, and always one generation in arrears.

Now if the selective disadvantage of an individual increases as its degree of character expression deviates more and more markedly from the mean, the very occurrence of genetical variation implies a reduction in the average fitness of the progeny of an individual. For the mean variability of the progeny of an individual will be highly correlated with the variability of the population as a whole, except in those rare cases where the population consists of a series of pure lines. Hence genetical variation round the mean is disadvantageous. Then one effect of natural selection will be to reduce this variability.

It is equally clear, however, that the variation cannot be wholly eliminated without danger to survival through failure to react to the changes in the environment, such as must be constantly occurring. Such invariable individuals would be at a disadvantage as compared with their more variable competitors.[1] Thus the organism is faced with a paradoxical situation in which variation results in an immediate loss of fitness while lack of variation means deferred but none the less serious disadvantage. In general then long term selection will effect a compromise between the two extremes. The mechanism of this compromise is linkage leading to balanced polygenic combinations.

[1] Environmental changes may be divided into two classes. There are directional trends continuing over long periods to which the organism must adapt itself or perish. There are also non-directional fluctuations to which the organism need not respond. In the case of ephemerals the seasonal changes are important in that the organism must not react strongly to the environmental cycle.

The way in which selection will act to build up these combinations can be seen from a simple example. Let us suppose that we have a population heterozygous for two genes (or groups of genes) **A** and **B**, cumulative in action, each of which is without dominance, and having gene frequencies of $\frac{1}{2}$. Thus **Aa** is intermediate between **AA** and **aa**, and **AaBb** will show the character in question to the same degree as **AAbb** and **aaBB**. Then we have nine genotypes which fall into five phenotypic groups, as follows:

| | |
|---|---|
| 2 | **AABB** |
| 1 | **AaBB, AABb** |
| 0 | **AAbb, aaBB, AaBb** |
| −1 | **Aabb, aaBb** |
| −2 | **aabb** |

The central group will be at the mean of the distribution and hence near the optimum. Now **AABB** and **aabb** are at a marked disadvantage as compared with the central class and **AaBB, AABb, Aabb** and **aaBb** are also at a disadvantage through not so great a one.

This means that selection will gradually reduce the numbers of **AB** and **ab** gametes produced as compared with the numbers of **Ab** and **aB** types. But **AB** and **ab** gametes are being continually replenished from the double heterozygotes which are formed from the fusion of **Ab** and **aB** gametes. Thus the action of selection in reducing variation and increasing immediate fitness is slowed down by random recombination of the two genes. If, however, **A** and **B** are linked this replenishment is reduced in rate and the number of zygotes in the disadvantageous classes, notably **AABB** and **aabb**, thereby diminished. So linkage carries an advantage in increasing the fitness of the population as a whole and selection will favour reduced recombination in doubly heterozygous individuals and hence ultimately in the population as a whole.

It is however important to note that although the proportion of **AB** and **ab** gametes is reduced by linkage in this way, some gametes of these types will be formed so long as any recombination does occur. Then if the environment changes in such a way as to cause a marked change in the optimum the corresponding change in the genetic constitution of the population can and will occur by means of the recombination. If **AB** or **ab** types are now favoured they are present and can be selected as a direct consequence of the action of crossing-over. Thus linkage of **A** and **B** results in the storage of variability in such a way that the immediate fitness of the population is maintained at a high level, while still gradually releasing variation, which makes possible a response to change in environment. **Ab** and **aB** are in fact balanced combinations.

Other causes of reduction of variability, such as homozygosity for **Ab** and **aB**, will be favoured immediately, but will eventually become disadvantageous when the environment changes. Such homozygosity must occur to some extent in nature. At this point it is important to remember that natural variation is preponderantly polygenic. In our simple illustration such homozygosity for one factor would change the whole set-up, but in the case of a polygenic character, dependent on possibly scores of genes, a much less erratic mean behaviour is expected. Some genes may become homozygous, but the remainder will still remain heterozygous and act in balanced combinations. Probably the proportions falling into the two classes will be a definite property of the genic organization and mutation rate of the species. Furthermore, in the simple digenic case considered above, only a proportion of the individuals were doubly heterozygous, and so open to the action of selection on recombination. With a polygenic character this proportion will be much higher and the action of selection correspondingly greater.

Thus it will be seen that, granted the occurrence of crossing-over and of some measure of outbreeding, the origin and behaviour of balanced combinations is an inevitable consequence of natural selection in a slowly changing environment. Heterozygosity for such combinations is the only mechanism which allows of maintenance of the variability necessary for future change in such a way that it does not constitute an immediate drawback. As such combinations are dependent for their working on linkage and recombination it is clear that this mechanism also allows one to see the real value of crossing-over to the organism. Such selective control of linkage is essentially similar to that postulated by Sturtevant & Mather (1939) but whereas they considered linkage of two genes, with its concomitant difficulties of maintenance of heterozygosity, it is now seen that the process depends on polygenic characters and so this difficulty vanishes.

### 9. The interaction of polygenic combinations

Balanced polygenic combinations can act as storing mechanisms for variation only if at least two occur in various representatives of one chromosome, i.e. only when some heterozygosity, real or potential, is present. Actually many more than two will exist.

Now any combination is balanced in that it contains + and − genes in a proportion which makes it, in combination with itself and the other combinations in the population, suitable to the environment. The fact that such a combination may be homozygous clearly implies close internal balancing if it is to survive. On the other hand, the fact that two combinations may occur together in one zygote means that they must be balanced one against the other so that their heterozygote is fitted for the environment. Thus two individuals, homozygous for different polygenic combinations, might both fall near the optimum in the variation range, yet their heterozygote might display heterosis, i.e. express the character to a markedly different and hence less favourable degree. In this case the heterozygote is at a disadvantage as compared with the homozygote and any mechanism tending to prevent intercrossing of the homozygotes will be favoured. Heterosis in such cases is disadvantageous and its occurrence will favour the origin of an isolation mechanism. Since heterogeneity for polygenic combinations is almost certain within populations, and hence inevitable between populations, it seems extremely likely that this may be a widespread cause of genetical isolation mechanisms (cf. Sturtevant, 1940). The effect of heterosis must however be clarified by further knowledge of the dominance relations of polygenes, before a final decision on its importance in this connexion can be reached.

Thus there are two balancing processes to which any polygenic combination is subject, (a) the internal one in homozygotes and (b) the relational balancing against other combinations. Both will occur in any population, but their relative effectiveness may vary according to the breeding behaviour of the organism. Where inbreeding is common, the opportunity for relational balancing will be rare and we may expect selection to act mainly on the internal constitution of the combinations. In such cases heterozygotes may be less well adapted than homozygotes and outcrossing will tend to be discouraged. The stored variability of the population then lies in its potential heterozygosity.

Where outcrossing is regular the occurrence of homozygotes may be rare and the relational balancing process becomes the more important. Good internal balance may then fail to be achieved and homozygotes will be poorly adapted. This will favour more certain outcrossing. Thus the breeding mechanism, whether in- or out-breeding in type, will tend to be strengthened in its action. It is not clear how far-reaching this process will be, and discussion is difficult without further knowledge of the organization of balanced combinations. It seems, however, highly probable that enforced hybridity of the type found in *Oenothera* and in the sex mechanism may have developed along these lines.

## 10. Polygenic combinations and polymorphism

A considerable number of cases are known where single gene characters regularly show heterogeneity in the wild. Species showing this behaviour are termed polymorphic. The occurrence of such regular polymorphism immediately raises the question of its maintenance; for normally if one allelomorph of a gene has a selective advantage over another the latter almost vanishes, its frequency becoming of the order of its mutation rate. Certain properties of polymorphism genes have been established which enable us to understand their occurrence better. Thus Fisher (1930), using the data of Nabours (1925), has shown in *Apotettix* that individuals heterozygous for these genes are more viable than their homozygous sibs. Since, however, this must be true of all such heterozygotes, if it is to provide an explanation for the existing polymorphism, it would seem unlikely that the viability differences are properties solely of the genes determining the visible polymorphism. It would seem more likely to be dependent on the complete linkage of these genes with, or their occurrence in, polygenic combinations, which had attained a particular relational balance.

As has been seen above, where outcrossing is the rule the relational balance may be more carefully adapted by selection than the internal balance. Then the homozygotes will show the character too much or too little and would be at a disadvantage as compared with the heterozygotes. This should tend to encourage mechanisms favouring outcrossing and in consequence we might expect a polymorphic animal such as *Apotettix* to show a preference for mating of unlike types. Unfortunately, no data are yet available for the verification of this point.

The question of the existence of various chromosome sequences ("inversions") in wild *Drosophila* probably falls into this category too. In fact, Sturtevant & Mather (1939) have suggested that such heterogeneity depends on interaction between the genic contents of such structurally variable chromosomes. It is notable that their argument, as translated into our present terminology, implies a better relational than internal balance, just as in *Apotettix*. In fact if each sequence in the third chromosome of *Drosophila pseudoobscura* carried a dominant marker gene, this species would simulate the behaviour of the grouse locusts. The structural variation in *D. pseudoobscura* is almost completely confined to one chromosome, viz. the third. It seems reasonable to expect that, in general, polygenic combinations will not be evenly distributed amongst all the chromosomes of the complement. The mechanism of variability storage is such that a new combination will frequently be more effective when in the same chromosome as those already existing, and so the *D. pseudoobscura* mechanism should often arise.

In a cross-breeding population any combination will occur most frequently heterozygous with other combinations in that same chromosome. Since there will be a number of such other combinations the selective advantage or disadvantage of the particular one in question seems likely to be capable of very delicate adjustment as a result of change in its frequency. Thus a combination when occurring in 10 % or 50 % of the gametes of a population in a given environment may be at a disadvantage, but when its frequency is say 30 % of the gametes it may be properly adjusted. In a somewhat different environment perhaps 10 % would be the proper frequency. That this will be so cannot be stated definitely until a full mathematical analysis has been undertaken, but it seems very likely to be true.

Granted the validity of this supposition, a very simple explanation of Huxley's "clines" becomes possible. Thus populations of Guillemots regularly contain a number of so-called bridled birds in addition to normal individuals. The frequency of bridling changes more or less regularly with the geographical location of the populations. Now if the bridling allelomorph is completely linked to one polygenic combination, just such a cline would be expected, because the environmental conditions are, broadly speaking, changing regularly with latitude and such environmental changes will result in correspondingly adjusted frequencies of the polygenic combination, and hence of the bridling character.

## 11. Conclusions

Experimental evidence has been advanced for the occurrence of internally balanced combinations of polygenes within chromosomes. Two of these when present in homologous chromosomes can cross over, lose their balance and so release variation, in the sense that the next generation will contain individuals showing an expression of the character outside the range previously encountered. It also appears that such combinations will be developed as a result of natural selection in the wild, since a population heterozygous for a number of these combinations on one or more chromosome pairs would show minimal immediate variation if effective crossing-over were low, and yet by virtue of this rare crossing-over would still include a very few of the extreme types which are necessary for adaptation to environmental changes. Each polygenic combination is subject to internal balancing by the action of selection in homozygotes and to relational balancing by selection operating on individuals whose homologous chromosomes are carrying unlike combinations.

The foregoing conclusions seem reasonably certain, but beyond seeing that the theory of polygenic combinations throws new light on such diverse questions as inbreeding and outbreeding mechanisms, polymorphism, heterosis and the origin of isolation mechanisms, as indicated above, it will be impossible to make much further progress until more information as to the organization of such combinations is available. We must have information as to the dominance relations of the individual polygenes, on the effect of selection on such dominance relations, of how thoroughly the $+$ and $-$ genes are mixed in the combinations and so on. Of even more importance, perhaps, is a knowledge of how a new combination, formed by crossing-over between two pre-existing balanced types, reacts with its progenitors. Can it differ sufficiently from them to constitute a new combination subject to balancing action, or will it in general be liable to reasonably free recombination with them and so tend to destroy both its own and their individuality? In other words, are the combinations of a population in a perpetual state of flux, with more or less fixed types only maintainable as a result of structural change, or are new combinations in general so unbalanced as rarely to establish themselves? Answers to these questions will do much to advance our knowledge of variation storage and its release in the wild, and the action of both natural and artificial selection.

## 12. SUMMARY

*Polygenic*, or "quantitative", variation is present in wild populations and is the raw material of species differentiation. Its properties may be investigated by suitable selection experiments.

The number of hairs on the ventral surface of the fourth and fifth abdominal segments of *Drosophila melanogaster* was chosen as the material for investigation. The *Drosophila* species differ in their hair numbers.

In one cross, $\underline{y} \times f\,B^lB^l$, where selection was exercised from $F_2$ onwards, the main advance was achieved in the first two selected generations. This was interpreted as being due to recombination of whole chromosomes.

In a second cross, $BB \times +$, following selection from $F_2$ onwards, an advance with selection was observed for two generations, followed by a period of stability. This was in turn followed by a second and larger advance with selection, which was interpreted as being due to the action of selection on variation released by recombination of genes in the same chromosome. This conclusion was tested and verified by a repetition of the experiment, by selection within one inbred line and by selection in certain backcross lines. The release of variation by intra-chromosome recombination implies the existence of balanced combinations of polygenes within the chromosomes.

It is shown that such *balanced polygenic combinations* will be developed by the action of natural selection. They effect the best compromise between the advantageous qualities of immediate stability and the ultimate variability necessary for a change in an altering environment.

Polygenic combinations will be subject to an *internal* balancing process in homozygotes and a *relational* balancing process in heterozygotes. The relative efficiency of these processes will affect the breeding mechanism of a population. Heterosis is a result of poor relational balancing and may be the cause of origin of some isolation mechanisms. Polymorphism in nature can be accounted for in terms of the relative efficiency of the internal and relational balances.

It seems likely that a given polygenic combination will have an optimum frequency in a population existing under given environmental conditions. This frequency will change with the environment. If a qualitative gene is completely linked with such a combination it will be expected to show a "cline" of frequencies over large geographical areas, as observed for example in bridled guillemots.

## REFERENCES

BAUR, E. (1924). "Untersuchungen über das Wesen, die Entstehung und die Vererbung von Rassenunterschieder bei *Antirrhinum majus*." *Bibl. genet., Lpz.*, **4**, 1–170.

DOBZHANSKY, TH. (1936). "Studies on hybrid sterility. II. Localisation of sterility factors in *Drosophila pseudoobscura* hybrids." *Genetics*, **21**, 113–35.

EAST, E. M. (1935). "Genetic reactions in *Nicotiana*." *Genetics*, **20**, 403–52.

FISHER, R. A. (1930). "The evolution of dominance in certain polymorphic species." *Amer. Nat.* **64**, 385–406.

FISHER, R. A., IMMER, F. R. & TEDIN, O. (1932). "The genetical interpretation of statistics of the third degree in the study of quantitative inheritance." *Genetics*, **17**, 107–24.

GOTTSCHEWSKI, G. & TAN, C. C. (1938). "The homology of eye colour genes in *Drosophila melanogaster* and *Drosophila pseudoobscura* as determined by transplantation. II." *Genetics*, **23**, 221–38.

MACDOWELL, E. C. (1917). "Bristle inheritance in *Drosophila*. II. Selection." *J. exp. Zool.* **23**, 109–46.

NABOURS, R. K. (1925). "Studies of inheritance and evolution in Orthoptera. V." *Tech. Bull. Kans. agric. Exp. Sta.* no. 17.

PAYNE, F. (1918). "An experiment to test the nature of the variations on which selection acts." *Indiana Univ. Stud.* **5**, 1–45.

—— (1920). "Selection for high and low bristle number in the mutant strain 'Reduced'." *Genetics*, **5**, 501–42.

STURTEVANT, A. H. (1918). "An analysis of the effects of selection." *Publ. Carneg. Instn*, no. 264, pp. 1–68.

—— (1938). "Essays on evolution. III. On the origin of interspecific sterility." *Quart. Rev. Biol.* **13**, 333–5.

—— (1939). "High mutation frequency induced by hybridisation." *Proc. nat. Acad. Sci., Wash.*, **25**, 308–10.

STURTEVANT, A. H. & MATHER, K. (1939). "Interrelations of inversions, heterosis and recombination." *Amer. Nat.* **72**, 447–52.

STURTEVANT, A. H. & TAN, C. C. (1937). "The comparative genetics of *Drosophila pseudoobscura* and *Drosophila melanogaster*." *J. Genet.* **34**, 415–39.

WATKINS, A. E. (1930). "The wheat species: a critique." *J. Genet.* **23**, 173–263.

ZELENY, C. (1922). "The effect of selection for eye facet number in the white Bar-eye race of *Drosophila melanogaster*." *Genetics*, **7**, 1–115.

ZELENY, C. & MATTOON, E. W. (1915). "The effect of selection upon the 'bar eye' mutant of *Drosophila*." *J. exp. Zool.* **19**, 515–29.

[*Editor's Note:* Plate 6 and its explanation have been omitted.]

# ATTENUATION OF GENETIC PROGRESS UNDER CONTINUED SELECTION IN POULTRY

I. MICHAEL LERNER and EVERETT R. DEMPSTER
*University of California, Berkeley*

Received 20.x.50

IN the absence of disturbing factors, a phenotypic trait whose genetic variance is determined by alleles at a single locus will respond to selection pressure until fixation of the most desirable allele is attained. Presumably similar exhaustion of genetic variation may account for deceleration and eventual cessation of gains from selection of a character under polygenic control. It is, however, also possible that other factors may interfere with the realisation of genetic gains long before the limitations imposed by the decrease in genetic variability enter the picture to a significant degree. Attention to a number of factors possibly responsible for this phenomenon has been drawn by Lush (1945), Dempster and Lerner (1949), Lerner (1950) and others. In general these factors represent mechanisms falling into one of two categories :—

(a) Negative correlation of the selected trait with fitness, and
(b) non-additive gene action (and possibly genotype-environment interaction).

The first of these may take the form of a negative correlation in individuals chosen as parents between genotypes for the selected trait and the numbers of surviving offspring, or of a negative correlation in their offspring between genotypes for the selected trait and the probabilities of survival to maturity.

Both (a) and (b) are basically varieties of what Lerner (1950) has called *genetic homeostasis*, or the property of a population at an adaptive peak to resist changes in its genetic composition. The more common and evolutionarily effective form of genetic homeostasis is, of course, selection for fitness which tends to maintain the population at an adaptive peak. Under artificial selection, however, differences in what may be called natural fitness are not necessarily any longer the sole determinants of selective advantage. Thus if artificial selection for a character not intrinsically connected with viability, rate of maturation, and fecundity is practiced, the individuals excelling in the desired trait and thus selected to be the parents of the next generation are not always those possessing the greatest fitness under natural selection. If they are actually less fit in the latter sense than individuals nearer the mean of the population with respect to the trait under artificial selection, it is clear that some of the artificial selection pressure applied will be dissipated in overcoming the pull of natural selection in the opposite direction from that of artificial selection. Under such circumstances, relaxation of artificial selection would result in the complete or partial return of the population to its original level with respect to the trait under selection.

Examples of this situation are provided in the long-range *Drosophila* selection experiments of Mather and Harrison (1949), of which only one needs to be cited here. From an $F_2$ cross between two strains of *Drosophila melanogaster* previously not subjected to artificial selection, a line selected for a high number of abdominal chæta was established. Reasonably regular gains were obtained for 20 generations, but concurrently there occurred a reduction in fertility of such severity as to threaten the extinction of the line. Selection was suspended at this point and the line reproduced *en masse*, with the result that chæta number reverted in the course of 5 generations to approximately the original level. With minor fluctuations this level was henceforth maintained for over 100 further generations.

At the 24th generation a selected line was again started from the mass culture. This line responded to selection without a loss in fertility and when selection was once more suspended 10 generations later no drop in chæta number was observed.

Mather and Harrison interpreted these results in terms of Mather's (1941, 1942, 1943) theory of polygenic balance. Briefly speaking, the gist of it as applied to this particular instance lies in the existence of linked combinations of genes some of which affect chæta number and others fitness. Under a high selection pressure for chæta number alleles with adverse effects on fitness were carried to high frequencies, being, so to speak, locked into blocks with desirable gross effects on chæta number. With suspension of the pressure for the latter trait, natural selection for fitness could once more operate to its full extent. In the meantime between the 24th and the 34th generation, blocks of different genic content arose as a result of crossing-over, so that a new adaptive peak at a higher level of chæta number came into existence.

Whether Mather's particular theory of polygenic balance or a broader basis such as suggested by Hazel (*see* p. 239 in Lerner, 1950) for the induction of negative genetic correlations between two traits under selection (chæta number and fitness in the *Drosophila* example described) be accepted, it seems that should mechanisms of this type be in operation in any given selection experiment, they might be detected by investigating rates of reproduction of individuals. Data for such an investigation are available from two experiments conducted with poultry. The purpose of the present communication is to examine them in the light of our previous remarks in order to determine, if possible, which of the three general factors * (exhaustion of variability,

---

* A fourth factor may be noted in the reduction of selection pressure. However, unless selection for other traits is introduced this factor must be in turn produced by one or more of the three listed ones.

balance between fitness and a trait under artificial selection, and non-additive variability) are of greatest significance in decreasing the efficiency of genetic gains under selection. In the first of these experiments, that on shank length, an actual plateau has been reached, although it is possible that the plateau is only a temporary one, such as obtained by Mather and Harrison in some of their selected lines. In the second experiment, on egg production, it is impossible to say at the present time whether progress has stopped. Nevertheless, the attrition of the artificial selection pressure by natural selection in the opposite direction is still subject to investigation even if only a reduction of the rate of gain per unit of pressure applied is obtained.

## SELECTION FOR SHANK LENGTH

Lerner (1943) has previously reported on the earlier phases of this experiment. It consisted of the establishment of a line of Single Comb White Leghorns selected for long shanks at maturity from the University of California production-bred flock which later served as an unselected control. Table 1 shows the progress attained from

TABLE 1

*Results of selection for increased shank length*

| Year | Production line | | Size line | | Difference in shank length, cm. |
|---|---|---|---|---|---|
| | Number of birds | Mean shank length, cm. | Number of birds | Mean shank length, cm. | |
| 1938 | 368 | 9·69 | ... | ... | ... |
| 1939 | 274 | 9·62 | 84 | 9·92 | 0·30 |
| 1940 | 137 | 9·55 | 89 * | 9·92 | 0·37 |
| 1941 | 346 | 9·37 | 106 | 10·20 | 0·83 |
| 1942 | 260 | 9·46 | 119 | 10·29 | 0·83 |
| 1943 | 543 | 9·42 | 191 | 10·46 | 1·04 |
| 1944 | 184 | 9·41 | 75 | 10·73 | 1·32 |
| 1945 | 382 | 9·48 | 61 | 10·84 | 1·36 |
| 1946 | 309 | 9·44 | 88 | 10·71 | 1·27 |
| 1947 | 377 | 9·77 † | 59 | 10·73 † | 0·96 |
| 1948 | 471 | 9·89 † | 53 | 10·96 † | 1·07 |
| 1949 | 473 | 9·53 | 92 | 10·78 | 1·25 |
| 1939-1944 | Change per year (least squares regression) | | | . . | 0·188 |
| 1943-1949 | Change per year (least squares regression) | | | . . | −0·007 |

\* Six birds inadvertently omitted in Lerner's (1943) report.
† Measured by different observer.

selection from 1938 to 1949. Only females were measured in this experiment, the selection of males being based on sister and progeny averages. All of the measurements of the birds up to 1946 were made by one of us (I. M. L.). In 1947 and 1948 a research assistant undertook this task. In 1949 the measurements were carried out jointly by the original observer and another research assistant. There is considerable reason to believe that the technique of measurement used in 1947-48 led to overestimates of 0·1-0·2 cm. Since this discrepancy was undoubtedly a constant one for all birds measured, the last column of table 1 is probably a more accurate indication of the gains attained than the column giving the mean shank lengths for the selected line. In any case it may be clearly seen that selection was effective for the first seven generations, after which little if any further gains were obtained. This is clearly demonstrated by the average gains per year obtained in the first seven years as compared with that in the second (based on the regression of shank length on year, and allowing a one-year overlap in the two periods), which appear at the bottom of the last column in the table.

Table 2 presents an analysis of variability of the successive generations of the selected line. The heritability figures given are estimates from intra-class half and full sister correlations. They represent a

TABLE 2

*Genetic parameters of size line*

(omitting families with only one offspring)

| Year | Number of birds | Coefficient of inbreeding ($F$) | Heritability | | Phenotypic variance | Genetic variance (raw) |
|---|---|---|---|---|---|---|
| | | | raw | corrected for $F$ | | |
| 1939 | 84 | 0 | 0·887 | 0·887 | 0·244 | 0·216 |
| 1940 | 87 | 0·029 | (−0·049) | (−0·050) | 0·168 | (−0·008) |
| 1941 | 99 | 0·072 | 0·262 | 0·288 | 0·174 | 0·046 |
| 1942 | 112 | 0·112 | 0·374 | 0·442 | 0·241 | 0·090 |
| 1943 | 182 | 0·145 | 0·053 | 0·063 | 0·186 | 0·010 |
| 1944 | 69 | 0·189 | 0·243 | 0·318 | 0·465 | 0·113 |
| 1945 | 52 | 0·222 | 0·649 | (1·023) | 0·280 | 0·182 |
| 1946 | 78 | 0·273 | 0·564 | 0·983 | 0·152 | 0·086 |
| 1947 | 45 | 0·299 | 0·105 | 0·157 | 0·187 | 0·020 |
| 1948 | 34 | 0·311 | 0·158 | 0·247 | 0·238 | 0·038 |
| 1949 | 90 | 0·344 | 0·289 | 0·519 | 0·232 | 0·067 |
| 1939-1944 Average | | 0·097 | 0·260 | 0·287 | 0·229 | 0·059 |
| 1943-1949 Average | | 0·231 | 0·269 | 0·410 | 0·236 | 0·062 |

considerable amount of variation from year to year, but the comparison between the average variability in the first and second halves of the period of selection indicates that the amount of genetic variance and the degree of heritability have not materially changed in the course of selection. This is substantiated by intra-sire daughter-dam regressions which lead to heritability estimates of $0·332 \pm 0·16$ for the first and $0·542 \pm 0·20$ for the second half of the selection period. It may be parenthetically noted that the heritability of shank length in the parent flock (the production line) was found to be 0·38, a figure comparable to the averages corrected for inbreeding in table 2.*

It seems that consanguineous mating practiced in the selection line has not reduced the genetic variance to the extent which might be indicated by the magnitude of the coefficient of inbreeding, paralleling the previous findings of Shoffner (1948). The significance of this observation cannot, however, be adequately assessed at this time.

We may thus see that in this experiment response to selection has ceased, although the genetic variability available for selection to operate on has not been exhausted. The question then arises as to which of the mechanisms noted as the possible explanations of such situations can be held responsible for the phenomenon observed.

To simplify the computations bearing on this question, we may confine our analysis to the offspring of dams used as parents for the first time, since the inclusion of re-mated dams would introduce selection criteria (in particular their reproductive performance and progeny tests) which would complicate the situation. In the material at hand, 82·5 per cent. (770 out of 933 birds) of the offspring in the years

TABLE 3

*Analysis of offspring from pullet dams*

| Year | Number of selected dams | Number of offspring | Index of fitness * | Average shank length of offspring, cm. | Selection differential | | Ratio realised : expected |
|---|---|---|---|---|---|---|---|
| | | | | | Expected cm. | Realised cm. | |
| 1940 | 15 | 70 | 4·67 | 9·91 | 0·480 | 0·476 | 0·99 |
| 1941 | 19 | 85 | 4·47 | 10·18 | 0·349 | 0·335 | 0·96 |
| 1942 | 18 | 64 | 3·56 | 10·22 | 0·194 | 0·109 | 0·56 |
| 1943 | 33 | 191 | 4·63 | 10·46 | 0·425 | 0·417 | 0·98 |
| 1944 | 23 | 57 | 2·48 | 10·73 | 0·558 | 0·505 | 0·91 |
| 1945 | 21 | 55 | 2·62 | 10·81 | 0·378 | 0·385 | 1·02 |
| 1946 | 23 | 71 | 3·09 | 10·66 | 0·273 | 0·132 | 0·48 |
| 1947 | 28 | 44 | 1·57 | 10·81 | 0·455 | 0·422 | 0·93 |
| 1948 | 23 | 51 | 2·22 | 10·94 | 0·339 | 0·190 | 0·56 |
| 1949 | 21 | 82 | 3·12 | 10·79 | 0·429 | 0·309 | 0·72 |
| 1940-1945 | ... | Average | 3·75 | ... | 0·404 | 0·380 | 0·94 |
| 1944-1949 | ... | Average | 2·47 | ... | 0·407 | 0·314 | 0·77 |

\* Number of offspring measured per dam, based on a standard 4-week hatching period.

1940-49 were from dams mated in their first year of life. As it may be seen from table 3 omission of the remaining 163 measured birds does not materially affect the trend of selection progress. Accordingly all of the computations referred to in table 3 were confined to 770 birds.

\* This value was computed for us by Mr Fred T. Schultz by pooling data from half-sister correlations for the years 1942-1949, involving measurements on 2983 birds. Mr Schultz also found that maternal effects account for 6-7 per cent. of the total variation, confirming the conclusions formulated by Hazel and Lamoreux (1947) and Lerner and Cruden (1951).

The first mechanism suggested lies in the lowered reproductive rate (as measured by the number of offspring reaching maturity) of parents with genotypes for longer shanks. It is quite apparent from the index of fitness in table 3 that this possibility exists, since the average number of offspring per dam has dropped from 3·75 in the period of gains to 2·47 in the period of the plateau. The question, however, is whether the elimination of offspring was differential or independent of the genotypes of the selected parents. We can attack this problem first on the phenotypic level, since the answer for the female parents at least does not present a difficult problem. If we consider the birds selected as parents of the next generation in any year we can readily compute the *expected selection differential* by subtracting from their average shank length the average of their own generation. If we weight the shank lengths of these females by the number of their offspring which survived until the time of measurement, the weighted average obtained would represent the *realised selection differential*. The ratio of the realised to the expected selection differential provides a measure of association between fitness and shank length. If it is less than unity, it means that among the selected birds those with longer shanks produce, on the average, less offspring than those with shorter shanks. This would provide a strong indication that natural selection for fitness does indeed operate in the opposite direction from artificial selection for longer shanks.

It may be seen that the selection pressure attempted, although variable from year to year, did not change much, if at all, between the period in which gains were realised and the one in which no further progress was obtained. On the other hand, the realised selection differential showed a considerable drop. The ratio between the two was 0·94 for the period ending 1945, and only 0·77 for the last six years. Although the gain realised in any year does not always reflect the variation in this ratio, it seems quite clear that by and large selection for fitness of dams has been in the opposite direction to selection for shank length, only mildly opposing it at first and fairly severely later.

It may be of interest to determine at what stage of the reproductive cycle the differential elimination of the offspring of the relatively longer shanked dams occurs. This can be readily done by computing the selection differentials realised at each of the successive stages. Thus the identical procedure used for computing the selection differential realised at the time of measurement can be applied to selection differentials at the time of incubation (based on the number of eggs laid by the dams), at the time of determination of fertility (based on the number of fertile eggs), and at hatching time (based on the number of chicks obtained). The results are summarised in the following table.

It may thus be seen, that differences in hatching power are the ones primarily responsible for the variation in the reproductive rate of dams having different shank length. Variation in egg production, fertility of eggs, and post-hatching mortality is not involved in the first period, although both egg production and chick survival differences contribute to the drop in the realised selection differential in the second.

*Selection differentials*

| Period | Expected | Based on eggs set | Based on fertile eggs | Based on chicks hatched | Realised |
|---|---|---|---|---|---|
| 1940-45 | 0·404 | 0·408 | 0·410 | 0·385 | 0·380 |
| 1944-49 | 0·407 | 0·382 | 0·382 | 0·337 | 0·314 |

The computations made so far strongly suggest that a polygenic balance of some sort may be in operation. It must, however, be realised that the negative correlation between shank length and reproductive fitness deduced to exist is on a phenotypic level. The attrition in the genotypic selection differential may be even greater than the ratio of 0·77 in table 2 indicates. To check this possibility computations were carried out for the expected and the realised selection differentials of family averages (including the dams) in a manner exactly similar to that described above for individual measurements. Family averages should provide somewhat better estimates of the genotypes of the dams than do their phenotypes. Indeed, the realised selection differential in the second period computed on this basis is an even smaller fraction of the expected selection differential. This is shown in the following table and suggests that the negative correlation between number of offspring and shank length may be greater along the genotypic than along the environmental pathway.

| Period | Expected selection differential, cm. | Realised selection differential, cm. | Ratio |
|---|---|---|---|
| 1941-45 | 0·137 | 0·134 | 0·98 |
| 1944-49 | 0·155 | 0·105 | 0·68 |

## THE ATTRITION OF SELECTION PRESSURE

The observed decrease in the phenotypic selection differential should lead to a greater reduction in gain if the correlation is entirely genetic than if it is due in part to an environmental component of like sign. It is possible to demonstrate, as shown in the appendix, that, on the extreme assumption of an entirely genetic correlation and on the basis of certain simplifying postulates (such as that all genetic variance is additive), the form of adverse natural selection being discussed would reduce the genotypes of offspring from their expected values by 0·03 cm. per generation in the early and by 0·11 cm. in the late periods of selection. Thus the additional loss in the second period as compared to the first would be about 0·11-0·03, or 0·08 cm. The assumptions underlying this estimate, which is nearly half of the gain per generation (0·188 cm.) attained in the early period, are likely to err in a direction leading to the attribution of too great an influence to differential fecundity, but bias in the reverse direction is not precluded. It may perhaps be agreed that the lower fecundity of dams with genotypes for higher shank length definitely results in some reduction in gains, that not improbably the reduction due to this cause may be nearly half the rate of gain achieved in the early period of selection, and that a still greater effect, though rather unlikely, is not disproved by the data.

In addition to the adverse natural selection *between* families due to larger dams having fewer offspring, there would also be some selection within families if the decrease in number of measured offspring per dam in the plateau period of selection, as compared to the early period, is due in part to the elimination within each family of genotypes for longer shank length. The intensity of this selection depends on unknown factors, such as the degree of possible negative correlation between genotype for hatchability and genotype for shank length. The following calculations based on postulates, very unfavourable for attaining gains, show that the within-families natural selection probably accounts for only a small portion of the decrease in rate of gain under artificial selection. The number of pullets per 100 fertile eggs set has been found to be 18·8 in the plateau period as compared to 24·3 in the early period, a drop of 22·6 per cent. Elimination of the 22·6 per cent. of the highest individuals in a normal distribution lowers the mean by 0·3885 standard deviations. However, it is extremely unlikely that the elimination is confined to the highest genotypes only, because the heritability of hatchability must be considerably lower than unity, as must also be the genetic correlation between hatchability and shank length. The value 0·5 is an overestimate of the product of the square root of the heritability and genetic correlation so that the decrease in genotype for shank length is almost certainly less than $0·5 \times 0·3885$ or 0·1943 standard deviations. This figure does not take into account the finite size of the families, which further decreases the shift in mean due to elimination of better genotypes. The genetic variance within families may be taken as approximately equal to one half the total genetic variance or (from table 2) as $0·062/2 = 0·031$ ; the genetic standard deviation is the square root of this or 0·176. Multiplying the latter figure by 0·1943 yields 0·034, which is less than a fifth of the gain per year achieved during the early years of selection. This component of natural selection for short shanks might then account for as much as, although probably less than, a fifth of the observed decrease in rate of gain.

It may be next argued that differential fertility of the males (on which measurements were not made largely because only a small percentage are normally raised to maturity) is responsible for a further reduction in the expected selection differential. This possibility can be disposed of by computations analogous to those made for the females, except for the fact that only family selection differentials can be utilised (the males selected as first-time sires being chosen on the basis of the average phenotypes of their sisters). The expected and realised selection differentials and their ratios for the periods 1941-44 and 1943-49 (there being only one male in 1940 not previously mated) are as follows :—

| Period | Per cent. of all offspring in period | Expected, cm. | Realised, cm. | Realised : expected |
|---|---|---|---|---|
| 1941-45 | 0·82 | 0·242 | 0·246 | 1·02 |
| 1944-49 | 0·77 | 0·197 | 0·217 | 1·10 |

The situation differs from that of the female parents in that the ratio of the realised to the expected selection differential in both periods exceeded unity, and in that the ratio shows a slight increase in the later period. In other words, while attrition of selection pressure applied on the maternal side can be demonstrated to have occurred, the paternal side seems to be free from this effect. This is also borne out indirectly by analyses of variance of the number of offspring (relative to the average in any given year) measured per dam. It may be seen from the following table that it is the dam rather than the sire that is mainly responsible for the negative correlation between genotype for shank length and the number of offspring, since the variance component due to differences between sires is in each of the periods analysed less than a twentieth of that due to the differences between dams, and not significant in terms of the latter.

| Period | Source of variance | Degrees of freedom | Mean square | Variance component |
|---|---|---|---|---|
| 1940-45 | Dams | 111 | 0·850 | 0·850 |
| | Sires | 12 | 1·147 | 0·041 |
| | Total | 123 | 0·879 | 0·891 |
| 1944-49 | Dams | 119 | 1·212 | 1·212 |
| | Sires | 14 | 1·083 | −0·019 |
| | Total | 133 | 1·198 | 1·193 |

It may thus be seen that with respect to shank length, where a plateau has been reached before genetic variability was exhausted, we have evidence that genetic homeostasis is mediated at least in part by natural selection operating on the reproductive fitness (hatchability and to some extent egg production and chick viability) of the dams chosen. Should this process and the possible intra-family offspring elimination be insufficient to account for cessation of progress—and their estimated magnitudes suggest that they are not likely to account for much more than half of the reduction in the rate of gain—non-additive gene action must be postulated.

This can take a number of different forms. For instance, if dominance of genes for long shanks is involved, it is possible that the frequency of the recessives has been reduced to such an extent that dominance deviations began to contribute significantly to non-additive genetic variance (in relative if not in absolute terms of magnitude). While there is no evidence for dominance in the crosses between the selected and the control lines reported by Lerner (1943), this possibility cannot be entirely dismissed.

Another type of non-additiveness may be found if overdominance exists, or, in other words, if selection has been directed towards heterozygotes. When sufficient data are available, this possibility can be checked by the technique suggested by Lerner (1950) of comparing the phenotypic variation of daughters of long and short-shanked dams mated to the same sire.

Finally, some form of epistasis may be present. The data at hand are inadequate to answer the question of non-additive gene action, but experiments involving relaxation of selection, further crosses between lines, and diallel matings are being initiated for this purpose.

## SELECTION FOR EGG PRODUCTION

We may next turn to the second of the experiments available for this study, that on the selection for a high hen-housed average egg production (production index). The details of this project have been elaborated in several previous publications, all of the pertinent details having been summarised by Lerner (1950). Table 4 gives the size of the population and the production index for each year. The latter is an average measure of egg production based on the total performance of an unculled group of pullets from the beginning of lay through 30th September of their second year of life. For reasons discussed in the section of this paper dealing with the shank-length selection experiment, the present study of selection differentials is confined to dams mated for the first time. Since within the period reported upon here, these in an overwhelming majority were two years old, the population studied has been restricted to dams of this age. Thus the performance of offspring reported in table 4 refers to birds hatched two years later than their dams.

Some of the columns in table 4 may require further comment. The production index of the selected dams (column 5) is, of course,

TABLE 4

Data on the production-bred flock of S.C.W. Leghorns

| 1 | 2 | 3 | 4 | 5 | 6 | 7 | 8 | 9 | 10 |
|---|---|---|---|---|---|---|---|---|---|
| Year of hatch of dams | Number of pullets in dams' generation | Production index of dams' generation | Number of selected dams | Production index of selected dams | Production index of dams' sister families (including dams) | Number of offspring | Per cent. of total number of pullets in generation | Production index of offspring | Production index of total flock |
| 1932 | 460 | 118.6 | 43 | 244.0 | 156.0 | 288 | 40.9 | 165.4 | 171.8 |
| 1933 | 535 | 125.6 | 48 | 249.2 | 176.6 | 237 | 48.9 | 191.9 | 186.5 |
| 1934 | 704 | 171.8 | 61 | 262.5 | 206.1 | 267 | 66.3 | 202.6 | 201.6 |
| 1935 | 485 | 186.5 | 65 | 275.3 | 216.4 | 362 | 61.1 | 184.5 | 184.0 |
| 1936 | 403 | 201.6 | 52 | 282.5 | 231.2 | 372 | 61.6 | 151.9 | 154.0 |
| 1937 | 592 | 184.0 | 48 | 277.7 | 219.9 | 228 | 49.0 | 177.0 | 179.2 |
| 1938 | 604 | 154.0 | 38 | 276.3 | 190.8 | 147 | 34.3 | 209.5 | 195.9 |
| 1939 | 465 | 179.2 | 68 | 279.0 | 233.3 | 376 | 58.9 | 221.4 | 213.5 |
| 1940 | 428 | 195.9 | 49 | 284.8 | 244.3 | 213 | 50.2 | 197.5 | 208.7 |
| 1941 | 638 | 213.5 | 93 | 289.6 | 252.6 | 596 | 91.7 | 190.5 | 190.9 * |
| 1942 | 424 | 208.7 | 52 | 283.0 | 239.7 | 142 | 54.6 | 219.4 | 218.7 |
| 1943 | 650 | 190.9 * | 67 | 268.7 | 216.6 | 302 | 67.9 | 218.0 | 214.8 |
| 1944 | 260 | 218.7 | 74 | 283.9 | 243.4 | 330 | 68.9 | 208.4 | 203.7 |
| 1945 | 445 | 214.8 | 68 | 294.9 | 247.4 | 348 | 73.4 | 208.2 | 207.3 |
| 1946 † | 479 | 203.7 | 77 | 282.1 | 228.0 | 400 | 72.9 | 227.5 | 224.7 |

\* The figure of 205.9 used in the computations by Lerner and Hazel (1947) was in error.

† There were 578 pullets alive at banding time in 1948; reproductive rates of dams were computed on this basis. At the time of banding 24 pullets constituting 11 complete families were discarded as a space-saving device, the only criterion of elimination being family size. The production index shown (224.7) is thus based on 549 pullets.

equivalent to the "survivors' production", since only birds living through the first laying year were available for selection in the second laying year. Since the primary criterion of selection was the family average, the production indexes of families selected are given in column 6. The number of female offspring in the next generation from the number of dams specified in column 4 is given in column 7, and the percentage that they form of the total number in the flock of their particular generation is shown in column 8. Finally columns 9 and 10 show the comparison between the production indexes of the group of offspring from 2-year-old dams and of the total population. It may be noted here in passing that the weighted average production index of the offspring of dams mated for the first time exceeds that of all other dams by approximately 4 eggs. The direction of this difference runs counter to the indications previously obtained from a restricted sample by Lerner and Hazel (1947), but tends to support the ideas advanced by Dempster and Lerner (1947) regarding the limited contribution progeny testing has made to the advances in egg production obtained in this flock.

It is impossible to say with certainty whether the flock has reached a plateau with respect to the production index. The inter-year variation appears to be so great as to preclude the possibility of establishing this point. It seems reasonably clear, however, that the rate of advance has tended to decline. Thus, the regression of gain on year in the period 1933-42 has been found to be 6·00 eggs, while that for the period 1939-48, only 2·95 eggs.* The question then still may be brought up whether the balance between artificial selection in one direction and natural selection in the other has played any role in the attenuation of the rate of gain.

It may first be noted without entering any details that unpublished data of Lerner and Cruden indicate no exhaustion of genetic variability in the production index (as judged from intra-year heritability estimates). The situation thus parallels that found in the shank-length selection experiment.

The reduction in selection pressure has similarly been found to be moderately small in extent. The figures appearing at the bottom of table 5 indicate that the average selection differential from 1934-42 (or 1932-40 if indicated by the year of hatch of the dams rather than of the offspring) was 99·7 eggs on the individual, and 39·0 on the family basis. For the later period (1940-48), the comparable figures were 82·9 and 35·6 eggs respectively. This reduction obviously cannot account for a 50 per cent. drop in the annual rate of gain on any hypothesis of exclusively additive gene action.

The computations leading to the answer to the question posed

* Lerner and Hazel (1947) gave the regression for the period of 1933-44 as 5·37 eggs. The error for 1943 noted in the footnote to table 4 calls for a revision of their figure to 5·15 eggs, which, of course, in no way affects either their or subsequent conclusions by Dempster and Lerner (1947) and Lerner (1950).

regarding the possible role of polygenic or other form of balance for egg production are analogous to those carried out for shank length. However, instead of the ratio between the realised and the expected selection differentials, it is more illuminating to compute the difference between them. As shown in table 5 the birds with higher egg

TABLE 5

*Selection differentials for egg production*

| Year of hatch of dams | Realised individual selection differential | Realised minus expected individual selection differential | | Realised family selection differential | Realised minus expected family selection differential |
|---|---|---|---|---|---|
| | | Total | Within sires | | |
| 1932 | 126·9 | 1·5 | −0·2 | 33·1 | −4·3 |
| 1933 | 126·1 | 2·5 | 4·5 | 49·5 | −1·5 |
| 1934 | 91·2 | 0·5 | 0·9 | 36·3 | 2·0 |
| 1935 | 89·4 | 0·6 | 0·8 | 29·6 | −0·3 |
| 1936 | 78·5 | −2·4 | −2·6 | 28·4 | −1·2 |
| 1937 | 93·5 | −0·2 | −1·1 | 39·4 | 3·5 |
| 1938 | 119·0 | −3·3 | −2·9 | 36·2 | −0·6 |
| 1939 | 99·4 | −0·4 | 0·9 | 52·4 | −1·7 |
| 1940 | 92·4 | 3·5 | 0·4 | 51·3 | 2·9 |
| 1941 | 77·0 | 0·9 | 2·1 | 39·9 | 0·8 |
| 1942 | 73·5 | −0·8 | −2·0 | 28·5 | −1·5 |
| 1943 | 79·2 | 1·4 | 0·6 | 28·4 | 2·7 |
| 1944 | 66·5 | 1·3 | 2·3 | 23·3 | −1·4 |
| 1945 | 80·4 | 0·3 | 3·1 | 33·9 | 1·3 |
| 1946 | 79·4 | 1·0 | 1·6 | 24·1 | −0·2 |
| 1932-1940 * | 99·7 | 0·22 | 0·13 | 39·0 | −0·37 |
| 1938-1946 * | 82·9 | 0·66 | 1·27 | 35·6 | 0·31 |

* Weighted according to number of offspring.

production actually produced somewhat more female offspring entering laying pens, since the realised selection differential exceeded the expected one by 0·22 eggs in the first, and by 0·66 eggs in the second period considered. Elimination of the possible effects of the sires on these figures, by computing the selection differentials within sires, leaves this conclusion unchanged, the excess of the realised over the expected selection differential being 0·13 and 1·27 eggs for the two respective periods.

Similar considerations of the family selection differentials confirms the lack of any balance mechanism. Undoubtedly, neither the figure of −0·37 eggs for the first period nor that of 0·31 eggs for the second for the differences between the realised and the expected differentials deviate significantly from zero. Furthermore, the change in the direction of the difference is opposite to the one expected should the balance mechanism have been involved in the reduction of obtained gains.

There seems to be but little doubt then that, in the case of selection for egg production, attrition of artificial selection pressure by natural selection for fitness has not played any significant role in this flock up to this time. It seems premature at this stage to enter speculations regarding the possible role of non-additive gene action in limiting the rate of genetic progress in the production index.

Before proceeding with a general discussion, comparable figures for two other economically important characters may be given. The traits considered are egg weight and sexual maturity, and from the data given in table 6 it should be apparent that once more there is

TABLE 6

*Selection differentials for egg weight and sexual maturity*

| Year of hatch of dams | March egg weight | | | Days to sexual maturity | | |
|---|---|---|---|---|---|---|
| | Mean of selected dams, gms. | Realised minus expected selection differential | | Mean of selected dams days | Realised minus expected selection differential | |
| | | Total | Within sires | | Total | Within sires |
| 1932 | 56·3 | −0·6 | −0·6 | 188·7 | 1·0 | 1·3 |
| 1933 | 57·5 | −0·1 | −0·4 | 184·2 | 0·9 | −0·1 |
| 1934 | 57·6 | −0·1 | −0·3 | 181·8 | −2·5 | −2·7 |
| 1935 | 56·9 | −0·1 | 0·0 | 175·3 | 0·1 | −0·2 |
| 1936 | 56·1 | 0·1 | 0·0 | 170·4 | 3·6 | 2·7 |
| 1937 | 57·6 | 0·3 | 0·4 | 167·6 | 0·3 | 0·9 |
| 1938 | 57·5 | 0·2 | −0·2 | 166·9 | −3·3 | −3·2 |
| 1939 | 57·7 | 0·0 | −0·8 | 170·4 | 0·9 | 0·4 |
| 1940 | 58·2 | −0·5 | −0·6 | 168·4 | 1·3 | 2·1 |
| 1941 | 56·3 | −0·6 | −0·3 | 169·9 | −1·1 | −1·1 |
| 1942 | 55·7 | −0·3 | −0·2 | 162·2 | 1·0 | 1·1 |
| 1943 | 57·9 | 0·0 | −0·1 | 177·9 | −0·1 | −0·6 |
| 1944 | 55·6 | −0·2 | 0·2 | 173·2 | 0·7 | 0·2 |
| 1945 | 56·5 | 0·1 | 0·3 | 157·8 | 1·1 | 0·8 |
| 1946 | 55·3 | −0·3 | 0·4 | 168·2 | −0·8 | −0·2 |
| 1932-1946 * | ... | −0·17 | −0·21 | ... | 0·26 | 0·11 |

* Weighted according to number of offspring.

no evidence that the balance between each of the traits and fitness plays a major role. It is, however, possible that here the balance may exist between these traits and egg production. Thus the selection pressure applied to the flock was primarily based on the number of eggs and viability. If birds laying too small or too large eggs do not lay or live as well as those with an intermediate size, the artificial selection pressure for egg weight may not have been unidirectionally applied and on the average was close to zero. The case of sexual maturity may fall in the same category, especially in view of the non-linear correlation between it and egg number observed by Knox (1930).

## GENERAL CONSIDERATIONS AND SUMMARY

It is an attractive hypothesis, and one that has met with a degree of success, that the construction of efficient selection indexes and the rough estimation of the rates of improvement to be achieved by selection of measurable characters of economic importance in domestic animals can be based on the postulates that the genetic variance is in large part additive in nature, and that such non-additive genetic variance as may exist has the same significance with respect to gains from selection as environmental variations from individual to individual. Although this simple model may be adequate in the early stages of a selection programme, it is certainly conceivable that the genetic and environmental relationships could become more complicated after sufficiently prolonged selection. Some theoretical possibilities are that the additive variance will approach exhaustion, that non-additive variation may increase to some extent, that it may no longer be neutral in its effects but may necessitate a positive selection intensity to prevent retrogression, and that negative correlations may appear between genotypes for the selected characters including components of fitness which, by definition, are subject to natural selection. All these possible hindrances may be studied on the genetic level in terms of variation and correlation; in some cases no doubt they may be effectively approached in terms of physiological limits and incompatibilities. Certainly, many attacks from many different standpoints will be required for a general understanding of the problems raised. The present paper attempts an analysis of some of the factors that may lead to a lowered response to selection in two poultry experiments of many years' duration.

It is shown above that selection for long shanks in a small population of White Leghorns over a span of eleven years (equivalent to eleven generations as far as the pullet and cockerel selections are concerned) led to a considerable increase in shank length during the first half and apparent cessation of gains during the latter half of the period. During this time there was no net decrease of phenotypic variability of shank length, nor are the estimates of heritability and of genetic variance, based on half and full sib correlations, any lower during the last half of the experiment than during the early period. Two possible reasons for the cessation of gains suggest themselves, namely, a change in the nature of the genetic variance and a counteracting tendency of natural selection. It is planned to test the first of these possibilities, which cannot be adequately evaluated from present data, in future experiments. The second possibility has been shown to exist to some degree. The data and the computations indicate that nearly one-half of the rate of gain achieved in the first period of selection may be counteracted in the second period by a negative correlation between genotypes of the dams for shank length and hatchability of their eggs, and that selective elimination among the offspring of individual dams might lead to some further reduction which, however, is probably considerably less than one-fifth of the early rate of gain. Some of the postulates on which the calculations are based may correspond only in a rough way with the actual situation. For example, it is unlikely that the regression of genotype for shank length on number of offspring of dams is strictly linear; it is entirely possible that the highest genotypes for shank length produce very few offspring and that the average and lower genotypes found among the dams produce approximately equal numbers. A non-linearity of this general type could result in greater attrition of gains than is indicated on the basis of the postulates adopted for computation. Due to uncertainties of this type, as well as to sampling errors, it cannot be positively affirmed that natural selection for shorter shanks does not account *in toto* for the cessation of gains; on the other hand, the data show with a high degree of probability that adverse natural selection is at least in part responsible. It is reasonable to suppose that the negative correlation between genotypes for fitness and for shank length has been caused by the prolonged selection for both, a possibility that has been predicted from theoretical considerations by a number of authors (*see*, for example, Lerner, 1950), and shown to occur to a high degree in the long-term selection experiments with *Drosophila* by Mather and Harrison (1949).

The selection for production index has continued for a longer period of time, although the number of generations involved is not necessarily different from that of the shank length experiment, largely because pullet selection was not used in the former. The size of the production flock is much greater and the inbreeding has been less as has also been the intensity of genetic selection due to the much lower heritability of production index as compared to shank length. In some ways, therefore, the production index experiment, may be considered to be in a somewhat earlier stage than the shank length experiment, although the possible physiological limit of an egg a day is approached by some birds, whereas no obvious physiological limit to shank length is as yet apparent. Although there are some indications that the rate of increase in production index in the later years was less than in the early years (or possibly gain has ceased entirely), the magnitude of the inter-seasonal variation precludes a definite judgment on this point. On the other hand, the dams with a high production index have as many offspring as (in fact, slightly more than) dams with a lower production index; such change in this respect as the data indicate, although probably statistically insignificant, is in the direction of an increased positive correlation between genotypes for production index and fitness. The notable difference between the two experiments in correlation between fitness and the character selected for may be due, in part at least, to the fact that production index includes one aspect of fitness, so that the component that these two characters have in common would favour a positive correlation between them. In addition, there has been some direct selection for number of offspring produced by a dam. However, neither the common component of fitness and production index, nor the direct selection for number of offspring, could preclude a *decrease* in correlation on continued selection. *A priori* it would be expected that some negative correlation should have been induced between production index and some aspect of fitness that does not directly enter into the production index, as, for example, the proportion of fertile eggs that hatch. It is planned to present an analysis of this relationship elsewhere.

[*Editor's Note:* Material has been omitted at this point.]

## REFERENCES

DEMPSTER, E. R., AND I. M. LERNER. 1947. The optimum structure of breeding flocks. *Genetics, 32,* 555-579.

DEMPSTER, E. R., AND I. M. LERNER. 1949. Selection problems in animal breeding. *Proc. Berkeley Symp. Math. Stat. Prob.,* pp. 481-483.

HAZEL, L. N., AND W. F. LAMOREUX. 1947. Heritability, maternal effects and nicking in relation to sexual maturity and body weight in White Leghorns. *Poultry Sci., 26,* 508-514.

KNOX, C. W. 1930. Factors influencing egg production. *Iowa Agr. Exp. Sta. Res. Bull., 119,* 311-332.

LERNER, I. M. 1943. Inheritance of size in Single-Comb White Leghorns. *J. Agr. Res., 67,* 447-457.

LERNER, I. M. 1950. *Population Genetics and Animal Improvement.* xviii+342 pp. Cambridge.

LERNER, I. M., AND CRUDEN, D. 1951. The heritability of egg weight : the advantages of mass selection and of early measurements. *Poultry Sci., 30,* (34-41).

LERNER, I. M., AND HAZEL, L. N. 1947. Population genetics of a poultry flock under artificial selection. *Genetics, 32,* 325-339.

LUSH, J. L. 1945. *Animal Breeding Plans.* viii+443 pp. Ames, Iowa : Collegiate Press.

MATHER, K. 1941. Variation and selection of polygenic characters. *J. Genet., 41,* 159-193.

MATHER, K. 1942. The balance of polygenic combinations. *J. Genet., 43,* 309-335.

MATHER, K. 1943. Polygenic inheritance and natural selection. *Biol. Rev., 18,* 32-64.

MATHER, K., AND HARRISON, B. J. 1949. The manifold effect of selection. *Heredity, 3,* 1-52, 131-162.

SHOFFNER, R. N. 1948. The variation within an inbred line of S.C.W. Leghorns. *Poultry Sci., 27,* 235-236.

# SELECTION RESPONSE AND THE PROPERTIES OF GENETIC VARIATION

FORBES W. ROBERTSON[1]

Institute of Animal Genetics, Edinburgh, Scotland

## Introduction

Body size in *Drosophila* provides suitable material for studying many of the central problems of quantitative inheritance. The distribution of size of wild stocks, reared under standard conditions, is continuous, while variable conditions, especially of nutrition, may cause striking variation. Unrelated wild stocks are similar in size, implying the existence of adaptive equilibrium; inbreeding causes a decline while crosses between inbred lines show heterosis. Naturally a variety of experimental methods must be used in the study of these problems and among these selection offers an especially fruitful line of enquiry. Thus the nature of the response to selection in either direction, the limits attained, the degree of stability at different levels, the consistency of response to parallel selection in different stocks, changes in variability, correlated change in other characters and so forth, provide data which are basic for further study of the inheritance of size. But the results of suitable selection experiments have a significance which extends beyond the confines of the particular material used. Thus comparison of response in different characters draws attention to possible relations between their genetic behaviour and their respective roles in the economy of the organism. Comparison of data from unrelated forms, for example, *Drosophila*, mice and poultry, broadens our theoretical basis and should lead us to more confident generalisations than we can attain at present.

The present account describes work carried out by Dr. E. C. R. Reeve and myself and deals chiefly with the results of two-way selection for body size in *Drosophila melanogaster*. A number of other methods have been used in our general study of the inheritance of size, especially the analysis of the effects of exchanging one or more chromosomes between lines which differ greatly in size or show heterosis when crossed. These experiments are too complex to describe in detail, but they will be referred to where they provide evidence which bears on the interpretation of selection results.

## Material

All the experiments described here have been carried out on stocks descended from individual, wild, impregnated females, caught at widely separated localities, and kept running in the laboratory in bottle populations. For purposes of experiment the flies are reared under favourable conditions of abundant food supply, while variable density of larvae in cultures is minimised by standardising the number of eggs which are collected and introduced into the cultures. To give an idea of the magnitude of environmental fluctuation within cultures, we may note that crosses between inbred lines, that is, flies of a constant genotype, have a coefficient of variation for size of about one per cent. Body size refers to the length of thorax or wing, recorded on the living fly, which is etherised and mounted on an adjustable platform fitted to the moving stage of a microscope—a quick and accurate method.

In each test or generation of selection a number of replicated cultures are set up and the selected flies are chosen according to their individual deviation from the mean of their own culture. Thus in one series of mass selection experiments 20 pairs of flies from each of five vials were measured, the extreme four pairs chosen from each vial, and combined to make a total of 20 pairs selected from altogether 100 pairs.

Records of phenotypic variation always refer to pooled within-culture effects. From time to time significant differences between cultures are encountered, due presumably to differences in the medium. Variation of this sort, together with minor fluctuations of temperature, contribute to the differences in mean size between successive generations, and tend to lower the precision of estimates of rate of change under selection. However, such difficulties have been largely overcome by rearing a sample of flies from the unselected stock along with the experimental flies, and this enables us to express the response to selection in terms of deviation from this control series. Complete details of medium and methods are given in an earlier publication (Robertson and Reeve, 1952a).

Choice of an appropriate scale with which to describe and analyse the data presents the usual problem. On theoretical grounds a logarithmic or multiplicative scale should be most satisfactory and it does eliminate the differences in variance between males and females. But the changes in thorax length produced by selection are rather small compared with average size, so that the analysis of size differences is not greatly affected by transformation to logarithms. However in the comparison of variance we have used a log scale, namely squared coefficients of variation, so that actual practice is something of a compromise between the competing claims of convenience and theory. Scale problems have been discussed in some detail in an earlier

[1] Member of Agricultural Research Council Scientific Staff.

publication (Robertson and Reeve, 1954), which provides some empirical justification for the ordinary linear scale in the comparison of means.

## THE EFFECTS OF SELECTION

### (1) Consistency between experiments

Since all wild stocks of *Drosophila* examined by us and other workers show great genetic variation, and for detailed work we have to deal with comparatively small populations with all the attendant hazards of sampling, it is natural to wonder how far any particular type of response to selection may be regarded as representative of similar experiments on other samples from the same or different populations. To throw some light on this problem, mass selection for both large and small size has been carried out simultaneously in three unrelated wild stocks, using thorax length as the measure of size. The wild stocks used are known as Crianlarich, Renfrew and Ischia and will be referred to as C, R and I; the first two are descended from flies caught in Scotland and the third is Italian in origin. Selection was continued in either direction until progress ceased, when the direction of selection was reversed. At the same time flies from the selected strains were allowed to mass mate under competitive bottle conditions and measured at intervals. Also, in each strain, selection was reversed after five or six generations of selection. The results are summarised in Figure 1. It should be noted that average thorax length, with sexes combined, of the unselected control stocks is close to 100 units in our scale of measurement (1/100 mm) so that a unit difference approximately corresponds to a one per cent change in thorax length. We shall deal first with the resemblance between the

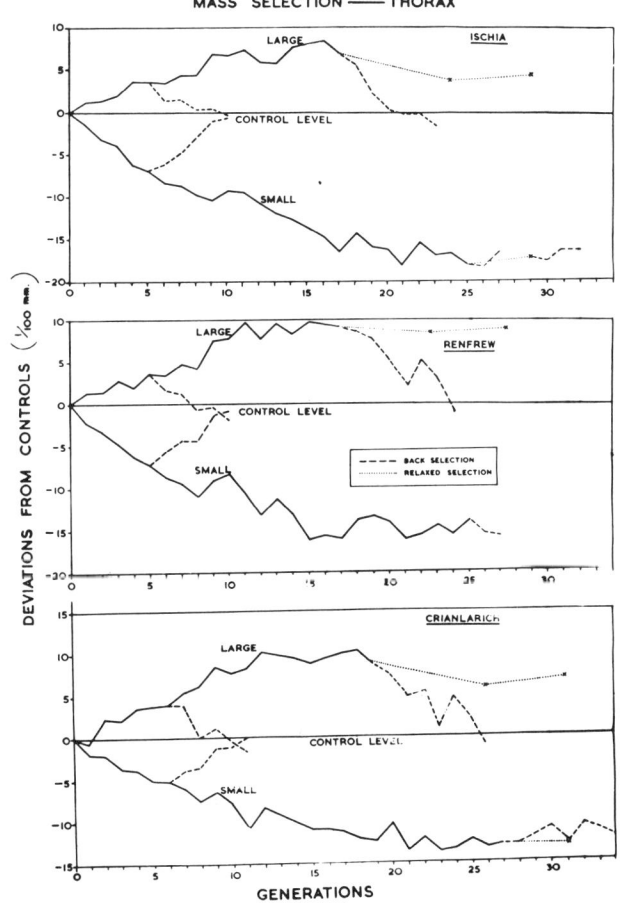

FIGURE 1

strains and then consider the respects in which they differ; the salient features are as follows:

(1) There is an immediate and sustained response to selection in either direction in all three stocks.

(2) The response tends to be asymmetrical since selection for small size produces a greater change than selection for large size. In the R and I series, the change per generation, based on the regression of total response on generation number over the first five generations of selection (Table 1) is about twice as great upward as downward. In the C stocks the response is approximately symmetrical in the early generations, but resembles the others in the longer period during which selection is effective in the direction of small size and in the greater deviation from the unselected stock at which the small strain stabilises.

(3) The actual levels of size at which corresponding strains stabilise are approximately the same, that is, 8-10 units up for the large strains, 14-17 units down for the small.

(4) Parallel strains behave in the same way when selection is reversed. In the first tests, at a comparatively early stage of the experiment, reversal of selection brings size back to the original level

TABLE 1. RATE OF CHANGE PER GENERATION DURING THE FIRST 5 GENERATIONS OF SELECTION (1/100 mm)

| Stock | Large | Small |
|---|---|---|
| C | .89 ± .18 | − .94 ± .10 |
| R | .59 ± .14 | −1.12 ± .09 |
| I | .74 ± .11 | −1.43 ± .09 |

in both large and small strains. There is perhaps a tendency for the large strains to return to the original level more quickly than they moved away from it. Also the reversed lines from large and small strains meet below the level of the unselected stocks, while selection up from the small strains goes a little faster than the initial selection for large size from the unselected stocks. At a later stage, after selection progress has ceased, large and small strains differ strikingly in behaviour. Selection down from the large strains returns size to the original level comparatively quickly; the change per generation is higher than in the earlier back selection tests. The sudden rise in the graph of the C and R responses is probably due to the controls fluctuating out of step with the selection lines. Evidently the failure to advance under selection is not due to loss of genetic variability since we can reduce size so easily by selection.

In the small strains, however, the situation is quite different since reversal of selection is ineffective. The different small strains behave as if they were homozygous, and when inbred lines were taken off the small strains, they did not change in size.

(5) Relaxation of selection in the large strains, after progress had ceased, was not accompanied by a return to the original level. In the R strain there was no evidence of change in size, while C and I showed a moderate decline. As might be expected, the small strains did not change under relaxed selection.

Thus, when quite unrelated stocks are selected under similar conditions, they respond to selection in much the same way—a valuable clue when we have to consider the general properties of the inheritance of size. These, of course, are not the only experiments which have been carried out. Thus, similar evidence of asymmetry is provided by a mass selection experiment carried out somewhat earlier on the Nettlebed stock. The selection procedure was identical except that wing, not thorax, length was the dimension selected. The responses to selection, are shown in Figure 2. The first test ran for 14 generations, and then a fresh sample was chosen from the Nettlebed stock and the experiment repeated for seven generations. The total difference between the selected strains after seven generations of selection was almost exactly the same in the two tests. In the first test selection was relaxed in the up and down lines after seven generations of selection and the flies mass mated under the favourable conditions of experimental culture; both relaxed lines showed little evidence of change in size.

We have also performed a number of experiments involving selection for long and short wing

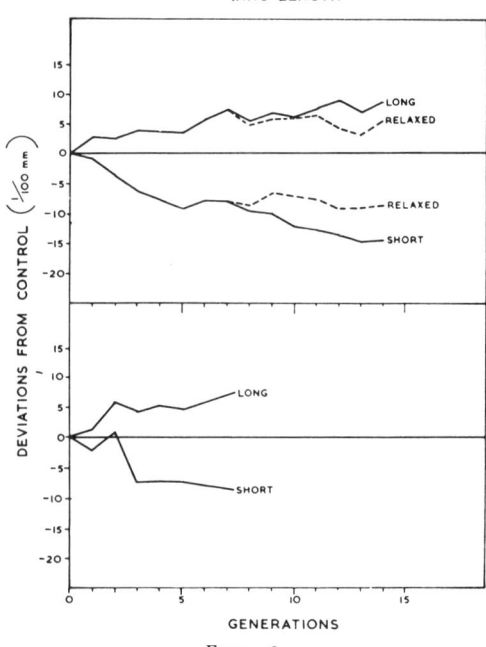

FIGURE 2

length on both the Nettlebed and Edinburgh stocks (Robertson and Reeve, 1952a). The detailed procedure differed from that applied in the later mass selection experiments, since the number of parents used per generation was smaller and the selection intensity was relatively greater. These selected lines showed less consistency in their type of response, and it appeared, at the time of these earlier experiments, as if there was not much evidence for regularity in the effects of selection. However with increasing experience and a greater variety of data to refer to, we can now fit the earlier results into the general pattern. It is likely that inbreeding effects and the smaller size of the foundation population are chiefly responsible for the more atypical types of response. However, these earlier experiments focussed attention on the importance of gene interaction, and a particularly interesting example of stability under selection was subjected to detailed analysis (Reeve and Robertson, 1953). This refers to a large strain, which, over a period of many generations, failed to respond to selection although reversal of selection quickly reduced size. Parent offspring correlations exceeded the value found in the unselected stock. It was concluded that correlated changes in fitness could not account for the behaviour of the line and it was necessary to seek an explanation in terms of gene behaviour. There are many points of resemblance between this case and the large strains in the mass selection experiments, although the latter have not been studied in the same detail.

*(2) Response and selection differential*

We may now return to the mass selection experiments and consider the nature of the responses to selection in more detail especially in relation to the intensity of selection. We can describe the situation either in terms of heritability, that is, the ratio of average change per generation to the average selection differential, or we can use the slightly different procedure proposed by Woolf and described by Falconer (1953). Here we calculate the regression of total response to selection for successive generations on the cumulated selection differential preceding any given generation to obtain what Falconer calls the "realised heritability." In the present case the general picture is much the same whichever measure we choose (Table 2). Now the selection differential is obviously related to the phenotypic variability; in general the more variable the sample the greater the differential for a constant number of parents per generation. Hence we must first look for any evidence of changes in variability. Figure 3 shows the level of the coefficient of variation throughout the selection experiment. The three unselected stocks, which are very similar in size, have almost the same coefficient of variation, which is close to two per cent, and since they did not show any systematic changes throughout the experiment, the records from the unselected stocks have been averaged and expressed as a horizontal line in Figure 3, which shows some interesting contrasts in the behaviour of the large and small lines.

In the R and I large strains, the variability fluctuates about the level of the unselected stock, being perhaps a little higher in the early generations of the I strain. In the C strain variability is generally lower than that of the controls. Hence, in the large strains, the changes in response to selection and the attainment of a selection limit have occurred without any appreciable change in the phenotypic variation. The small strains provide quite a contrast to this picture. In the R and L strains there is an immediate rise in variance above the control level; while the C strain shows little change, until the later stages of selection, when there is a dramatic rise in variance. But all three groups agree in that the small strains are more variable than the large —a difference which is established right from the beginning of selection. The distribution of size has been examined in the generations which show a particularly high variance and there is a slight suggestion of asymmetry, with the longer tail pointing in the direction of small size, although such departures from symmetry are unimportant except for the last generations in the small C strain, where the high variance is due to a few particularly small individuals in an otherwise apparently symmetrical

TABLE 2. AVERAGE VARIABILITY AND RESPONSE TO SELECTION OVER THE FIRST 10 GENERATIONS OF SELECTION.

| Stock | $C^2$ | Average selection differential | Response per generation (1/100 mm) | Estimates of heritability 1 | 2 |
|---|---|---|---|---|---|
| | | | Large | | |
| C | 3.2 | 2.23 | .85 ± .07 | .38 ± .03 | .39 ± .04 |
| R | 3.9 | 2.32 | .71 ± .08 | .30 ± .03 | .32 ± .05 |
| I | 4.5 | 2.51 | .63 ± .06 | .25 ± .02 | .25 ± .03 |
| | | | Small | | |
| C | 4.5 | −2.65 | −.73 ± .06 | .28 ± .02 | .27 ± .02 |
| R | 6.1 | −2.96 | −.91 ± .14 | .31 ± .05 | .26 ± .05 |
| I | 6.1 | −2.81 | −1.03 ± .10 | .37 ± .04 | .34 ± .03 |

Estimates of heritability refer (1) — to the ratio of change per generation to average selection differential and (2) — to the "realized heritability."

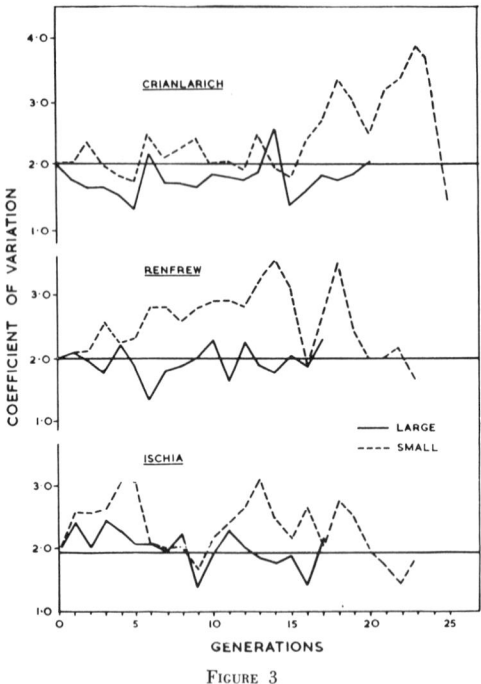

FIGURE 3

distribution. In view of the evidence implying loss of genetic variance in this strain, it is possible that they are due to greater environmental sensitivity rather than genetic segregation.

The results of the first 10 generations of mass selection of C R and I are shown in Table 2. The average response per generation is calculated as the regression of response on generation number and the first estimate of heritability is the ratio of this average response to the average selection differential per generation. Heritability estimate (2) is the "realised heritability" of Falconer referred to above. The two estimates generally differ little. Average phenotypic variability is expressed in terms of squared coefficients of variation. As expected, the selection differentials are relatively greater in the more variable small strains. The strains selected in opposite directions show an interesting contrast with respect to the response per generation and phenotypic variance. Thus in large lines selection response is inversely related to phenotypic variance, so that most response is made by the least variable line. But in the small lines the position is reversed, and the more variable lines show the greater response. Apparently, when selection is for large size, an increase in phenotypic variance is a handicap, whereas in selection for

small size it is an aid. It will be remembered that the three foundation stocks are almost equally variable; evidently selection for respectively large and small size produces qualitatively different effects, in that the latter very quickly leads to increased evidence of genetic segregation. It seems likely that this contrast in the variability of large and small strains is related to the genetic causes of the asymmetrical response to selection.

*(3) Changes in fitness*

Response to selection may be influenced by the properties of the genetic variation which contributes to phenotypic variation and also by changes in fitness which accompany selection. We might wonder whether the latter could account for the failure of large strains to advance under selection although they retain so much heterozygosity, and also whether it could contribute to the asymmetry of selection response. However, records of various components of fitness provide no grounds for attributing differing rates of change in the earlier stages of selection to such a cause. Also when selection was reversed after five to six generations of selection there was no change in viability. In the present experiments we have, as a by-product of the method of culturing a standard number of eggs per culture, a record of the proportion of these eggs which produce adults and the data for successive generations of selection are recorded in Figure 4. The figures quoted cannot be regarded as very reliable since there is considerable fluctuation in successive generations due partly to changes in the medium and partly to variation in the age of female flies from which eggs were collected; older females tend to produce a higher proportion of infertile eggs (Robertson and Sang, 1944). However the general trend is reasonably clear. In the C and R stocks selection in either direction leads to a decline in viability especially in later generations, but the large strains are consistently more viable than the small. In the I experiment there was little evidence of consistent difference between the strains nor of much decline with selection. Hence, when lines selected in opposite directions differ, it is the small line which has the lower viability and which might be expected to encounter the stronger resistance to progress—and yet such lines respond most to selection and apparently reach fixation.

It might be thought that pressure of natural selection would serve as an adequate explanation of the selection limit in the large strains. As we have seen, viability declines as the experiment continues and when selection was reversed after the response had ceased, there was some evidence of a rise in viability. But this explanation encounters serious difficulties. Thus when selection was relaxed and the flies cultured under much more competitive conditions than occur in our experimental cultures, there was no change in the R strain while the others showed only a moderate decline—a result

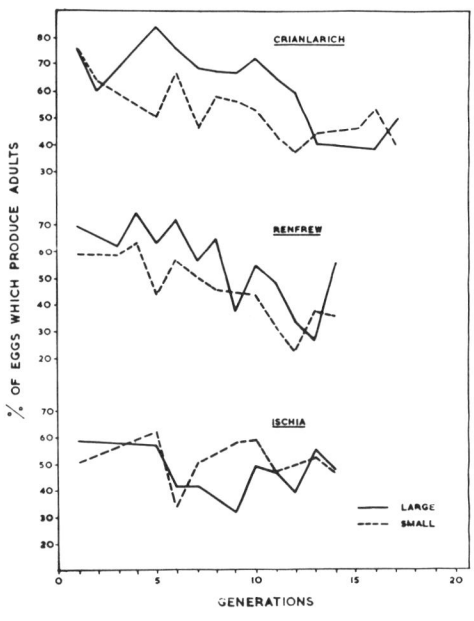

FIGURE 4

which strongly suggests that correlated changes in fitness are not the cause of the behaviour of the large strains. And we have to remember too, the investigation of an apparently similar situation in the line selected for long wings in the Nettlebed stock, referred to earlier. This indicated that the behaviour of the line under selection could not be accounted for in terms of fitness changes. Hence, it appears likely, that although fitness changes may be involved in the response to selection, they are not primarily responsible for the salient contrasts of two-way selection in the present experiments.

These considerations are relevant to the effects of reversing selection in the large strains after a number of generations of stability under selection. Here the rate of return to the unselected level is particularly high (Table 3). Phenotypic variability rises above the level found in the forward selected strains; the comparisons are based on the average of estimates for the last five generations of the latter and the average for the generations of reversed selection. There is evidence of the same tendency which has been noted earlier in the selec-

TABLE 3. EFFECTS OF REVERSING SELECTION IN THE LARGE STRAINS

| Stock | $C^2$ | Response | Heritability |
|---|---|---|---|
| C | 4.8 | $-1.00 \pm .26$ | $.38 \pm .10$ |
| R | 6.3 | $-1.19 \pm .27$ | $.35 \pm .08$ |
| I | 6.7 | $-1.42 \pm .18$ | $.51 \pm .06$ |

tion for small size, namely the response tends to be greatest in the strain which shows the highest phenotypic variance, while the three strains fall into the same order as previously with respect to their level of variability. Although the particularly high rate of change per generation might be attributed to fitness changes which tend to re-enforce the direction of selection, the parallelism between this present instance and the original selection for small size, in which fitness changes were almost certainly unimportant, re-enforces the argument in favour of explanations in terms of particular types of gene effects and interactions as a possible clue to the characteristic responses to selection for body size.

*(4) Selection from crosses*

When mass selection in either direction reached a level beyond which it was difficult, if not impossible, to proceed, strains selected in the same direction were inter-crossed and the selection continued from the $F_2$. Dealing with large strains first, Figure 5 shows that all crosses show heterosis in the $F_1$, a decline in the $F_2$, followed by a response to selection which takes size well beyond that of the cross between the parent lines. Indeed the new lines advance beyond the average size of the parent strains by about 50 per cent of the deviation of the latter from the control stocks. The success with which crossing, followed by continued selection, takes the selected character well beyond the original selection limits may well have practical

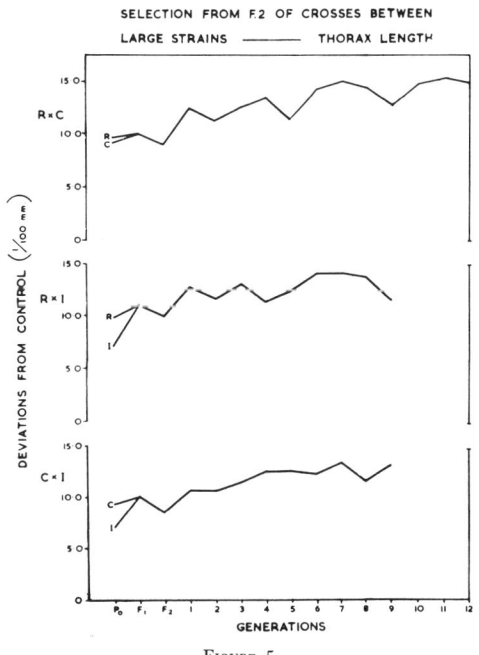

FIGURE 5

implications. Selection downward from the $F_2$ between small strains offers some contrasts (Fig. 6). There is heterosis in the $F_1$, sufficiently striking in the crosses with the I strain to take size about one third the way back to the original level. Only in the $F_2$ of the C × I cross is there a decline in size; the others show either no change or an increase in size. Also, selection is ineffective at first. Thus the progeny of flies selected from the $F_2$ are either as big as or bigger than their parents. Thereafter however, selection becomes effective and leads apparently to stability after five or six generations; the line selected from the C × R cross had to be discontinued after six generations due to contamination. It is worth noting that in the lines selected from crosses to the I small strain, stability is apparently reached at a level which exceeds that of the smaller parent strain. Thus, unlike selection for large size, continued selection from crosses between small strains has not succeeded in increasing the total response to selection, and encounters initial difficulty in making progress from the $F_2$.

Such differences in relation to the direction of selection, especially in the early stages, are reflected in the variability of the large and small strains (Fig. 7). In the former there is little evidence of any systematic change in variance during the experiment and the value fluctuates about the average for the unselected stocks of two per cent. Only in the cross between the I and C strains is there evidence of a decline in $F_1$ variance below that of parents; there is little difference between parents and progeny in the other cases. In the small strains the position is rather different. The $F_1$ variability shows a striking decline below the level of the parents, recalling the behaviour of crosses between

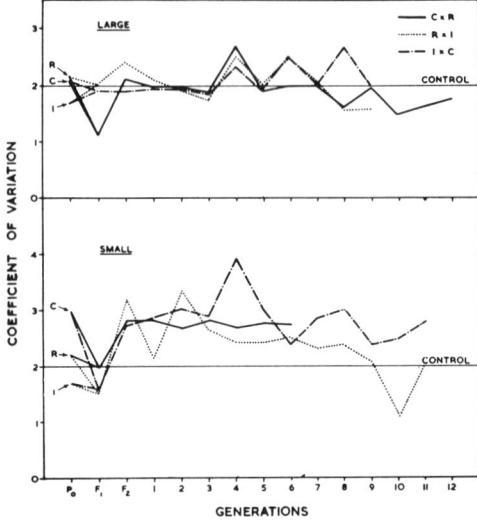

FIGURE 7

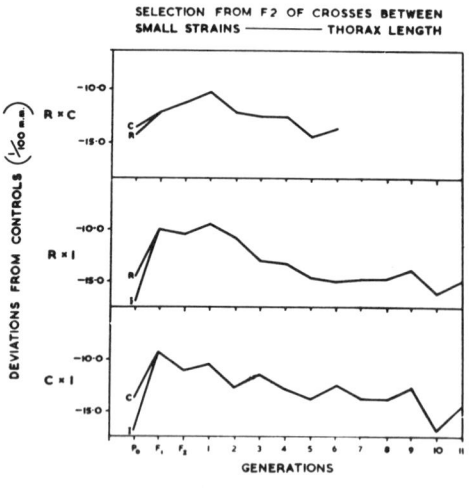

FIGURE 6

inbred lines which almost invariably show a reduction of variance of size, due presumably to the greater resistance to environmental variation of the more heterozygous types (Robertson and Reeve, 1952b; Reeve and Robertson, 1953b). In the $F_2$, there is good evidence of genetic segregation in the jump in coefficient of variation to a value of about three per cent. The variability remains high throughout the experiment, except that in the R × I line there is some evidence of a decline in later generations. Thus we find here further examples of what appears to be a general rule, namely that selection for small size is accompanied by higher phenotypic variance than occurs with selection for large size, and there are strong indications that this high variance is attributable to genetic segregation.

The interesting demonstration that selection for large size can make a good deal further progress when stable lines are crossed is supported by parallel evidence from earlier experiments on the Nettlebed stock. Here two lines, which had long ceased to respond to selection for respectively long wing and long thorax, were crossed and the $F_2$ split into two groups, in which either large wing or long thorax selection was carried out. The results are shown in Figure 8 for thorax length. As in the mass selection experiments, there is heterosis in the $F_1$, a decline in the $F_2$ followed by a striking response, which was comparatively modest in the line selected for long wings, but very striking in the long thorax line, which continued steadily for about six generations before falling off, comparatively suddenly, to be followed by a period of sta-

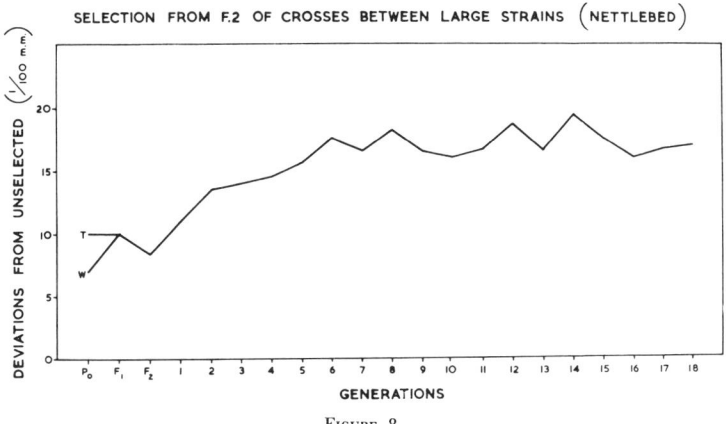

FIGURE 8

bility during which other tests demonstrated that the line remained heterozygous. Thus, there appears little doubt that the intercrossing of large lines enables us to increase body size up to a level considerably higher than would otherwise be the case.

### INTERPRETATION

In general these experiments demonstrate a fair degree of consistency in the effects of selecting for large and small size in wild stocks. Among the more regular features we may note the immediate response to selection in either direction; the tendency to asymmetry with the greater response downward; the unstable equilibrium in large strains which can often be brought down to the original level by reversing selection; the eventual complete stability of small strains; the higher phenotypic variability which developed during selection for small size; the inverse relation between response and phenotypic variance in different large lines compared with the positive relation between them in small strains; the heterosis in crosses between large and also between small strains, in which it is more striking; the further increase in size caused by selection from crosses between large strains which have reached a limit; the initial failure to make progress in down selection from crosses between small strains; and finally the inference that these characteristic responses to selection are to be interpreted chiefly in terms of gene behaviour rather than correlated changes in fitness.

Demonstration of consistency in response to selection is a first step to understanding the underlying genetic situation. By its very nature, however, selection is ill fitted to discriminate between alternative genetic mechanisms which have to be examined by other methods. Fortunately we can draw on evidence from a number of other experiments designed to throw light on the general properties of the inheritance of size. But before dealing with this, a few comments on the "character" body size seem called for. Different wild stocks of *Drosophila melanogaster* have about the same body size, which may be regarded as part of the adaptive response to environmental conditions. Hence selection away from the normal level is likely to lower adaptation, but it does not follow that movement in either direction will be equally disadvantageous. Thus inbreeding leads to a decline in size and there are strong indications that the genetic phenomena associated with selection for small size and inbreeding have a great deal in common and present a contrast to the effects of selection for large size. Hence in studying the reasons for the responses to selection we are at the same time dealing with aspects of heterosis.

The more general evidence on these problems is derived from an analysis of the effects of exchanging chromosomes between lines inbred without selection or descended from selected strains. With the aid of dominant marked inversions and methods of backcrossing which minimise error due to recombination, many different genotypes have been prepared and the comparison of their size and particular genetic constitution enables us to test various hypotheses about gene and chromosome behaviour (Robertson and Reeve, 1953; Robertson 1954; Robertson and Reeve, 1955). Among the principal findings we may note the following:

(1) The exchange of chromosomes between lines reveals widespread interaction between non-homologous chromosomes and a non-additive relation between homologues. Since this type of analysis, in which we deal with whole chromosomes, must grossly under-estimate the extent and variety of gene interaction, it is unwise to take apparently additive chromosome effects at their face value. The demonstration of widespread gene interaction falls into line with comparable evidence, relating to various aspects of fitness in wild or laboratory pop-

ulations, which has been presented by other workers (Dobzhansky and Wallace, 1953; Brncic, 1954; Cordeiro and Dobzhansky, 1954; Levene, Pavlovsky and Dobzhansky, 1954).

(2) It is impossible to account generally for the heterosis in crosses between lines simply in terms of the sum of the dominance effects of individual chromosomes nor of over-dominance relations between homologous pairs, and gene interaction has to be recognised as the essential clue to both heterosis and inbreeding decline.

(3) The gene interaction detected in these experiments is not of a purely haphazard nature and certain general features are apparent. Thus gene combinations which favour more normal size tend to be epistatic to those which favour smaller size to that in heterozygous combinations, a set of genes from a more normal line may more or less completely suppress the expression of gene combinations which favour smaller size. The term epistasis as used here is more akin to the original Batesonian sense of the term than the current one, met with in statistical studies, where it is more or less synonymous with interaction between non-homologous genes. The substitution in one line of a single chromosome from another, that is, the presence of a single pair of heterozygous chromosomes, may do much to overcome the effects of inbreeding, and examples have been encountered in which a single substitution, in an otherwise homozygous background, has increased size up to the level of the cross between the parent lines.

(4) The evidence for this sort of epistasis is greatest in the effects of chromosome substitutions in homozygous backgrounds, especially of the smaller lines, and becomes less with an increase in the level of heterozygosity and is least evident in the exchange of chromosomes between large lines.

(5) Environmental variation is not constant for different genotypes and tends to be greatest in homozygous lines and least in crosses between them. Increases in the level of heterozygosity due to progressive chromosome substitution are accompanied by a decline in variance. But since the level of size rises, decline in environmental variance and increase in size may be regarded as dual aspects of the substitution of heterozygous for homozygous combinations in this sort of analysis. But it is possible to obtain very striking size differences which are independent of the level of heterozygosity, by selecting in both directions and then inbreeding (Robertson and Reeve, 1955a); so that size is far from being simply a function of the level of heterozygosity, and depends more, in general, on the nature of the gene combinations present.

These experiments, together with the selection data, point to certain general conclusions. The high level of variability found in adapted wild stocks is consistent with stability of response to environmental conditions. This variation is presumably co-adapted (Wallace, 1953) in the sense that mutual compatibility of the array of genotypes arising by segregation is favoured by natural selection. When this situation is interfered with by inbreeding or selection, not only are we almost bound to lower fitness but we may, at the same time, bring to light properties of the genetic variation, which can hardly be demonstrated on the original stocks, but which nevertheless give us some insight into the genetic basis of their stability of phenotype.

Our chromosome analysis and other sources of evidence, notably the studies of variation within wild populations by the workers noted above, emphasise the conditional nature of gene effects. In general, the effects of segregation at a given locus on a character like size or viability is determined by the genetic background and hence attempts to describe phenomena in terms of additive genes, dominance and gene frequency and so forth represent a considerable act of faith, unless we can be quite certain that, within the array of genotypes under study, such segregation effects are comparatively unchanged. We have eventually to determine the conditions and situations in which static models are relevant. However certain general features may be suggested. In the normal, unselected wild stocks, under favourable conditions, individual gene effects both within and between loci appear more additive with respect to size, than when inbreeding occurs. In the outbred stock, under our environmental conditions, it appears as if there were extensive duplication of gene function. This does not necessarily imply, of course, comparable similarity in primary effects, but may be attributable to the number of steps intervening between primary action and the phenotypic changes we measure. Inbreeding leads to a loss of genes which can play a similar role with respect to size, so that remaining genes of this sort develop greater segregation effects. Also interaction between different loci will become more important since it will matter more which particular genes are present together, and, in general such interaction will become increasingly important as inbreeding proceeds. At the same time, gene-environment interactions are likely to become more evident, since it will also matter more which particular environmental conditions are experienced. The sort of dynamic biochemical system which could be visualised as responsible for such different kinds of gene behaviour have been considered by Wright (1941) and more recently by Stone, Alexander and Clayton (1954), while Waddington (1952) has discussed these problems in relation to the canalisation of development. Evidently we need deeper analysis of the properties of gene interaction in relation to particular characters and particular environmental conditions if we wish to interpret gene behaviour in suitable biological terms.

Inbreeding leads to a reduction in size and it is reasonable to regard this as a symptom of the reduced metabolic and assimilatory efficiency of the more inbred organism. Now selection for small size appears largely equivalent to a directed form of inbreeding in which we systematically reject or break up gene combinations which tend to cover up potential genetic tendencies for small size. Selection generally leads to an increase in variance which appears to be largely due to the increased effects of genetic segregation and this constitutes an aid to selection progress. It is possible that linkage plays a part here, since selection may build up groups of linked genes which tend to segregate as a unit (Mather, 1954). We might expect such effects to be relatively more important in a species like Drosophila melanogaster, than in others where recombination is higher. However, although we cannot be certain as to the precise origin of the increased variance, the observed situation is quite compatible with the evidence from chromosome assay tests which suggested that gene combinations which favour smaller size tend to be hypostatic to those which favour more normal size, Although selection may depend, in the earlier stages, on more or less additive effects, continued progress is likely to rely to increasing extent on the rejection of epistatic genes or gene combinations which are brought into greater prominence. And there is no reason why the process should not continue until fixation is reached, as appears to be the case in all the small strains which have been created. As long as the strain retains a fair level of heterozygosity, segregation in each generation will tend to offset the effects of selection, since selected parents will differ somewhat in genetic constitution and contribute different epistatic genes or gene combinations to their progeny. The clearest evidence of this is seen in the initial ineffectiveness of selection from the $F_2$ of the crosses between small strains.

Selection for large size introduces a new phenomenon, since it carries us well above the adapted optimum for size, while presumably reducing the general adaptive level. There are obviously many gene combinations which are able to produce larger size, but these are not normally selected for. The chief problem concerns the fall-off in selection progress when heterozygosity remains high and it is possible to bring size back to the original level by reversing selection. Since we appear to be selecting heterozygotes, the question turns on whether we are doing so because of their joint effects in increasing size or because correlated changes in viability limit further progress. As we have pointed out earlier, it is improbable that the latter provides the main key to the situation. If some sort of over-dominance accounts for the behaviour of the large strains when they reach an upper limit, then this form of gene interaction contributes more to the phenotypic variance than was originally the case in the unselected stock, although the total variance remains much the same. As the general level of heterozygosity declines, due to the combination of selection pressure and comparatively small population size, the relative segregation effects of remaining heterozygous combinations may increase so that they dominate the situation. The evidence from the chromosome assay experiments could be reconciled with this since effects of particular heterozygous combinations increase as the background becomes more homozygous. Or it could be maintained that the particular sort of genetic changes produced by selection is a more important reason for the increased importance of heterozygous combinations. But, whatever explanation is preferred, it is likely that particular overdominance relations are conditional on the genetic situation in the strain. And this is consistent with the effects of selection from the $F_2$ of crosses between large lines which leads to a substantial increase beyond the level of the $F_1$ of the cross. Here too there is further evidence of genetic variability when selection response has ceased due probably to the emergence of different heterozygous combinations which mark the end of selection progress.

Thus the responses to selection for body size present a reasonably intelligible picture in terms of the properties of the genetic variation which have been detected in more analytical experiments, and these in turn are clearly related to the nature of the "character" we are dealing with. The increasing evidence of gene interaction as size declines, whether by inbreeding or selection, suggests progressive interference with the dynamic processes which are responsible for the normal phenotype. In the wild stock the integrity of the developmental system which underlies normal development is sustained by a margin of safety, of which we have some indication in the epistatic gene combinations which are stripped away or broken up by selection for small size. If this is so then these effects provide a suitable point of attack for further study.

Selection for large size suggests changes of rather a different sort probably because larger size is only possible if the processes which control growth and assimilation are not interfered with to the same extent or in the same way and this contrast would appear primarily responsible for the asymmetrical response to selection. The evidence for the selection of heterozygotes in the later stages of selection for large size could be taken as evidence of the difficulty of securing maximum performance without at least some degree of heterozygosity.

It is obvious that metrical characters are not to be regarded as a class with common characteristics about which we can generalise with equanimity. Each must be considered in relation to the physiology of development and the immediacy of its contribution to fitness. In general, reaction to inbreeding and general sensitivity to environmental variation — especially of nutrition — are correlated with relative importance of variation in relation to fitness. In an adapted population such characters are near an upper limit of performance. We might expect asymmetry of response would be greatest in characters like egg production and viability, since, to put it rather crudely, there are many more ways of destroying the integrated response than of improving it. We might guess too, that the sort of gene interaction we have detected in the inheritance of body size would appear even more clearly in characters like egg production and viability, and chromosome assays support that view. It is conceivable, in such cases, that the genetic situation in the wild stock may have something in common with that in the large strains, when selection progress is no longer effective and when there is complete asymmetry in the response to two-way selection. If so, estimates of heritability may be a poor guide to selection responses, since, although progeny tests give high estimates, selection response is nil in one direction—the way we want to go.

Emphasis on the attributes of gene behaviour, as a likely key to many types of selection response, should not be taken to imply that selection limits, due to correlated change in fitness, are not also important, as Lerner (1954) has stressed. Each case has to be examined on its own merits to see which explanation is more relevant. With increasing understanding of the genetic control of different characters and the physiology of development, it may be possible to inter-relate these causes, and assess their relative importance in particular instances.

We have not so far discussed the significance of environmental variation to which body size is particularly sensitive. It must be remembered that all the experiments described here have been carried out under favourable conditions of abundant food supply and also, of course, a certain quality of diet determined by the species of yeast used and the type of medium on which it is grown. Also, although environmental conditions have been kept as uniform as possible, they are certainly heterogeneous both within and between cultures. Recent experiments suggest that such environmental variation cannot be regarded as independent of the expression of genetic differences even in the unselected stock, and probably less so in strains undergoing selection or inbreeding. Hence gene-environment interactions, especially in relation to nutritional variables, may contribute appreciably to the phenotypic variation. Here we enter a field of great subtlety but cardinal importance, since it is likely to reveal the genetic variation in wild stocks in a fresh light. The interrelations between controlled changes in the environment and the behaviour of genetic variation affecting particular characters provides an important experimental approach to the study of genetic variation and the processes which control normal development. And this has to be related to the type of selection pressures which are most important in the variable environment to which the stocks are adapted.

Thus the study and interpretation of the effects of selection open up many avenues of exploration. The nature of the response to selection, the behaviour of the genetic variation in relation to the direction of selection and the part played by the selected character in the economy of the organism, together with the influence of environmental conditions on the expression of genetic variation would appear as aspects of an underlying unity which the experimentalist must seek to expose if we are to understand the origin and significance of the genetic variation with which selection deals.

## References

BRNCIC, D., 1954, Heterosis and the integration of the genotype in geographic populations of *Drosophila willistoni*. Genetics 39: 77-88.

CORDEIRO, A. R., and DOBZHANSKY, TH., 1954, Combining ability of certain chromosomes in *D. willistoni* and invalidation of the wild type concept. Amer. Nat. 88: 75-86.

DOBZHANSKY, TH., and WALLACE, B., 1953, The genetics of homeostasis in *Drosophila*. Proc. Nat. Acad. Sci. Wash. 39: 162-171.

FALCONER, D. S., 1954, Asymmetrical responses in selection experiments. Intern. Union Biol. Sci., Naples, Symp. Genetics of Population Structure.

LERNER, I. M., 1954, Genetic homeostasis. Edinburgh: Oliver & Boyd.

LEVENE, H., PAVLOVSKY, O., and DOBZHANSKY, TH., 1954, Interactions of the adaptive values in polymorphic experimental populations of *Drosophila pseudoobscura*. Evolution 8: 335-349.

MATHER, K., 1954, The genetical units of continuous variation. Proc. IX Intern. Cong. Genet. Part I. Caryologia 6: 106-123.

REEVE, E. C. R., and ROBERTSON, F. W., 1953a, Studies in quantitative inheritance. II. Analysis of a strain of *Drosophila melanogaster* selected for long wings. J. Genet. 51: 276-316.

1953b, The analysis of environmental variability in quantitative inheritance. Nature, Lond. 171: 874-876.

ROBERTSON, F. W., 1954, Studies in quantitative inheritance. V. Chromosome analysis of crosses between selected and unselected lines of different body size in *Drosophila*. J. Genet. 52: 494-520.

ROBERTSON, F. W, and REEVE, E. C. R., 1952a, Studies in quantitative inheritance. I. The effects of selection of wing and thorax length in *Drosophila melanogaster*. J. Genet. 50: 414-448.

1952b, Heterozygosity, environmental variation and heterosis. Nature, Lond. 170: 296.

1953, Studies in quantitative inheritance. IV. The effects of substituting chromosomes from selected strains in different genetic backgrounds in *Drosophila melanogaster*. J. Genet. 51: 586-610.

1955a, Studies in quantitative inheritance. VII. Crosses between strains of different body size in *Drosophila melanogaster*. Z. indukt. Abstamm.- u. Vererblehre 86: 424-438.

1955b, Studies in quantitative inheritance. VIII. Further analysis of heterosis in crosses between inbred lines of *Drosophila melanogaster*. Z. indukt. Abstamm.- u. Vererblehre 86: 439-458.

ROBERTSON, F. W., and SANG, J. H., 1944, The ecological determinants of population growth in a *Drosophila* culture. II. Circumstances affecting egg viability. Proc. Roy. Soc. B, *132*: 277-291.

STONE, W. S., ALEXANDER, H. L., and CLAYTON, F. E., 1954, Heterosis studies with species of *Drosophila* living in small populations. Univ. Texas Publ. *5422*: 272-307.

WADDINGTON, C. H., 1952, Canalization of the development of quantitative characters. Quantitative Inheritance. H.M.S.O.: London.

WALLACE, B., 1953, On coadaptation in *Drosophila*. Amer. Nat. *87*: 343-358.

WRIGHT, S., 1941, The physiology of the gene. Physiol. Rev. *21*: 487-527.

## DISCUSSION

DOBZHANSKY: Americans are sometimes reproached by their transatlantic colleagues for, allegedly, confusing the concepts of "big" with "good" and "great." Yet, not only some Americans but also some European geneticists have contended that an increase in size in hybrids compared with the strains crossed must be called heterosis by definition. The difficulties to which such definitions would lead are shown most clearly by Dr. Robertson's work. When lines of *Drosophila* selected or unselected for size are intercrossed, the hybrids are either larger than the parents, or, at least, as large as the parental strains. And yet, Dr. Robertson shows that large flies are not necessarily the most highly fit flies; high fitness is rather a property most often met with in strains of a certain intermediate size. Here, then, is a remarkably clear case of what may be regarded as luxuriance of hybrids in size, not accompanied by increase in fitness, and, indeed, sometimes accompanied by deterioration of fitness. Why, then, should heterozygosis be usually accompanied by increased body size, even at the expense of fitness? Clearly, here is a problem of developmental genetics, and, moreover, one of the problems the solution of which may lead to a cooperation of developmental and of population geneticists, and even to emergence of a synthesis of these disciplines.

## ERRATA

Page 174, col. 1, line 16 should read: "... which favour smaller size to ...."
Page 174, col. 2, line 9 from the bottom should read: "... behaviour has been considered ...."

# The Genetics of Litter Size in Mice

### D. S. FALCONER
*Agricultural Research Council Unit of Animal Genetics, Institute of Animal Genetics, Edinburgh*

Litter size is a quantitative character of some considerable complexity, and the title of this paper should more properly have been "Some aspects of the genetics of litter size." The complexity arises mainly from the fact that the character belongs partly to the parental generation and partly to the filial generation; that is to say, the number of young born in a litter depends partly on the fertility of the parents—chiefly, as we shall see, that of the female—and partly on the viability of the embryos that will constitute the litter. There is also an interesting maternal effect, but its interest hardly compensates for the difficulties it introduces. For these reasons a complete description of the genetics of litter size is a goal for the future rather than a present achievement.

In this paper I shall give an outline of a series of investigations made by R. C. Roberts, J. C. Bowman, and myself, which were concerned principally with the reactions of litter size to inbreeding and to artificial selection, and with the nature of the changes produced by these two procedures. From the reaction to inbreeding we can learn something about the dominance relations of the genes that influence litter size, and from the response to selection we can determine the proportionate amount of additive genetic variance. The total amount of genetic variance can be discovered only from a comparison of the variances of genetically uniform and genetically heterogeneous groups, and this has not yet been done. In addition to the inbreeding and selection, some studies were also made on an unselected control population, from which information was obtained about the influence of male fertility, the maternal effect, and the parent-offspring correlation. The investigations of the nature of the changes produced by inbreeding and by selection were aimed at discovering the extent to which ovulation rate, implantation rate, and fetal mortality were involved. A preliminary account of the selection experiment was published some time ago (Falconer, '55). Descriptions of the inbreeding experiments will be found in three papers, by Roberts ('60), Bowman and Falconer ('60), and Falconer and Roberts ('60). The remaining work summarized here will be fully described elsewhere.

For the purposes of measurement in all the experiments, litter size was taken to be the number of live young found in the first litters of females aged between 6 and about 9 weeks. All the experiments were done on the same basic stock, known in the laboratory as the J stock. It originated in crosses between several different non-inbred strains and had subsequently been maintained by random mating for some ten generations.

### INBREEDING

*Inbreeding depression.* Litter size, as everybody knows, is reduced by inbreeding. The conclusion to be drawn is that the genes that reduce litter size are on the average recessive to their alleles that increase it. Figure 1 shows the rate of decline found with intense inbreeding, and very little selection, in two experiments. In the first experiment (upper graph) there were thirty lines inbred by full-sib matings with no artificial selection. In the second experiment (lower graph) there were 20 lines inbred by a double-first-cousin mating followed by consecutive full-sib matings. Selection within lines was applied to ten of the lines, but the intensity of selection was low and the rate of decline was not affected. The two groups of lines are not shown separately on the graph. The decline of litter size in both experiments was linear with re-

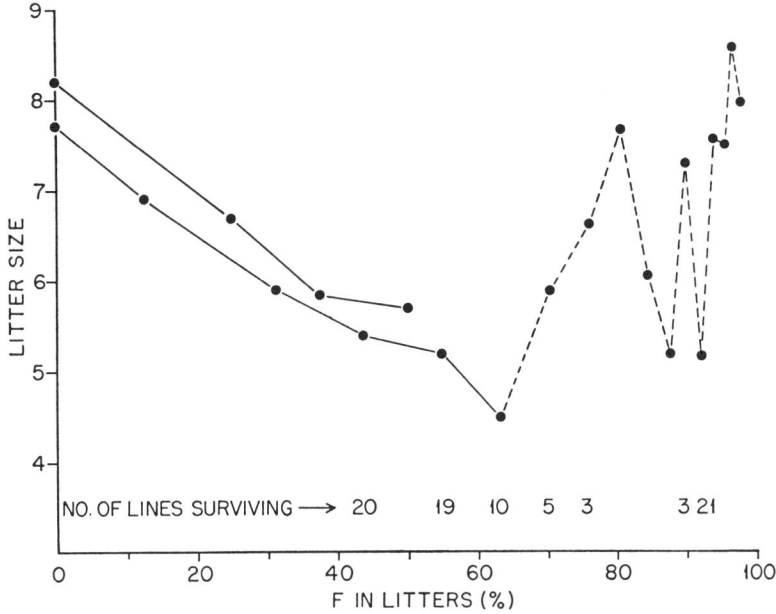

Fig. 1 Reaction of litter size to rapid inbreeding. Mean litter size plotted against the inbreeding coefficient of the litters.

spect to the coefficient of inbreeding, and the rate was 0.49 young per 10% increase of inbreeding in the first experiment, and 0.56 young in the second experiment. If the linear decline had continued indefinitely the litter size would have been reduced to two young at 100% inbreeding. The first experiment was stopped after three sib matings (F = 0.5), and nothing further will be said about it here. The second experiment was continued for as long as the lines survived. Each line was propagated from the first litter of one female. All the offspring in this litter were mated and the line was continued from one of them. Any line became extinct when there was no litter containing at least one of each sex. All of the 20 lines survived to an inbreeding coefficient of 44%. Three-quarters of the lines were lost in the next three generations, and by the time the inbreeding coefficient had reached 76% only three lines survived. The loss of lines resulted in an increase of the mean litter size, because, of course, the surviving lines were those with the higher litter sizes. Two of the three surviving lines dropped out at the eleventh and twelfth generations when the inbreeding coefficient was about 90%. The remaining one out of the original 20 lines survived indefinitely and its mean litter size was equal to, or a little above, the non-inbred controls. (This line, now in its twenty-eighth generation, has the official status of an "inbred strain" and is known as JU.)

The records of the three lines that survived longest showed that these lines were not particularly good ones at the beginning, and their long survival was due to the fact that they did not decline in litter size. This absence of inbreeding depression in some lines, and the fact that one line reached very high levels of inbreeding without any decline of litter size proves, I think, that overdominance cannot have been a major factor in the inbreeding depression of this population. In other words there cannot have been any overdominant locus with more than a trivial effect on litter size. Simple dominance—or deleterious recessives—is a perfectly adequate explanation of the situation. From the practical point of view the results of this experiment show that selection between lines is effective in counteracting the inbreeding depression of litter size, and that if one aims to end up with a certain number of highly inbred lines one must have somewhere about 20 times this number at the start.

*Relative importance of inbreeding in mother and young.* The effect of inbreeding on litter size is complicated by the fact that, under continuous inbreeding, the inbreeding coefficient of the young in the litters is always one step ahead of that of the mothers. The reduction of litter size may be due partly to the reduced fertility of the females and partly to the reduced viability of the embryos. In order to separate the effects of inbreeding on the young from those on the mother, crosses were made between partly inbred lines and the litter sizes of the inbred mothers with crossbred young were compared with those of equally inbred mothers with inbred young. The mean litter sizes found are given in table 1. The three comparisons in the first row show the effect of inbreeding on the fertility of the mothers, and they give a mean value of 0.175 for the reduction of litter size per 10% of inbreeding. The comparisons in the second and third columns show the effect of inbreeding on the viability of the young, and they give a mean value of 0.245 for the reduction per 10% of inbreeding. Adding the two contributions together gives about 0.42 young which agrees well enough with the rate of decline under continuous inbreeding. About 40% of the total inbreeding depression of litter size is thus attributable to reduced fertility of the females and about 60% to reduced viability of the young. The inbreeding of the father, it should be added, did not influence the size of the litter sired.

*Cause of reduced fertility.* The reduced viability of the inbred embryos was not further investigated, though for the sake of completeness it would be interesting to know the developmental stage at which death most frequently occurs. The cause of the reduced fertility of inbred mothers was, however, investigated by dissections of pregnant females. Inbred females were mated to males of another line so that the embryos were noninbred. Dissections were made at 16 days of gestation, counted from the finding of a vaginal plug. Counts

TABLE 1

*Mean litter sizes according to the inbreeding coefficients of the mothers and of the young*

| | | Inbreeding coefficient of mothers | | |
|---|---|---|---|---|
| | | 0% | 37½% | 50% |
| | | *mean no. of young per litter* | | |
| Inbreeding coefficient of young | 0% | 8.2 | 7.5 | 7.3 |
| | 50% | — | 6.3 | — |
| | 59% | — | — | 5.8 |
| Reduction of litter size per 10% inbreeding of | mother: 0.19, 0.18, 0.16. | | | |
| | young: 0.24, 0.25. | | | |

TABLE 2

*Numbers of corpora lutea and percentage losses in inbred and noninbred females*

| Females dissected | | | Mean no. of corpora lutea | Mean loss | | |
|---|---|---|---|---|---|---|
| Series | Inbreeding coefficient | No. | | Preimplantation: % of corpora lutea | Postimplantation: % of implants | Total: % of corpora lutea |
| | % | | | | | |
| I | 50–59 | 86 | 10.0 | 17.6 | 12.9 | 28.4 |
| | 0 | 58 | 10.1 | 11.1 | 11.5 | 21.4 |
| II | 50 | 13 | 10.9 | 22.5 | 13.6 | 33.1 |
| | 0 | 15 | 11.7 | 4.0 | 13.0 | 16.6 |
| III | 63 | 17 | 12.5 | 37.1 | 18.7 | 49.0 |
| | 0 | 59 | 10.3 | 12.0 | 10.2 | 21.0 |

were made of the corpora lutea (as a measure of the ovulation rate), the numbers of implantation sites, and the numbers of live embryos. These counts were then compared with similar counts made on comparable noninbred females. The results are summarized in table 2. Series I and III refer to the first and second inbreeding experiments described here, and series II refers to another inbreeding experiment with the same stock, which also provided the data for table 1. The results show clearly that the reduced fertility of the inbred females was due almost entirely to a greater preimplantation loss of eggs or embryos. The ovulation rate was not reduced, and the postimplantation loss was only a little, and non-significantly, increased. The preimplantation loss was, however, much greater in inbred than noninbred females, and the differences are significant at the 5% level in series I and II and at the 1% level in series III. Three possible causes of the greater preimplantation losses in inbred females may be postulated, but I do not know which is the right one. One cause might be a higher proportion of abnormal eggs; another might be failure of fertilization through impaired transport of the sperm; and the third might be failure of implantation from endocrine malfunction.

The fact that the ovulation rate was not affected by inbreeding calls for some comment. Ovulation rate is correlated with body size; the regression of the number of corpora lutea on the weight of the female at 6 weeks was $0.24 \pm 0.06$ corpora lutea per gram. Body size might well be expected to decline on inbreeding and, so to speak, carry the ovulation rate with it. But in fact the body size of the mice in these experiments did not decline on inbreeding, because the reduction of litter size led to an improved maternal environment which compensated for any decline of intrinsic growth rate that there may have been. Thus the conclusion that the ovulation rate is independent of inbreeding is valid only if there is no change of body size. The conclusion about gene action that we can draw is that the genes that affect ovulation rate independently of body size do not show directional dominance, though the genes that affect it through their effects on body size may do so.

THE CONTROL LINE

An unselected control line was maintained with minimal inbreeding over the whole course of the inbreeding and selection experiments, and there are some conclusions to be drawn from it that should be described before we consider the selection.

*Inbreeding in the control line.* The control line was maintained by ten pairs of parents in each generation, with equal representation among the parents of the next. The effective population size was therefore 40, and the rate of inbreeding was 1.25% per generation. The litter size in the control line did not change systematically and, apart from irregular fluctuations, it remained at about 7.5 young over the whole course of the experiment. This fact (which may be seen from figure 2), besides being very convenient for the analysis of the selection responses, is also interesting in connection with the inbreeding. The inbreeding coefficient computed from the effective population size works

TABLE 3

*Analysis of variance of litter size in the control line up to generation 28*

| Source of variation | d.f. | M.S. | Variance component |
| --- | --- | --- | --- |
| Between generations | 27 | 7.64[a] | 0.11 |
| Within generations | 882 | 4.11 | 4.11 |

| Variance of observed generation means | | | |
| --- | --- | --- | --- |
| Expected | | | Observed |
| Real | Sampling | Total | |
| 0.11 | 0.13 | 0.24 | 0.23 |

[a] $F = 1.86$; $P < 0.01$. Mean number of litters per generation, 32.5.

out to be 32% at generation 31. If the litter size had declined at the same rate as it did with rapid inbreeding the control line would have dropped to a mean of about six young by the end. A decline of this amount would certainly have been detected. It therefore looks as if natural selection has been effective in counteracting the inbreeding. If natural selection is indeed the explanation, then it must have worked chiefly through its action on the viability of the young because, on account of the breeding system, there was very little opportunity for it to act on the fertility of the female. There may, however, be no need to invoke natural selection because, it will be remembered, three of the 20 rapidly inbred lines reached much higher levels of inbreeding without showing any decline of litter size. Whatever may be its real explanation, the constancy of the control line suggests that an effective population size of 40 may be large enough to allow a strain to be maintained for many generations without any deterioration of litter size.

*Variation between generations.* From the graph of the control line in figure 2 it will be seen that the mean litter size fluctuated eratically between the limits of 6.9 and 8.6. The variance of the observed generation means is 0.23. How much of this variation between the generations was real and how much due to sampling? This question was answered by a simple analysis of variance between and within generations up to generation 28, which is shown in table 3. The variation between generations is significant at the 1% level and therefore without doubt is real. But it is rather small in amount: the component between generations is only some 2½% of the variance within generations. When the possible sources of variation between generations are considered—quality of the food, temperature, light, and other seasonal effects, it does seem surprising that these have so little influence on litter size in comparison with the differences between contemporaneous individuals. This, however, does not answer the question of how much of the observed variation is real. The variance of the observed generation means is the sum of the real variance, which is 0.11 (from table 3), and the sampling variance. The expected sampling variance is $1/n$ times the within-generation variance, where $n$ is the number of litters per generation, and this works out to be 0.13. Thus about half of the variance that appears as fluctuations of the generation means is attributable to sampling and half to real differences between the generations.

*Variation attributable to male fertility.* The records of the control line provide information about the effect of the male on the size of the litter he sires. It is necessary first to explain what the records consisted of. Each generation consisted of the first-litter progeny of ten pairs of parents. These ten full-sib families contained, on the average, about three or four females. Each of these females was test-mated to a male from another family, and her litter size was recorded from the litter subsequently born. Before the test litters were born one female of each sibship was chosen to be a parent, so that her litter was retained for testing in the next generation and the other test litters were discarded. One male from each of the ten sibships was used in the test matings, and the females were arranged in harems of three or four per male. The females of a harem all came from different sibships, so that each male was mated to a set of females unrelated to each other and to him. Any difference between the mean litter sizes of harems, in excess of what would be expected from sampling, would therefore be attributable to differences of male fertility. Only the first 16 generations have so far been analyzed in this way. The analysis of variance (table 4) refers to the variation within generations, pooled over the generations. Because of the non-random, but also irregular, distribution of the sibships among the harems, this analysis is an awkward one and the compositions of the mean squares have not yet been worked out. The best thing to do meantime seems to be to remove the variation attributable to sibships and use the residual mean square as the error variance. This makes the mean square between harems significant at the 1% level and, though the significance is certainly overestimated, there seems to be little room for doubt that males do influence the size of the litter they sire. The component of variance attributable to males works out to be about 10% of the total variance. This, however, is an upper limit because the mean square for harems contains some variation attributable to sibships which has not been recognized in the analysis. Whatever the precise figure may be, we can conclude that the influence of the male parent in determining litter size is small compared with that of the female parent and the litter itself.

*Parent-offspring correlation.* The control line provides data for the estimation of the parent-offspring correlation, though the data have not yet been fully analyzed in this respect. The data, of course, accumulated as the experiment went on and the information was not available for the prediction of the expected response to selection. If it had been, the selection would probably not have been attempted because the correlation is virtually zero. An analysis of the first 16 generations yielded a daughter-dam regression of $-0.066 \pm 0.053$. A graphical representation of the relationship between daughters' and mothers' litter sizes based on the whole experiment shows clearly the absence of any correlation (see figure 5).

The correlation between mothers' and daughters' litter sizes is, however, complicated by an interesting maternal effect. The litters in this experiment were not adjusted to a standard size at birth. Under this system mothers who have large litters rear their daughters in a large litter. The daughters are consequently retarded in growth and this tends to make them have small litters. In this way the maternal effect contributes negatively to the correlation between mothers and daughters, and the correlation observed is the combination of this negative environmental correlation balanced against any positive genetic correlation that there may be. The two component parts of the correlation can be separated to some extent by taking account of the daughters' body weights, which provide a measure of at least part of the maternal effect. If weight is held constant, then the (partial) regression of daughters' on dams' litter size becomes $+ 0.058 \pm 0.053$. Doubling the regression gives an estimate of 11.6% for the heritability of litter size, but the large standard error renders any value between 0 and about 30% compatible. Apart from the distressing magnitude of its standard error, this is not a completely satisfactory estimate of the heritability because the standardization of body size eliminates not only the unwanted maternal effect but also the variation of litter size that is associated with genetic differences of body size. The heritability will therefore be underestimated, and the responses to selection do indeed indicate a higher value.

## SELECTION

Selection for increased litter size was made in one line, referred to as the "high line," and for decreased litter size in another line, referred to as the "low line." The selection in both lines was carried out in the following way. As in the control line, each generation consisted of ten full-sib families. All the females in each family were test-mated but, unlike the control line, sisters were mated to the same male. When the test litters had been born, selection was made within the sibships. That is to say, in each sibship the female with the best litter was selected. Her litter was

TABLE 4

*Analysis of variance of litter size within generations of the control line up to generation 16*

The F ratios, both significant at the 1% level, and variance components are only approximate for reasons explained in the text.

| Source of variation | d.f. | M.S. | F | Variance component | |
|---|---|---|---|---|---|
| | | | | | % |
| Harems | 138 | 4.49 | 1.47 | 0.45 | 10.6 |
| Sibships | 139 | 5.41 | 1.77 | 0.73 | 17.3 |
| Residual | 204 | 3.06 | | 3.06 | 72.1 |
| Total | | | | 4.24 | 100.0 |

Mean number of litters per harem, 3.48
Mean number of litters per sibship, 3.46

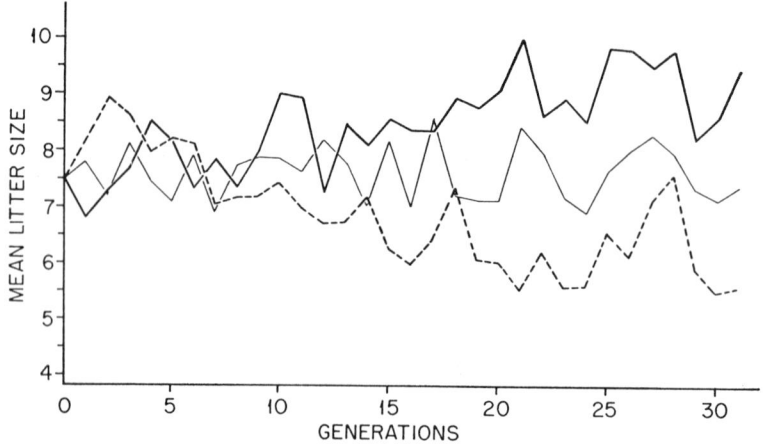

Fig. 2. Response of litter size to selection. Heavy line, selection for large litters; light line, unselected control; broken line, selection for small litters.

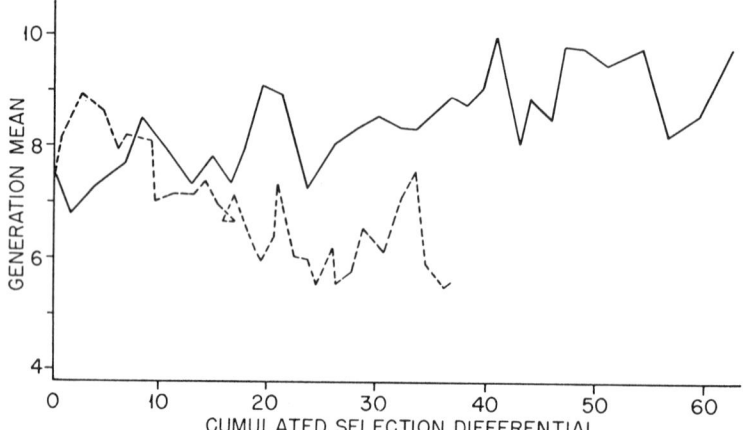

Fig. 3 Response of litter size to upward and downward selection.

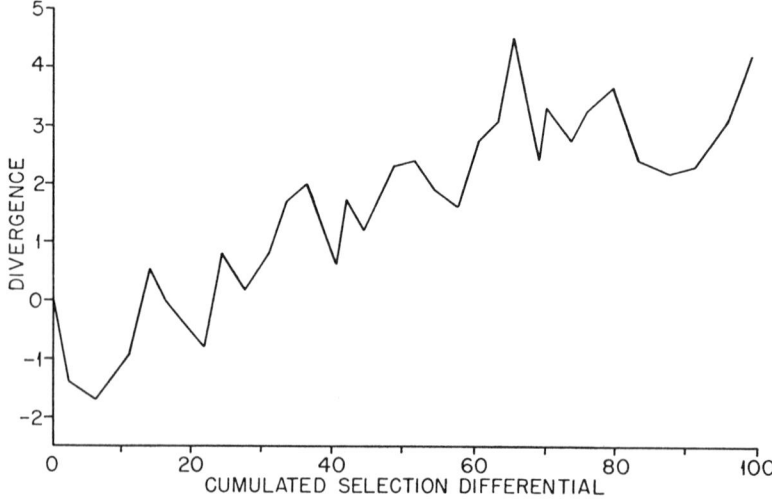

Fig. 4 Divergence between upward and downward selected lines.

reared for testing in the next generation and the other test litters were discarded. This procedure amounted to within-family selection applied to females, males being taken at random. Because the selection was made within families the negative maternal effect was circumvented, each group of females among which selection was made having been subjected to the same maternal environment.

*Response to selection.* The responses to selection are shown in figure 2 plotted against the generation number, and in figures 3 and 4 plotted against the cumulated selection differential. The results are fairly straightforward and need not be discussed in detail. There was a contrary response in the first two generations, attributable to the negative maternal effect mentioned earlier. Thereafter, the responses went in the right directions and continued till about generation 20 when both lines ceased to respond. The rates of response shown by figures 3 and 4 must be doubled to give the realized heritabilities because only one sex was selected. The realized heritabilities, estimated roughly from the graphs and discounting the contrary response in the first two generations, are 8.3% for upward selection, 22.9% for downward selection, and 12.6% for the divergence between the high and low lines. Thus there was a marked asymmetry in the responses, downward selection responding at nearly three times the rate of upward selection. I do not know the reason for this. These realized heritabilities refer, of course, to within-family selection, and are not directly comparable with the estimate from the daughter-dam regression. Conversion to individual heritabilities gives values of 14.5% from the upward selection, 40% from the downward selection, and 22% from the divergence. These values are considerably higher than the value of 11.6% obtained from the daughter-dam regression. But the daughter-dam regression, as noted before, referred to variation in litter size not associated with genetic variation in body size, whereas the selection could make use of any such variation. I therefore do not regard this discrepancy as a serious one.

The final levels reached, after the responses had ceased, were 9.2 young in the high line, and 6.0 in the low, compared with a mean of 7.6 in the control. Thus an improvement of about 1.5 young per litter was made in both directions, and the final difference between the high and low lines was 3.2 young. This difference amounts to 1.6 times the original phenotypic standard deviation and 3.3 times the additive genetic standard deviation. This total response to selection is very small compared with responses in other experiments, which commonly yield some 10–20 phenotypic standard deviations of response, or 20–30 additive genetic standard deviations. The conclusion to be drawn may be either that relatively few genes are concerned with the variation of litter size, or the limits to selection do not represent fixation at all relevant loci. The latter seems the more probable because lethal and semilethal genes which may cause variation of litter size through their effects on embryonic viability could not be brought to fixation.

*Nature of the changes made by selection.* One circumstance has so far been omitted from the consideration of the responses to selection, and that is the fact that the females of the high line were reared in large litters and the females of the low line in small litters. Should not some adjustment be made for the differential maternal effect so produced? In order to explore this problem I calculated, over generations 20-31 in the high and low lines, the mean daughters' litter size for each maternal litter size, hoping thereby to make a comparison between the lines at a standard parental litter size. The resulting regressions of daughters' on dams' litter sizes are depicted in figure 5 along with a similar calculation for the control line over the whole experiment. The situation revealed is very striking. The control line, as mentioned earlier, shows no correlation at all between daughters and dams, but the high line shows a negative correlation and the low line a positive one. It is not clear to me how this information should be used to make a just comparison between the lines because, in the first place, the genetic properties of the three lines are clearly different, and in the sec-

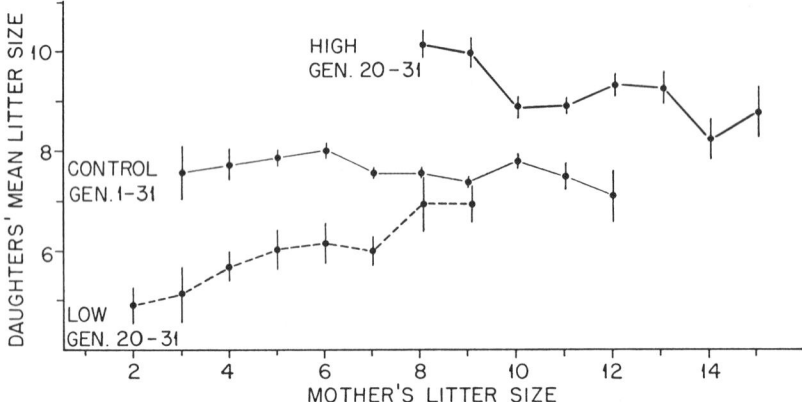

Fig. 5 Mean daughters' litter sizes plotted against the mother's litter size, showing the regression of daughters' on dam's litter sizes. Vertical lines extend to ± one standard error.

ond place, the difference between each of the selected lines and the control will vary according to the parental litter size taken as standard. Perhaps they should be compared at a standard parental litter size of 7.5, which is the initial level. Then the high line is about 2.6 young above the control and the low line about 1.0 below. The asymmetry of the unadjusted responses might be accounted for by invoking the maternal effect in this way, but I do not feel confident enough to pursue the matter here.

The different genetic properties of the selected lines brought to light by the daughter-dam regressions depicted in figure 5 lead to another line of thought, which, in conjunction with the facts to be mentioned later, leads to a hypothesis about the nature of the genes that have been responsible for the responses to selection. A plausible interpretation of the difference between the daughter-dam regressions is that selection for increased litter size had exhausted the additive genetic variance, so that what is left at the end is the negative maternal effect; whereas selection for reduced litter size had increased the genetic variance, so that the genetic correlation overweighs the environmental and the daughter-dam regression becomes positive. An increase of genetic variance in the low line would be compatible with the hypothesis that low litter size is due to lethal and semilethal genes in the embryos. These genes, presumably at low frequencies initially, could be brought by selection to intermediate frequencies, but not beyond; and at intermediate frequencies they would make their maximum contribution to the variance of litter size. In this way the response to downward selection would cease when the daughter-dam regression was at its maximum.

This idea of lethal genes in the low line was suggested by the distributions of litter size which differ strikingly between the lines. These distributions are shown in figure 6A, where the high and low line distributions refer to generations 20–31 and the control line distribution to the whole experiment. The distributions of the high line and control have a small "tail" at low litter sizes, whereas the distribution in the low line looks as if it were a compound of two distributions — one with a mean at the mean of the control and the other with a mean corresponding to the tail of the control distribution. If we suppose that the tail represents litters that are segregating a lethal, then an increase of the frequency of these litters in the low line would satisfactorily account for the altered form of the distribution in the later generations of the low line.

The final investigation to be described here concerns the ovulation rate in the selected lines. The ovulation rates were determined by egg counts after natural mating. Two sets of counts were made, one at generations 16 and 17, when only the selected lines were counted, and the other at generation 31, when all three lines

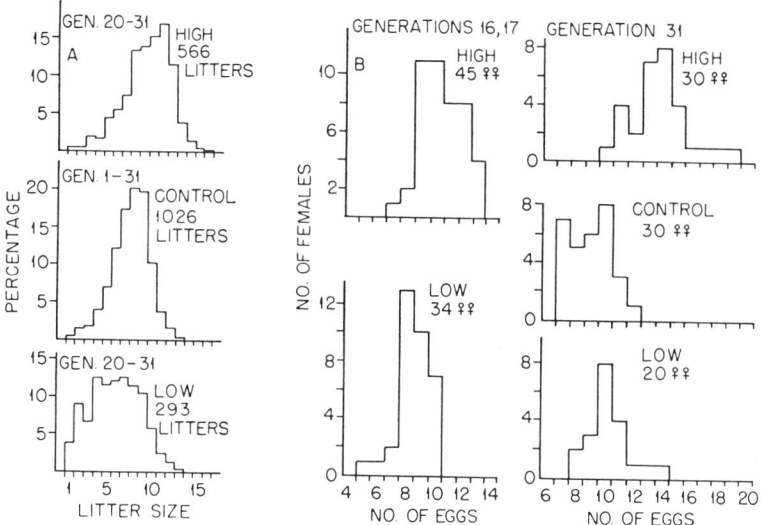

Fig. 6 A. Frequency distributions of litter size. B. Frequency distributions of numbers of eggs ovulated in natural matings.

were counted. The results are given in table 5, and the distributions of the egg numbers are shown in figure 6B. It is immediately clear that selection, unlike inbreeding, has changed the ovulation rate. If the ovulation rates are compared with the litter sizes in the high and low lines, as shown in table 5, then it appears at first sight that the difference in ovulation rate alone is enough to account for the difference in litter size. But if we deduce the loss of eggs and embryos from the difference between the ovulation rate and the corresponding litter size, we find that this conclusion is not fully justified. The losses must be compared on a proportionate, or percentage, basis, and then the loss is greater in the low line than in the high line. This fits in well with the lethal hypothesis. The ovulation rate in the low line at the end of the experiment is not lower than that of the control, and this also agrees with the lethal hypothesis, because the distributions of litter size suggest that the low line has a mode equal to the mode of the control line, which suggests in turn that the size of undepleted litters is the same in the two lines. It should be mentioned, however, that the ovulation rate in the low line is in fact significantly higher than that of the control; the reason for this is not clear.

My tentative conclusion about the nature of the changes produced by selection is this: that selection for increased litter size has acted chiefly on the fertility of the females by increasing ovulation rate, though there has at the same time been an increase in the proportion of eggs or embryos lost; selection for decreased litter

TABLE 5

*Ovulation rates and deduced losses*

|  | Generations 16 and 17 | | Generation 31 | | |
|---|---|---|---|---|---|
|  | High line | Low line | High line | Low line | Control |
| Number of eggs[a] | 10.4 | 8.5 | 13.7 | 10.3 | 8.9 |
| Difference (high–low) | 1.9 | | 3.4 | | |
| Litter size | 8.4 | 6.2 | 9.2 | 6.0 | 7.6 |
| Difference (high–low) | 2.2 | | 3.2 | | |
| Number lost | 2.0 | 2.3 | 4.5 | 4.3 | 1.3 |
| Percentage loss | 19 | 24 | 33 | 42 | 15 |

[a] All differences significant.

size, in contrast, has acted chiefly on the viability of the embryos, and has resulted in little or no decrease of ovulation rate but a marked increase of embryonic mortality.

## DISCUSSION

We have tried, in the investigations reviewed here, to break down the character "litter size" into its component characters, and have found that these components have different genetic properties. In particular, ovulation rate is influenced by genes with predominantly additive effects on it (more precisely, without directional dominance), whereas implantation rate and embryonic viability are probably influenced more by deleterious recessive alleles at low frequencies. To understand the reasons for these differences, we shall have to discover the relationships of litter size and its components with natural fitness. About the nature of these relationships, however, I can at present offer little more than conjecture.

On the analogy of clutch size in birds, one would expect an intermediate litter size to confer maximal fitness—that is, to yield the greatest number of adult offspring—and the existence of a fair amount of additive genetic variance of litter size is in accord with this expectation. I have, however, been unable to find any evidence of an intermediate optimum in laboratory mice: the number weaned begins to drop off only when the litter size exceeds about 13 born alive, and there are very few deaths after weaning. The situation in the wild, with a limited food supply, is probably very different, and the genetic properties of litter size in laboratory mice may perhaps reflect an adaptation to conditions in the wild. But it is difficult, nevertheless, to believe that this adaptation could survive so long under domestication when the pressure of natural selection was shifted toward higher litter sizes.

If we accept the postulate that an intermediate litter size is optimal, or has been in the past, then the different genetic properties of the component characters are readily understandable. It is inconceivable that the implantation rate or embryonic viability should have intermediate optima; any loss is wasteful of maternal effort and the upper extreme values must represent maximal fitness. Depression on inbreeding is just what would be expected of such characters and what was in fact found. Adjustment of the litter size to an intermediate optimal value would then be achieved best by an appropriate ovulation rate. It is therefore in the ovulation rate that we should expect to find the genetic properties of a character with an intermediate optimum. Such a character should respond to selection and not change on inbreeding, and this is what was found.

## SUMMARY

Litter size, measured as the number of live young born in first litters, was studied as a quantitative character by inbreeding and by selection; the changes produced were investigated by dissections of pregnant females.

Inbreeding led to a decline of about 0.5 young per 10% increase of the inbreeding coefficient, and the decline was linear with respect to the inbreeding coefficient. Three out of 20 lines survived to about 90% inbreeding and one of these survived indefinitely. These lines reached high levels of inbreeding without any decline of litter size. This suggests that overdominance cannot have been a major cause of the inbreeding depression. About 40% of the reduction of litter size on inbreeding was attributable to reduced fertility of the females and the remaining 60% to reduced viability of the young. The reduced fertility of inbred females was due almost entirely to an increased pre-implantation loss of eggs or embryos. The ovulation rate was not influenced by inbreeding.

An unselected control line, maintained with minimal inbreeding, did not decline in litter size though by the end (generation 31) its computed inbreeding coefficient was 32%. Environmental differences between generations contributed only about 2½% of the total variation of litter size.

Differences of fertility between males in the control line contributed, at the most, 10% of the variation of litter size. The correlation between parents and offspring was virtually zero, but this is complicated by a maternal effect that contributes negatively to the correlation and counterbalances the positive genetic component of the correlation.

Selection applied to females within families yielded progress with a realized heritability of 8% for upward selection and 23% for downward selection. Progress ceased after about 20 generations of selection, when the mean litter sizes were 9.2 young in the high line, 6.0 in the low, and 7.6 in the control. Selection, unlike inbreeding, affected the ovulation rates, the mean ovulation rate in the high line being 3.4 eggs greater than that of the low line. But the low line females ovulated more, not fewer, eggs than the control.

Comparisons of the properties of the lines after the response to selection had ceased suggested the tentative hypothesis that, whereas the response to upward selection had been achieved through an increased ovulation rate, the response to downward selection resulted from a reduced viability of the embryos.

[*Editor's Note:* The Discussion has been omitted.]

## LITERATURE CITED

Bowman, J. C., and D. S. Falconer 1960 Inbreeding depression and heterosis of litter size in mice. Genet. Research, 1: 262–274.

Brumby, P. J. 1960 The influence of the maternal environment on growth in mice. Heredity, 14: 1–18.

Dickerson, G. E., C. T. Blunn, A. B. Chapman, R. M. Kottman, J. L. Krider, E. J. Warwick, J. A. Whatley, M. L. Baker, J. L. Lush, and L. M. Winters 1954 Evaluation of selection in developing inbred lines of swine. Missouri Agr. Expt. Station Research Bull., No. 551, pp. 1–60.

Falconer, D. S. 1955 Patterns of response in selection experiments with mice. Cold Spring Harbor Symposia Quant. Biol., 20: 178–196.

Falconer, D. S., and R. C. Roberts 1960 Effect of inbreeding on ovulation rate and foetal mortality in mice. Genet. Research, 1: 422–430.

Roberts, R. C. 1960 The effect on litter size of crossing lines of mice inbred without selection Genet. Research, 1: 239–252.

## AN EXPERIMENTAL CHECK ON QUANTITATIVE GENETICAL THEORY

### I. SHORT-TERM RESPONSES TO SELECTION

By G. A. CLAYTON, J. A. MORRIS* AND ALAN ROBERTSON†

*Institute of Animal Genetics, Edinburgh*

There is a theoretical approach to the problems of animal breeding which is founded upon the work of Sewall Wright and developed in the main by Lush. This approach is essentially statistical, and the main integrating concept is the additive genetic variance between individuals and its related parameter, the heritability. These concepts appear in the estimation of the breeding value from performance in individuals, in considerations of the accuracy of progeny testing, in the prediction of rates of improvement using different selection techniques, and in the evaluation of selection indices. Estimates of these parameters relate specifically to the particular population (under its particular set of environmental conditions) from which they were derived.

The value of this approach in the improvement of economic traits in farm animals is difficult to check because of the large numbers required for an adequate analysis. In most animals the generation interval—e.g. $4\frac{1}{2}$–5 years in cattle—is so great as to make it impossible to secure conclusive results within a reasonable time. This means that very little experimental check of the theory exists, and the experiments to be described have been designed in part to try to meet this need.

The basic premises of the theory to which we have referred earlier seem not to have been explicitly discussed in any publication. Before we describe our experimental work, we should perhaps put down what conditions we consider necessary for the predictions of the theory to be adequately borne out. First of all, it must be made clear that we can separate the applications into two parts, which we can loosely describe as static and dynamic. In the static part we are concerned to describe phenomena in the existing population, mostly in terms of the observed similarity between relatives. In the dynamic part we are concerned to describe the response of the population to external stress, either by selection or a change in the mating system practised.

The static description of the existing population is more satisfactorily treated than the dynamic, and the recent work of Anderson & Kempthorne (1954) and Cockerham (1954) has almost completed the making of statistical models to describe the situation. The only basic premises, in dealing with a diploid organism, are

(i) that the inheritance of quantitative characters is particulate and diploid,
(ii) that the Mendelian laws of segregation hold,
(iii) that there is no natural selection between different genotypes.

---

\* Present address: C.S.I.R.O. Poultry Research Station, Werribee, Australia.
† Member of the Scientific Staff, Agricultural Research Council, Great Britain.

Within this framework, a satisfactory description of the observed phenomena can be given. It will be noted that there are no assumptions about the kind of gene action involved.

The moment we turn from a static to a dynamic model the problem becomes more complex. We are faced with the fact that we are dealing with an essentially discontinuous process (involving the segregation of individual genes) as a continuous one. Because of the essential discontinuity, we cannot hope that response to selection can be predicted exactly, and have therefore to make a subjective judgement as to whether we shall consider any discrepancies important or not. The effect of selection is to change gene frequencies, and these will again alter the parameters involved in the 'static' description of the population. But although the parameters be used to predict the change in mean of the population, they cannot in any way be used to predict their own changes. To do this, we have to deal with the effects and frequencies of the individual genes. *A priori*, we can therefore do no more than lay down conditions under which the basic parameters may be expected to give 'adequate' predictions of response to selection over, say, the first five generations. It is to be expected that such predictions will not hold good if the following additional conditions are grossly violated:

(iv) That all genes concerned have small effects. Under our conditions of selection (20% of the population selected), this implies effects due to individual genes of less than one-half the observed phenotypic standard deviation.

(v) That only a small proportion of the hereditary variance is due to interactions between genes at different loci.

(vi) That the environmental variance of different genotypes is not regularly connected with their mean value. (v) and (vi), taken together, imply that scale effects are not important. In general, we may expect scale effects to assume importance only when the coefficient of variation of the character is large, say over 40%.

(vii) For *a priori* predictions of improvement by selection, a normal distribution is often assumed. If the coefficient of variation is small, then normality seems not to be important, and all we require is that the distribution can be transformed into one approaching normality.

This final list of conditions may appear to be a little dogmatic, but to discuss them in detail would require more space than would be worth while in this essentially experimental paper. They should, however, be regarded not as necessary conditions for the adequacy of the theory, but rather conditions which, if grossly violated, will cause its breakdown. It will be noted that they do not contain any reference to dominance. Additive action at individual loci (in the sense that the heterozygote is intermediate between the two homozygotes) is *not* a necessary condition.

We had in this laboratory a large random-breeding population of *Drosophila melanogaster*, which had been maintained under constant conditions for several years previously. This served as a source of material and could be resampled at any time. Our approach in this work was first to measure similarities between relatives of different kinds in the base population, in a sense to investigate first the validity of the static model. Having obtained results coherent within themselves, we then proceeded to select lines from the population by different methods—by individual selection of different intensities and by different kinds of family selection. The responses to selection for the first five generations could then be compared with those expected from the parameters derived from the base population. In addition, observations were made wherever possible on characters which were secondary—in the sense of not being selected—with the intention of analysing the correlated responses. The approach was therefore operational in nature, although fundamental in the sense that if things went wrong it is possible with *D. melanogaster* to use 'trick' methods to find out why. The choice of main character, abdominal sternital bristles, was dictated solely by ease of measurement. We were therefore not basically concerned with the inheritance of abdominal bristles as such; we were using them merely as a convenient tool for analysing the response of a population to various external stresses.

## MATERIAL AND METHODS

The stock of *D. melanogaster* employed derived from a capture in Kaduna, West Africa, early in 1949. The Kaduna stock from a few generations after its capture was kept in a population cage with an average population size of about 5000, in a constant-temperature room maintained at $25 \pm 0.5°$ C. It is not known whether the original capture was of a single inseminated female or of several individuals. The population had been kept in this cage for about three years (sixty generations) before these experiments began. Repeated sampling has shown no change in bristle number since this work started. In all experiments the standard agar food of this laboratory was used. Overcrowding of bottles was avoided by inspection, parents being removed before laying too many eggs. An optimal egg density of about 400–500 was aimed at, though it proved difficult to attain in the later stages of the experiment when serious fecundity problems arose. In the case of the full-sib selection experiments parents were shaken over into fresh vials at 24-hourly intervals. The relative insensitivity of this character to environmental variation was amply evidenced by the rarely significant and, in all cases, small contribution of the variation between cultures to the total. All flies developed at 25° C.

The character selected for in these experiments was the sum of the bristles on the fourth and fifth abdominal sternites, the character used by Mather & Harrison (1949) and Rasmuson (1955) in their selection experiments. The means and standard deviations of bristle number in the cage population, based on a very large number of observations, are given in Table 1.

Table 1. *Statistical parameters of the base population*

|  | Males | Females |
|---|---|---|
| Mean | 31·4 | 39·2 |
| S.D. | 3·03 | 3·54 |

Miller (1950) summarizes what little is known of the anatomy and function of the bristles of *Drosophila*. Variously modified bristles serve gustatory, olfactory, chemotactile, auditory and tactile functions. It is not surprising then that the bristles are intimately associated with the peripheral nervous system. The precise function of the ventral abdominal bristles is not at all apparent. It is possible that the micro-chaetae serve to prevent the fly from sticking to surfaces under moist conditions.

## THE HERITABILITY DETERMINATIONS

The additive genetic variance and its related parameter, the heritability, describe the genetic situation in the population as it is at a particular time. The heritability can be used to predict the response of the mean to selection, but not that of the heritability itself. In other words, the value of the prediction of changes in the mean after several generations of selection is doubtful. One of the objects of this work was to investigate the number of generations during which the response did proceed according to initial predictions. Accordingly we have regarded the response to selection as part of the process of verification of prediction and have considered as measures of heritability only those parameters derived from the base population itself.

The estimates obtained from half-sib and full-sib correlations came directly from the first generation of the selection programmes described later, but the offspring-parent regression analysis required an experiment specifically designed for this purpose.

### (1) Offspring-parent regression

Samples of eggs were obtained from the control population and placed in food bottles under conditions of optimum density. Samples of males and females were taken from the bottles at hatching, and each male was allowed to mate with a large number of virgin females. The females were then scored for abdominal and sternopleural bristles and transferred to separate vials for egg-laying. When the progeny emerged, two males and two females were scored from each dam. In all, 506 females mated to 27 males produced offspring.

An analysis was then made of the offspring-parent regression within sire groups. In the case of female offspring, this is equal to half the heritability. For male offspring, the regression was corrected by the ratio of the standard deviation of counts in females to that in males and this then equated to half the heritability. The daughter-dam regression was equal to 0·269 and the son-dam regression was 0·206, which after correction became 0·241. Estimates of the standard errors were obtained from the analysis of covariance. The estimates of heritability became:

| | |
|---|---|
| From daughter-dam regression | 0·54 + 0·11 |
| From son-dam regression | 0·48 ± 0·11 |

The two estimates were then pooled to give a final value of 0·51 ± 0·07.

### (2) The half-sib correlation

The half-sib family selection experiment consisted of three high and three low lines, in each of which were ten families. Each family was sired by one male which had been mated to ten females. The flies were allowed to mate for 72 hr. in order to ensure that all or nearly all females would be inseminated. Before this, the females had been kept in isolation for 72 hr. It was found that this treatment was adequate to ensure insemination. After the mating period the parent flies were shaken into fresh bottles at 12-hourly intervals. Five males and five females were scored from the second and third bottles of each family, a total of 200 individuals in each line. Male and female scores were analysed jointly. This was done within lines and the results pooled, with the results shown in

Table 2. It will be noted that significant differences were found between bottles, but that they made up only 5% of the total variance. To calculate the heritability of bristle score, we must allow for the fact that the average relationship between members of a family is slightly higher than 0·25 because there is a chance of one-tenth that any two will in fact be full-sibs. For an estimate of within bottle heritability, we can then take

$$h^2 = \frac{1 \cdot 256}{0 \cdot 275 \, (1 \cdot 256 + 9 \cdot 12)} = 0 \cdot 484.$$

This has a standard error of 0·11.

Table 2. *Half-sib analysis of variance—base population*

| Source of variation | D.F. | Mean squares | Components |
|---|---|---|---|
| Between families within lines | 54 | 39·51*** | 1·256 |
| Bottles within families | 60 | 14·39** | 0·527 |
| Residual | 1020 | 9·12 | |

### (3) *The full-sib correlation*

The full-sib selection group again consisted of three replicates in each direction, and in each line twenty families were scored. The parents were shaken over into fresh vials at 24-hourly intervals. Offspring were scored from the second and third vials of each line— three males and three females from each vial. In each line 240 individuals were thus scored. The analysis of variance gave the results shown in Table 3. In this case, the variation between vials within families was not significant. The within-vial heritability estimate thus obtained is $\frac{3 \cdot 08}{\frac{1}{2} \times (3 \cdot 08 + 8 \cdot 54)} = 0 \cdot 53$. The standard error, by the usual formula, is 0·07.

We have now three estimates of the heritability

(i) from the parent-offspring regression   $0 \cdot 51 \pm 0 \cdot 07$
(ii) from the half-sib correlation   $0 \cdot 48 \pm 0 \cdot 11$
(iii) from the full-sib correlation   $0 \cdot 53 \pm 0 \cdot 07$

We can combine these to give an overall estimate of 0·52.

Table 3. *Full-sib analysis of variance—parental generation*

| Source of variation | D.F. | Mean squares | Components |
|---|---|---|---|
| Families within lines | 114 | 46·60*** | 3·08 |
| Vials within families | 120 | 9·71 | 0·20 |
| Residual | 1080 | 8·54 | |

There is, however, a further source of information as to the nature of the observed variation. We are operating on a total score, the sum of two separate observations on the fourth and fifth sternites. The separate scores have similar means and variances. We can then separate the causes of variation into those common to both sternites and those specific to each. Those readers accustomed to the genetics of farm animals will have recognized the situation of repeated measurements of the same character. We can then use the variance of the difference between the two scores, $\sigma_D^2$, as a measure of that part of the variance of total score which is specific to each sternite.

From the observations of Reeve & Robertson (1954) and from our own half-sib and full-sib analyses, it appears that the genetic correlation between scores on the two sternites is close to unity. Thus it seems that the genes controlling this character have the same effect on both sternites, and that the difference between sternites is therefore mostly non-genetic in character. This we can simply call 'developmental error', some of which will be due to the unit nature of the score. The variance of the difference, $\sigma_D^2$, thus measures a part of the total variance not utilizable in the selection of total score. The remainder, $\sigma_P^2 - \sigma_D^2$, contains the genetic variance of both sternites together and the common environmental variance. We shall frequently make use of the ratio $\sigma_D^2/\sigma_P^2$ as an indication of the probable make-up of $\sigma_P^2$. If the genetic variation is being used up by selection, then this ratio should approach unity. If we end in a balanced situation in which considerable hereditary variation is maintained, this may remain low. To use the ratio in later generations involves assumptions that the 'common' environmental variance remains small, but, bearing in mind this and one or two other reservations, the ratio $\sigma_D^2/\sigma_P^2$ is a useful guide to the situation.

In our initial population, we found values of $\sigma_D^2$ of 4·97 in females and 3·40 in males. Thus about 35% of the variance in total score is due to 'developmental error' specific to each sternite. From our estimates of heritability, we find that 52% of the variance is of the additive genetic kind. Reeve & F. Robertson (1954) presented evidence from analysis of inbred lines and of crosses between them that the environmental variance common to segments contributed only a small percentage of the total. We can thus make up our variance balance sheet as in Table 4. Thus about 80% of the hereditary variance is due to the average effects of genes. The consequences of this have been elaborated elsewhere (Robertson, 1955), but it seems difficult to avoid the conclusion that we are here dealing with a situation in which non-additive genetic variation is slight and in which most of the genes are really acting additively in the sense that the heterozygote is intermediate between the two homozygotes.

Table 4. *The balance sheet of the initial variation*

| | |
|---|---|
| Additive genetic | 0·52 |
| Genetic complexities | 0·09 |
| True environmental error | 0·04 |
| Developmental error | 0·35 |
| | 1·00 |

### THE EFFECT OF INBREEDING

Two series of inbred lines were made from this population by continual full-sib mating. In the first series, ten lines were started with four pair-matings in each generation to minimize loss of lines, and of these seven were still in existence at the 10th generation of inbreeding and five at the 22nd. In the second series, 100 lines were started without reserves, and when they had been reduced to fourteen at the 8th generation, reserve matings were made to minimize further loss. Both series of lines were measured at the 10th generation when the expected inbreeding coefficient, $F$, was 0·87. The scores of the individual lines had diverged widely, the highest having a mean value (averaged over the sexes) of 45·5 and the lowest of 28·6 compared to the initial value of 35·3. The first series averaged 32·2 for seven lines and the second series 36·4 for twelve lines. These two values differ significantly from one another and the former differs significantly from the initial value. However, whatever may be the cause of the difference between the series, we seem to be justified in saying that inbreeding has little average effect on the mean.

The variance component between lines within series was 12·94 units based on 17 degrees of freedom. If we assume that all genetic variance is due to additively acting genes, then the expected value of this is $2F\sigma_g^2$, in the present case 10·25 units, which does not differ significantly from the observed figure. At generation 22 of the first series, further measurements were made on five lines. The average variance within lines was 4·7 units in males and 5·3 units in females. Assuming that little genetic variance remained, we would expect this variance to be due to developmental error and to environment common to both segments. The variance in the inbreds was in fact slightly higher than the variance of the difference between sternites in the outbred material. The common environmental effects may account for part of this, as will any differential susceptibility to environmental variation in inbreds, which Tebb & Thoday (1954) have shown to operate to some extent for bristle characters.

## THE SELECTION PROGRAMME

The different selection programmes can be briefly summarized as follows:

(i) *Individual selection* at different intensities, but with the same number of selected parents in each generation which were mass mated.

In these, twenty males and twenty females were selected on their own performance from different numbers of scored flies. In all cases the selection was replicated as follows:

(a) 20/100 (20 selected out of 100)—five replicates in each direction. These were continued for more than thirty generations and their later behaviour will be the subject of further papers (Fig. 1).

(b) 20/75—three replicates in each direction. Relaxed at generation 7—discontinued at generation 15 (Fig. 2).

(c) 20/50—as (b).

(d) 20/25—three replicates in each direction—discontinued at generation 7 (Fig. 2).

(e) In addition, a series of five controls were kept under identical environmental conditions to the above lines (Fig. 8).

(ii) *Family selection* by different systems but with identical rates of inbreeding. The inbreeding was, however, greater than in the individual selection lines.

(a) *Half-sib selection.* Ten single sire families—ten dams per sire—were made in each generation. Ten males and ten females were measured in each family and the next generation bred from equal numbers of *unscored* members of the two extreme families. The females were alloted at random to each male. There were three replicates in each direction for seven generations (Fig. 3).

(b) *Full-sib selection.* In each line 240 offspring (six males and six females from each of twenty matings) were scored in each generation. The matings for the next generation were made up from equal numbers of unscored members of the extreme four families, mated at random. There were three replicates in each direction for seven generations, followed by relaxation (Fig. 4).

(c) *Family selection with inbreeding.* A small experiment was carried out using full-sib family selection, scoring four families in the same way as in the other full-sib family experiment, but selecting parents from the best family so that they were full-sibs. There were two replicates in each direction for nine generations, though one line did not survive beyond the sixth generation (Fig. 6).

### SELECTION RESULTS

The results of the first five generations of individual selection are shown in Figs. 1 and 2. As would be expected from the high heritability estimates the response to selection was very marked. A notable feature is the lack of randomness among replicates. Within each group of replicates an order was established within one or two generations which was

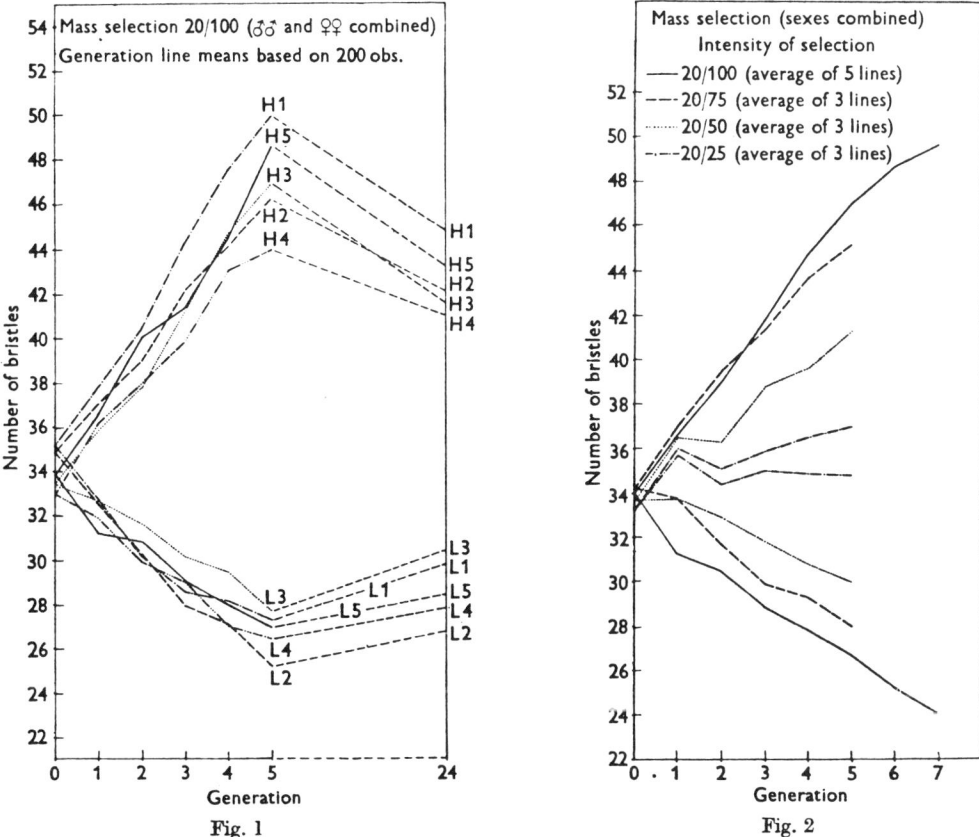

Fig. 1. Response to individual selection. Selection intensity 20/100 of each sex. Generation means based on average of male and female scores. H 1...H 5 and L 1...L 5 are individual high and low lines respectively, all lines relaxed at generation five for nineteen generations.

Fig. 2. Response to individual selection of differing intensities but identical inbreeding rate.

largely maintained from generation to generation. At the 5th generation of the most intense selection, the average response in the 'up' lines was 12·6 bristles, but this varied from 16·0 to 9·6 in the separate lines. This is an important observation from the point of view of experiments set up to compare different types of selection programmes. Suppose we were comparing 20/100 individual selections with some methods of selection thought to be superior to it by 20%. It is quite obvious that if we relied on just one line of each type, the answer would be quite meaningless, as amongst our replicates we can get a deviation on either side of the mean of 25 %.

To what extent can this variation among lines be due to sampling? In these individual selection experiments, we chose twenty males and twenty females as parents. In theory, if all parents have an equal chance of leaving offspring, this would give us an increase in the inbreeding coefficient of 1·2 % each generation. However, as Crow (1954) has shown, unequal egg-laying ability of females and mating ability of males will increase this figure. In experiments with *Drosophila* he finds that the ratio of the effective number to the actual number is 0·71 for females and 0·48 for males. In our case, this would lead to an increase in inbreeding to 2·2 % each generation. The expected drift variance is equal to $2F\sigma_g^2$ (Wright, 1921), where $\sigma_g^2$ is the additive genetic variance. To use this formula, we must assume additive action, reasonable in this instance.

The problem is now complicated by scale. It will be seen later that we have good evidence that the genetic variance declines rapidly in the first few generations of selection downwards. From the last three generations of the family selection experiments, when the mean values had approached those at the 5th generation in the 20/100 lines, we found average values of $\sigma_g^2$ of 5·07 and 2·45 in the two directions. Assuming an $F$ value of 0·11 at this time, we obtain expected variances due to drift of 1·12 and 0·54 respectively. The observed variances between replicates at generation 5 (based on the 4 degrees of freedom between replicates in each group) were 5·40 and 0·75. The former figure is significantly greater than expected, the latter is not. It seems therefore that genetic sampling could account for divergences between the down-lines but not between the up-lines. However, whatever may be the reason, these results show that *Drosophila* may not be a suitable organism for experiments designed to evaluate different selection programmes.

## THE ACCURACY OF PREDICTION

The extent to which the average responses of the different sets of lines correspond to expectation is shown in Fig. 5. The expected response to selection of each generation is given by the expression
$$\Delta G = \bar{i} h^2 \sigma_P,$$
where $\bar{i}$ is the selection differential (in standard units), $h^2$ is the heritability and $\sigma_P$ is the phenotypic standard deviation. In the parent populations we found a value of 0·52 for the heritability and values of 3·03 and 3·67 for $\sigma_P$ in males and females respectively, giving a mean value for the two sexes of 3·35. The predictions and observations may be summarized as shown in Table 5 for the first five generations of the individual selection.

Table 5. *Predicted and observed response to individual selection*

| Intensity | $\bar{i}$ | Expected change/generation | Observed change/generation | |
|---|---|---|---|---|
| | | | Up | Down |
| 20/100 | 1·40 | 2·42 | 2·62 | 1·48 |
| 20/75 | 1·24 | 2·14 | 2·20 | 1·26 |
| 20/50 | 0·97 | 1·68 | 1·46 | 0·79 |
| 20/25 | 0·35 | 0·61 | 0·28 | −0·08 |

The lines selected upwards are in fair agreement with expectation, bearing in mind the divergence between replicates. But the lines selected downwards have in all cases responded much less than expected. In fact, the lines with very weak selection applied

# 140  An experimental check on quantitative genetical theory

have gone in the wrong direction. In the other three groups of replicates, the response to down selection has been about 60% of that expected. One cause of this is that the phenotypic variance has changed as a result of selection. The expected selection differential in the 20/100 lines, from the expression $\bar{i}\sigma_p$, is 4·69 units. The average over the first five generations of selection is 4·96 in the up-lines and 4·05 in the down-lines. This reflects the decrease in variance in the down-lines, but is not a sufficient explanation of the smaller response. Dividing the response by the observed selection differential, we can calculate the 'realized heritability' (Falconer, 1953) in the two directions. The results obtained are 0·53 and 0·37 respectively. Thus there seems to have been a real change in the heritability and the genetic variance in the down-lines after two or three generations of selection. A possible explanation of this would be that natural selection was opposing artificial selection. We shall see later that relaxation of selection gives no support to this.

There is an odd consistency about the discrepancies. In both directions the ratio of observed to predicted response is highest in the 20/100 lines and declines continually to the 20/25 lines. It is difficult to suggest a reason for this, and perhaps the first thing to do is to check that this trend is real by further experiments. The adaptation of the population to the now uncrowded situation cannot have been responsible, for the controls, set up to test this point, did not show any change in mean value.

### Family selection

The results of family selection are shown in Figs. 3 and 4. The same features are shown as in the individual selection. There is an early establishment of an order between replicates and a smaller response in the down-lines than in the up-lines. The expected response to selection can be calculated from the formula.

$$\Delta G = \bar{i} h_f^2 \sigma_f,$$

where the subscript now refers to families treated as units and not to individuals. Although the proportion selected is the same as in the more intense individual selection, the value of $\bar{i}$ is slightly less because we are now dealing with selection from a small number of units. Thus $\bar{i}$ is 1·40 for 20/100, but 1·27 for 2/10 and 1·33 for 4/20.

We can then make up the balance sheet (Table 1) of variance for the two systems of selection (assuming an individual heritability of 0·52), bearing in mind that each half-sib family consisted of twenty individuals in two cultures and each full-sib family twelve individuals in two cultures. The between-culture component is derived from the first generation of selection.

Table 6. *The calculation of response to family selection*

|  | Half-sib ($r=0.275$) | Full-sib ($r=0.50$) |
|---|---|---|
| Between families | 1·60 units | 2·93 units |
| Between cultures within families | 0·53 units | 0·20 units |
| Within cultures | 9·12 units | 7·96 units |
| $\sigma_f$ obs. | 1·52 bristles | 1·92 bristles |
| $h_f^2$ | 0·69 | 0·79 |
| $\bar{i}$ | 1·27 | 1·33 |
| $\Delta G$ | 1·33 bristles/generation | 2·02 bristles/generation |

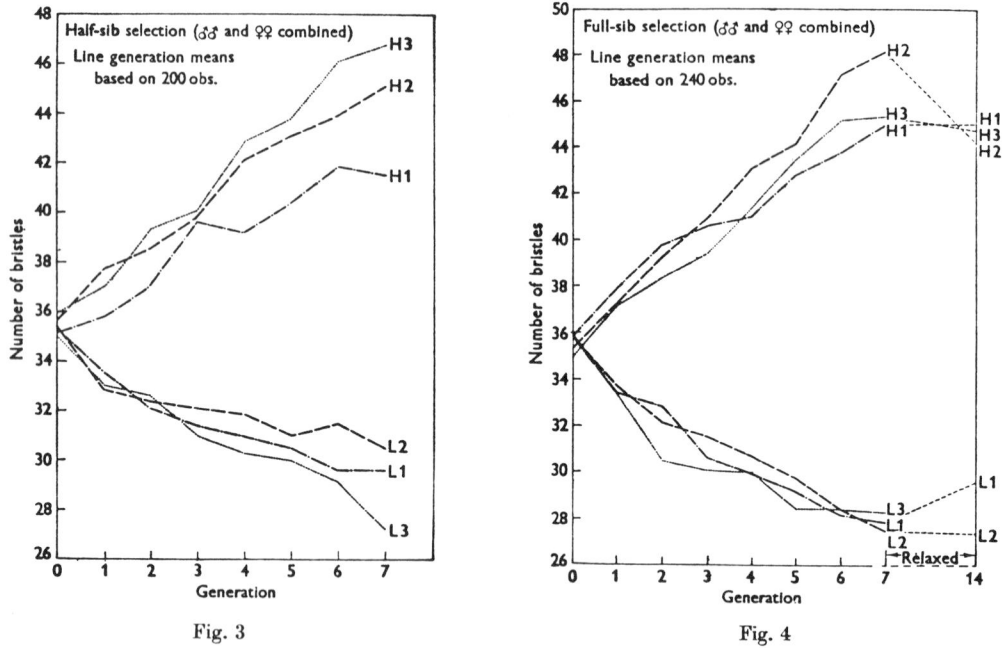

Fig. 3. Response to half-sib family selection (two families chosen from ten, twenty flies scored per family).

Fig. 4. Response to full-sib family selection (four families chosen from twenty, twelve flies scored per family).

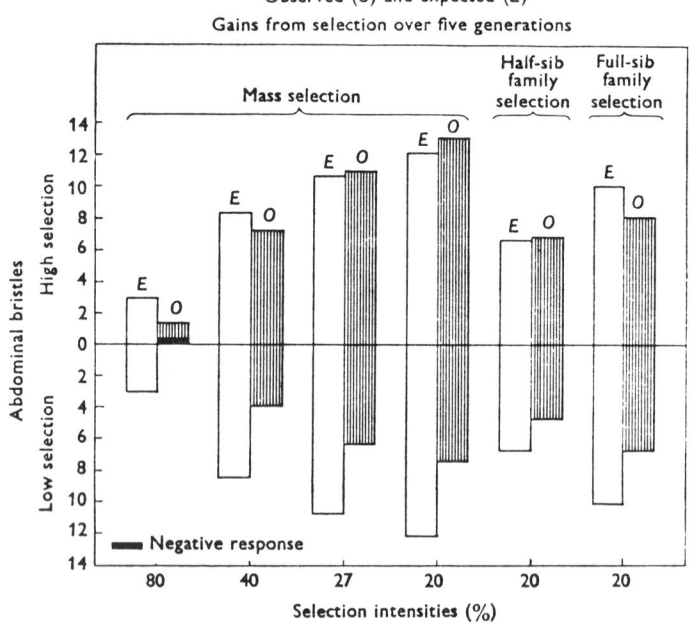

Fig. 5. Comparison of observed and predicted response for the first five generations of different selection programmes.

We can now compare these predictions with the average response in the first five generations of selection (Table 7).

In family selection we have the advantage that we can estimate the heritability in each generation from the variation between family groups. Table 8 shows the estimates, averaging over the 2nd, 3rd and 4th selected generations. These results confirm the suggestions in the individual selection experiments that the essential reason for the lower response in the downward direction is a decline in the additive genetic variance and consequently in the heritability in the early generations of selection. There is no doubt that this is superimposed on a scale effect. But if it were simply a problem of scale, the heritability would not be expected to change. The scale effect can therefore only supply a partial answer to the problem of the early asymmetry.

Table 7. *Predicted and observed response to family selection*

|  | Expected | Observed Up | Observed Down |
|---|---|---|---|
| Half-sib | 1·33 | 1·38 | 0·94 |
| Full-sib | 2·02 | 1·62 | 1·36 |

Table 8. *The change of genetic variance during selection*

|  | Base | Half-sib Up | Half-sib Down | Full-sib Up | Full-sib Down |
|---|---|---|---|---|---|
| $\sigma_g^2$ | 5·86 | 5·63 | 2·13 | 4·25 | 2·34 |
| $\sigma_P^2$ | 11·25 | 12.76 | 7·23 | 11·16 | 7·90 |
| $h^2$ | 0·52 | 0·44 | 0·28 | 0·38 | 0·30 |

### FAMILY SELECTION WITH INBREEDING

While the full-scale full-sib family selection was being carried on, it was decided to do a small experiment with the same intensity of selection, but with full-sib inbreeding every generation. The details were the same as in the full-scale experiment except that the parents from the next generation were chosen from the best family out of five. The selection was carried on for nine generations with the exception of one line, which could not be carried beyond the 6th because of sterility. In order to try to retain this line, many more pair matings than necessary had been made up in each generation. In the final and crucial generation, only one pair of flies out of sixteen produced offspring and in the previous generation only three out of fifteen.

Somewhat to our surprise, the lines responded well and were only slightly behind the full-sib lines selected without deliberate inbreeding. Calculation of the expected response, on the assumption that the additive genetic variance was all due to genes acting additively and was being reduced in the expected manner by the inbreeding, gave an expected response in five generations of 5·77 bristles compared to 10·10 in the other full-sib selection lines. The observed responses are shown in Fig. 6 and in Table 9.

Thus the response to the selection with maximum inbreeding was in moderate agreement with prediction, while that without intense inbreeding was not. However, it must be remembered that in the former we had only two replicates, and the number of flies scored each generation was only sixty.

The variation declined as expected in the low lines, but it is interesting that the high line H 3 which showed infertility also showed high variability. Table 10 shows the average phenotypic variation over the 3rd–6th generations. The values of the 2nd–4th generations in the other full-sib-lines (when the mean values were at the same level) are included for comparison.

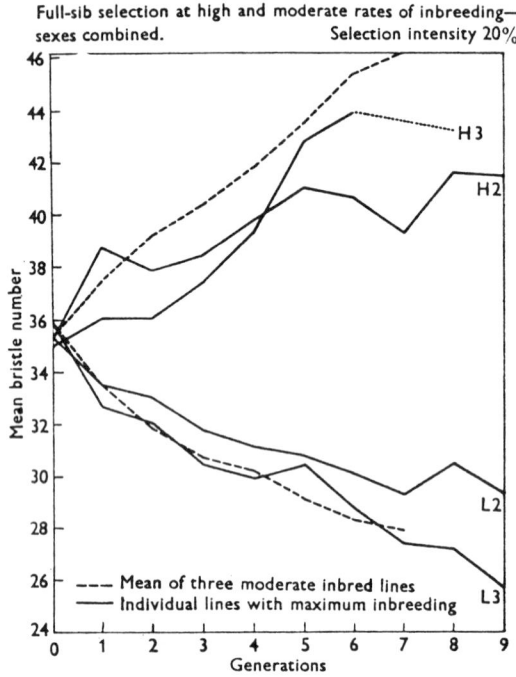

Fig. 6. Comparison of response to full-sib selection at moderate and maximum rates of inbreeding.

Table 9. *Predicted and observed response to family selection with maximum inbreeding*

|  | Expected | Observed | |
|---|---|---|---|
|  |  | Up | Down |
| Full-sib inbred | 5·77 | 6·7 | 4·5 |
| Mildly inbred | 10·10 | 8·1 | 6·8 |

Table 10. *Variation in inbred selected lines*

|  |  | D.F. | $\sigma_P^2$ |
|---|---|---|---|
| Full-sib-inbred | H 2 | 196 | 8·27 |
|  | H 3 | 179 | 13·15 |
| Mildly inbred | Up | 2161 | 11·16 |
| Full-sib inbred | L 2 | 196 | 3·77 |
|  | L 3 | 196 | 5·54 |
| Mildly inbred | Down | 2151 | 7·90 |

In the inbred lines we were surprised by the extent of response. But the above analysis suggests that the discrepancy really lay in the mildly inbred full-sib lines. We may write our expression for expected response as $\bar{\imath} h_f \sigma_{gf}$, where $\sigma_{gf}$ is the genetic standard deviation between families. In family selection for a character with as high a heritability as we find, the important effect of the early stages of inbreeding is on $\sigma_{gf}$, as $h_f$ is fairly close to unity and will change little. Now $\sigma_{gf}$ will decline as $(1-F)^{\frac{1}{2}}$ and not as $1-F$ (making the assumption, of course, that all the genetic variation is additive), so that the rate of response may be higher than we might have expected.

### THE RELAXATION OF SELECTION

The behaviour of selected lines on the cessation of selection is of interest from two points of view: it gives a measure of the natural selection which is opposing the artificial selection and is therefore relevant to the problem of asymmetrical response; and, if done in the early generations before the fixation of many segregating genes has taken place, shows the strength of the evolutionary forces holding the mean in its original position.

The behaviour of all these lines on relaxation of selection was observed, with the exception of the half-sib family selection. Measurements do not need to be made at each generation, and relaxed lines can therefore be carried with little labour. The lines in the 20/100 mass selection were relaxed at generation 5. In this case only, the lines were kept under crowded conditions, and no attempt was made to restrict the number of parents. It was realized that, although this intensifies the forces of natural selection, it did not give a measure of their effect under the conditions of artificial selection, in which larval competition was reduced as much as possible. The other relaxed lines were kept under the same conditions as during selection. The results are summarized in Table 11 and Figs. 1 and 4. Although all the lines did not behave in the same way, there seems little point in giving all the details. The effect of the relaxation is given in terms of the proportion of the original response which was retained.

Table 11. *The proportion of response retained on relaxing selection*

| Lines | Generation selected | Response | Generation relaxed | Response retained | Generation relaxed | Response retained |
|---|---|---|---|---|---|---|
| 20/100 H | 5 | 13·1 | 6 (crowded) | 0·65 | 19 | 0·64 |
| 20/100 L | 5 | 7·4 | 6 (crowded) | 0·65 | 19 | 0·72 |
| 20/75 H | 5 | 11·1 | 11 (optimum) | 0·91 | — | — |
| 20/75 L | 5 | 6·1 | 11 (optimum) | 0·98 | — | — |
| 20/50 H | 5 | 6·3 | 10 (optimum) | 0·77 | — | — |
| 20/50 L | 5 | 4·1 | 10 (optimum) | 1·05 | — | — |
| Full-sib H | 7 | 10·5 | 7 (optimum) | 0·86 | — | — |
| Full-sib L | 7 | 8·0 | 7 (optimum) | 0·90 | — | — |

One fact is immediately obvious—that the degree of return to the original mean is rather small. The 20/100 lines, for instance, had only returned about a third of the way in nineteen generations of relaxation under crowded conditions. Evidently the forces governing the maintenance of the mean in the base population are not very strong. As might be expected, the return under optimum conditions is slower. The average return of the high lines is 15 % and that of low lines 2 % of the original response in eleven generations. In fact, one of the sets of low lines showed an increase in response after ten generations of relaxation, but this could well be due to sampling. In all cases we find that the percentage return of the low lines is less than that of the high lines. The opposition of natural selection cannot then be invoked as a reason for the smaller original response of low lines to artificial selection.

In order to check the maintenance of genetic variation in lines which had not reverted to any extent on relaxation, back selection was carried out for two generations for two of the lines in the 20/75 experiment. The results are given in Table 12. It is apparent that there is a response to the back selection indicating the maintenance of genetic variation, in spite of the stability under relaxation.

Table 12. *Back selection after relaxation*

| Line | Mean before relaxation | Mean after relaxation (11 generations) | After back selection (2 generations) |
|---|---|---|---|
| L3 20/75 | 28·6 | 27·7 | 30·0 |
| H3 20/75 | 44·0 | 44·0 | 40·7 |

It seems from these results that many of the genes controlling bristle numbers must have little or no effect on natural fitness. These last two lines, for instance, had shown considerable response to selection over five generations, responded to back selection after relaxation and would no doubt have responded once more to selection in the original direction. In these and other relaxed lines, the gene frequencies must have been considerably disturbed from their values in the base population, but nevertheless the lines show only a slight tendency to return to their original mean value. What forces govern the array of gene frequencies in the base population? On this evidence, these forces must be very weak in the case of many genes, suggesting that perhaps mutation plays an important part.

### The crossing of selected lines

The behaviour of lines on crossing is to some extent informative as to their genetic situation. If, for instance, a complexly integrated genetic situation had been produced in the lines so that each line had a different complex, we might expect crosses between two high lines to show a regression to the mean and the same for the low lines.

Only the lines of the 20/100 experiment were crossed and, though this was done at generation 12, the results are included here because of their relevance to the situation in the base population. The following crosses were made:

(i) five between high lines and low lines, reciprocally,
(ii) five between high and high, reciprocally,
(iii) five between low and low, reciprocally,
(iv) crosses of all lines back to the base population, using the males of the latter. In these crosses, females alone were scored.

The crosses between high and low lines give in the males an indication of the effect of the $X$-chromosome. Although the difference between the groups of lines was at the time thirty bristles, the difference in the male score of the two groups of reciprocal crosses was 2·6, showing that at most only a small fraction of the response could have been due to genes in the $X$-chromosome, in agreement with the results of Mather & Harrison, and with the offspring-parent regressions presented earlier.

The results for the female score are best appreciated in Fig. 7. On the whole, the crosses are not widely different from the means of the parental values. The observed mean value is generally closer to that of the control population, but the extent of this regression is not great. Perhaps the $F_1$ values would not be expected to reveal the full effect of the breakdown of the epistatic situation between the lines, but in the one cross much later in the selection in which an $F_2$ was taken the mean of the latter was not closer to the control mean than the $F_1$ had been. It must be noted that the crosses are made from unselected members of the lines and that the effect of one generation of relaxation is added to that of crossing. On the whole the conclusion that would be reached from the behaviour of these crosses is that different epistatic combinations have not been built into the lines, a result again in agreement with Mather & Harrison who found that interaction between chromosomes from selected lines was of little importance.

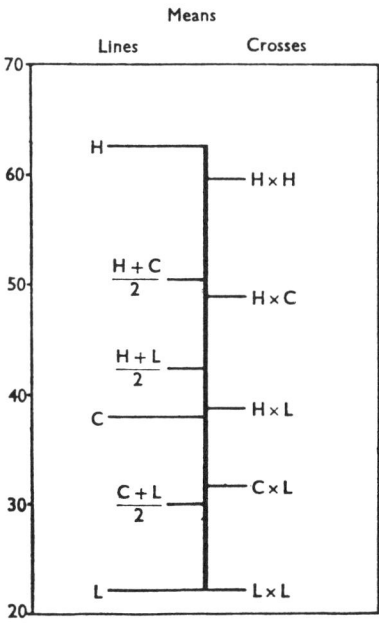

Fig. 7. The average results of crosses between lines at generation 12 of the 20/100 selection experiment (female scores only). H, high lines, L, low lines, C, base population.

## Discussion

On the whole, the experimental observations of the initial population are consistent with each other in the light of existing theory. The different estimates of heritability were in good agreement with one another, bearing in mind their various standard errors. The full-sib estimate was insignificantly larger than the other two, suggesting that a proportion of the variance might be due to dominance; but the standard error of any such estimate of the variance due to dominance deviations would be so large as to make the estimate almost worthless. The fact that the character is made up of two separate measurements, allows us to infer from the observed variance between measurements that dominance deviations cannot be very important, and this is supported by the fact that inbreeding had no effect on the mean of the character.

In the main, selection for bristles in this population has given results in fair agreement with theory, though this is in some measure a subjective judgement. We can, however, move forward from this purely operational point of view and inquire what further information can be obtained from the results. The results throw some light on the nature of the genetic variation in the initial population and on the forces which maintain it. It is difficult to avoid the conclusion that the genes controlling abdominal bristles show additive action or something close to it, in the sense that the heterozygote must be roughly intermediate between the two homozygotes on the average. We observe that hereditary variation cannot make up more than 61 % of the total variance, while 52 % is due to the average effects of genes, i.e. the latter variance makes up 85 % of the genetic variance. It is possible to account for this on the assumption of complete dominance if all the dominants are present at low frequencies, but this seems to be rather an improbable hypothesis from an evolutionary aspect. The suggestion of additive action is borne out by the crosses between selected lines which indicate little in the way of genetic complications, by the effect of inbreeding, and quite independently by Mather & Harrison's (1949) chromosome assays of their selected lines. The significance of this finding has been discussed elsewhere (Robertson, 1955), and it has been suggested that there is little direct connexion between bristle number and fitness. The modification of dominance properties of the genes will take place primarily along the causal pathways leading from primary gene action to reproductive fitness or selective advantage. 'Peripheral' characters lying off this pathway will therefore not be affected by the evolution of dominance and will show the additive action which we often observe when we can get close to primary gene action, as in the case of blood groups. It may be argued that if bristle number has no direct relationship with fitness, or, to put it in more general terms, is not 'useful' to the animal, then there is little point in investigating its inheritance. It would be a more reasonable view to regard the results as being of interest because they refer to a character of an extreme type with, of course, fitness itself at the other extreme. Its apparently simple genetic control does not make bristle number any less interesting. From the animal breeder's point of view, some characters of economic importance may be of this type, such as the fat content of milk in cattle and the thickness of backfat in pigs.

This experiment and others recently reported by Rasmuson (1955) show that considerable genetic variation in this character exists in wild populations. We may reasonably ask what light our results throw on the forces controlling this variation. The results of the relaxation of selection in lines selected for five generations, which showed that the line means had returned only about one-third of the way back to the original mean after nineteen generations of relaxation (Fig. 1), suggest that these forces were not very strong, and that many of the genes controlling the character must be nearly neutral in their effects on reproductive fitness. We have sufficient evidence on the effects of relaxation to feel justified in relying on the results. Somewhat contradictory evidence comes from the parent-offspring regression experiment in which it was observed that a higher proportion of the extreme females were sterile than were the intermediates. The females put into vials had a variance of abdominal bristle score of 14·08, and those which produced offspring had a variance of 12·53 units. Rather oddly, the same effect was not observed with sterno-pleural bristles in the same females in which the variances were 1·36 and 1·46 respectively. The greater sterility of the females extreme for abdominal count could be explained either as an effect of the bristle deviation itself or as a consequence of the fact that extremes will show a higher level of homozygosity (see Robertson, 1956). At first sight these results are not in agreement with the results on relaxation, but further evidence is needed before the problem can be discussed in detail, though the relaxation evidence must at the moment be given greater weight.

Recent work on the origin of new variation in this character (see Durrant & Mather, 1954; Clayton & Robertson, 1955; and Paxman, unpublished) suggests that new variance

148  *An experimental check on quantitative genetical theory*

is arising by spontaneous mutation at the rate of roughly 0·01 unit each generation. This is too small to affect to any extent experiments on the time scale that we have been using, but indicates that mutation must be considered seriously in any discussion of the maintenance of genetic variation in this character in wild populations. More experimental evidence is needed on the role of natural selection in maintaining variability before we can come to any definite conclusion on these points.

The view which one takes of the agreement between response to selection and prediction seems to be a personal one, judging by the reaction of people with whom we have discussed this work. There are those who have been surprised by the uniformity of response and those who remark immediately on the individual behaviour of the replicate lines.

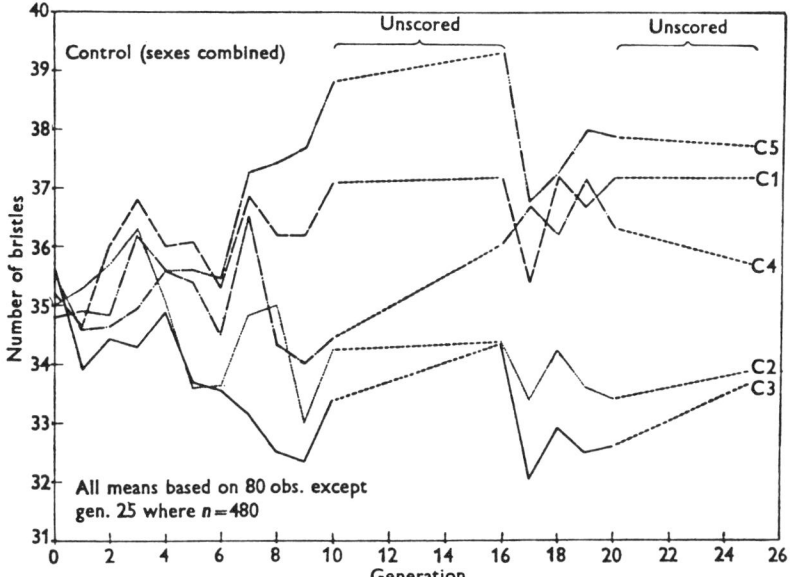

Fig. 8. Genetic divergence between control (unselected) lines (twenty pairs of parents in each line). Maintained under conditions identical with the mass selected lines.

One aspect of this is undoubtedly that the magnitude of chance deviations between lines due to genetic sampling is not usually realized. Our control lines (Fig. 8) established a definite order within themselves in the absence of selection, which analysis has shown to be in agreement with that expected on inbreeding theories. The divergence between replicates is, as would be expected, greatest in the family selection where the increase in inbreeding coefficient was $6\frac{1}{4}\%$ each generation. By the 5th generation, the expected standard deviation between replicates is 1·9 bristles, compared to the expected response of 6·65 and 11·10 in the half-sib and full-sib selection. It is then not surprising in this case that there is not agreement between replicates.

The differences among replicates were considered in some detail in the 20/100 individual selection experiment. In the up-lines the deviations were considerably greater than would be expected on the basis of genetic sampling alone. There are several other possibilities which may play a part in this. As well as affecting the mean of a character, genetic drift would be expected also to change the genetic variance in the different lines. If these lines were being subjected to selection, the effect of drift on the variance would be superimposed on the drift in the mean. Furthermore, one would expect the effect of drift in variance to be to some extent persistent from generation to generation. In other words, if the genetic variance were higher than average for a certain line in one generation, it would be expected to remain higher in the next generation. A theoretical consideration of this problem in the case of genes with additive action and two alleles at each locus suggests that

(i) The effect of genetic sampling on variance will be at its minimum when allele frequencies are close to one half.

(ii) The variation between lines in genetic variance will depend on the number of genes concerned and will be higher the smaller the number of genes.

(iii) If gene frequencies are close to one extreme, there will be a correlation between the drift effect in the mean and its effect on variance. If selection is tending to increase genes from low to intermediate frequencies, there will be a positive correlation between the two, i.e. a sampling change in gene frequency which increases the mean will also increase the variance. In the opposite direction there would be a negative correlation. This might explain the apparently greater drift in the up-lines than in the down.

(iv) If we include the possibility of having many alleles segregating at each locus or, in rather wider terms, several alternative segregating units, then the possibility of divergence between replicates becomes even greater. The presence of a rare allele or combination of genes in one line may then make the variance and response to selection of that line quite different from other lines.

The effect of the number of genes concerned is important because *D. melanogaster* has effectively only three chromosomes, one of which is only half the length of the others, and there is no crossing-over in the male. The effect of the selection may be to increase the frequency of whole sections of the chromosome, so that the effective number of genes is much below its actual value. The discrepancies between replicates may then be a property of *Drosophila* selection experiments and therefore perhaps need not cause general concern. The observation remains that if *Drosophila* are to be used in model experiments for the comparison of selection programmes, a considerable number of replicates are necessary.

Of the two other main discrepancies with theory the first concerns the smaller response in the lines selected downwards. Here we must first of all deal with the problem of scale. From the point of view of the variance, the most satisfactory scale to use would be the logarithmic one in that the coefficient of variation on the natural scale remains fairly constant in the early generations of selection. If all effects, both genetic and environmental, are to alter in the same manner on shifting the mean expression, then it would be expected that the change in response would be paralleled by a change in the total variance and not by a change in the heritability. In fact, we find that the heritability, whether determined by the ratio of response to selection differential or from half-sib and full-sib analyses, has declined by the 2nd or 3rd generation in all these down selection experiments. The results of relaxation show that opposing natural selection plays no part in this asymmetry. Closer examination of the variance suggests that the environmenta variance has in fact behaved as would have been expected in a log scale, but that the genetic variance has been considerably reduced. The cause of this remains obscure. We may interpret the facts rather differently if we say that the effects of gene substitutions become smaller as we go down the scale, but this is hardly an explanation. From the genetic point of view, we might suggest that a single gene of fairly large effect has been fixed in the low line but not in the high line. The fall-off in response to selection did not occur until the 2nd or 3rd generation, and the response in the 1st generation was fairly large. This early response might be the effect of the fixation, and the fact that the gene had been fixed would then cause the variability to be reduced. However, we have no proof of this suggestion.

Finally, we have the surprising and perhaps important fact that in the individual selection lines the ratio of observed to expected response shows a regular decline from the high intensities to the low. What this means is difficult to say. What it suggests is that the low intensity work should be repeated and this we propose to do.

## Summary

1. The paper describes the response, in the early generations, of a large random-breeding population of *Drosophila melanogaster* to selection for abdominal bristle score. The main purpose of the work was an examination of the adequacy of existing theory in describing these phenomena.

2. Estimates of the heritability of the character by parent-offspring, full-sib and half-sib and half-sibs correlations were in good agreement, the mean value being 0·52.

3. As a further 35% of the variance in total score could be ascribed to causes affecting sternites separately, it seems very probable that the genes affecting this character act additively or nearly so. This is in agreement with the results from crosses between lines.

4. The responses to selection over seven generations, based either on individual or family score, were in fair agreement with predictions from these estimates with three reservations:

(a) There was considerable divergence between replicate lines selected upwards, even when the degree of inbreeding was low. This may be peculiar to *Drosophila*, which should therefore be used with care in experiments intended for extrapolation to other species.

(b) The responses to downward selection were less than expected, as a consequence of a decline in genetic variation within two or three generations.

(c) The agreement with prediction was best at high intensities of selection, the response at lower intensities being below expectation.

5. Various of these observations are relevant to the problem of the maintenance of such variation in a population. It seems probable that the forces controlling any genetic equilibrium cannot be strong and that in this character mutation may play an important role.

## References

ANDERSON, V. L. & KEMPTHORNE, D. (1954). A model for the study of quantitative inheritance. *Genetics*, **39**, 883–98.
CLAYTON, G. A. & ROBERTSON, A. (1955). Mutation and quantitative variation. *Amer. Nat.* **89**, 150–8.
COCKERHAM, C. C. (1954). An extension of the concept of partitioning hereditary variance for analysis of covariances among relatives when epistasis is present. *Genetics*, **39**, 859–82.
CROW, J. F. (1954). Breeding structure of populations, II. Effective population number. *Statistics in Mathematics and Biology.* Ames, U.S.A.: Iowa State Coll. Press.
DURRANT, A. & MATHER, K. (1954). Heritable variation in a long inbred line of *Drosophila*. *Genetica*, **27**, 97–119.

FALCONER, D. S. (1953). Asymmetrical response in selection experiments. *I.U.B.S. Symposium on Genetics of Population Structure.*
MATHER, K. & HARRISON, B. J. (1949). The manifold effect of selection. *Heredity*, 3, 1–52 and 131–62.
MILLER, A. (1950). The internal anatomy and histology of the imago of *Drosophila melanogaster*. *Biology of* Drosophila, ed. M. Demerec. New York: John Wiley and Sons.
RASMUSON, M. (1955). Selection for bristle number in some unrelated strains of *Drosophila melanogaster*. *Acta Zool., Stockh.*, 36, 1–49.
REEVE, E. C. R. & ROBERTSON, F. W. (1954). Studies in quantitative inheritance. VI. Sternite chaeta number in *Drosophila*: a metameric quantitative character. *Z. Vererbungslehre*, 86, 269–88.
ROBERTSON, A. (1955). Selection in animals, synthesis; *Symposia on Quantitative Biology*, XX. Cold Spring Harbor.
ROBERTSON, A. (1956). The effect of selection against extreme deviants based on deviation or on homozygosis. *Genetics* (in the Press).
TEBB, G. & THODAY, J. M. (1954). Stability in development and relational balance of X chromosomes in *Drosophila melanogaster*. *Nature, Lond.*, 174, 1109–10.
WRIGHT, S. (1921). Systems of mating. I–V. *Genetics*, 6, 111–78.

# THE EVALUATION OF NEW METHODS FOR THE IMPROVEMENT OF QUANTITATIVE CHARACTERISTICS[1]

A. E. BELL, C. H. MOORE[2] and D. C. WARREN[2]

Purdue University, Lafayette, Indiana

## Introduction

Rapid and extensive changes are occurring in the applied as well as the experimental approach to improvement of quantitative characteristics in animal populations. Three major developments which tend to intensify these changes of viewpoint are: (1) The widespread acceptance of population genetics as an effective method of approach to the problems of breeding for improvement of quantitative traits; (2) the realization by many breeders that selection within closed populations was yielding diminishing or no returns as higher performance levels were reached; and (3) the need for an evaluation of breeding techniques designed to utilize heterosis or hybrid vigor.

### Population Genetics Approach

While the basic principles of population genetics were early defined largely by Fisher, Haldane and Wright, application in the applied fields lagged some 20 years. Due largely to the persistent efforts of Lush (1945) and his students, the principles of population genetics were first applied in the field of animal breeding. The most intensive application to poultry breeding has been the work of Lerner (1950). Other investigators also have contributed to our present day wealth of information on the heritability of various quantitative traits, the genetic correlations between these traits and the combination of these traits into selection indices. An important phase of this approach concerns the accuracy of these statistical tools of population genetics in predicting subsequent performance. In the final analysis, the practical utility of these genetic concepts will depend upon the effectiveness of this approach in obtaining desired improvement in animals and plants. While testing these concepts in each of our economic species is highly desirable, other genetic problems are equally demanding on the available experimental facilities. Rapidly reproducing laboratory animals are proving efficient in testing the reliability of many of the concepts such as coefficient of inbreeding, heritability, genetic correlations, etc. Notable examples of such studies are those of Kyle and Chapman (1953) with rats, Falconer (1953) with mice, and Robertson and Reeve (1952) with *Drosophila*.

### Plateaued Populations

The genetic structure of a population which fails to respond to selection, even though genetic variation is apparently present, poses another challenging problem to the geneticist. This problem is of immediate practical significance to the applied breeder. Some breeders, after a careful analysis of performance records over several generations, are convinced that they are making little or no progress in improving certain important quantitative traits. Will continued selection eventually break through these performance plateaus? Will further improvement await the introduction of new genetic variation into the closed population, or is a new method of breeding or selection required for further progress? Could these plateaus possibly be due to physiological ceilings for the species or as a result of negative genetic correlations between selected characteristics? Does irradiation provide an avenue for further improvement in plateaued populations as suggested by the studies of Scossiroli (1954) with *Drosophila*? Answers to these questions are not readily apparent on the basis of today's knowledge. Additional experimental evidence is needed.

### Breeding for Heterosis

The classical contributions of Shull, East, Jones and others in the early experimental exploration of inbreeding and hybridization in maize, followed by its successful application by commercial corn breeders, have served as a challenge to animal breeders. Extensive investigations into the feasibility of inbreeding and hybridization of economic species of animals have been made over the past quarter of a century by both experiment stations and commercial establishments. While some favorable results have been reported especially in swine and poultry, few will deny that the success of this method of breeding has hardly measured up to the example of hybrid corn in America. More efficient methods of developing and identifying inbred lines of superior combining ability would facilitate this program.

The practical utilization of this phenomenon of heterosis, undoubtedly, would be simplified by understanding its genetic base. A frank appraisal of our knowledge to date on this subject indicates that many of the manifestations of heterosis have been unearthed; yet, as to the underlying basic gene action, we are scarcely more enlightened than were

---

[1] Journal Paper No. 885 of the Purdue University Agricultural Experiment Station. Supported in part by grants from the Rockefeller Foundation and the National Science Foundation.

[2] Animal and Poultry Husbandry Research Branch, Agricultural Research Service, U.S.D.A., Lafayette, Indiana.

the pioneers in this field. In fact, recent investigations are reviving the early theory of heterosis expounded in principle by Shull and East. They explain heterosis largely on the basis of heterozygosity *per se* in contrast to the more widely accepted hypothesis of linked dominant factors initiated by Jones. Enlarging on the concepts of Shull and East, Hull (1946) concluded that overdominance must be an important factor in heterosis of yield in corn. His efforts should be recognized as a major force in stimulating animal and plant geneticists to search for new approaches to a maximum expression of heterosis. From theoretical calculations, Crow (1948) added considerable support to the overdominance concept. He suggested that heterosis might be accounted for by overdominance at only a relatively few loci. Numerous other important contributions could be cited.

The above developments have lead to the realization that more efficient methods of selection for improving quantitative traits are theoretically possible. Two new methods have been proposed as being especially efficient in exploiting the maximum heterosis in traits influenced to a large degree by overdominance. These are: (1) Recurrent Selection to a tester line for specific combining ability as proposed by Hull (1945); and a logical modification of Hull's scheme, (2) Reciprocal Recurrent Selection for specific combining ability as developed by Comstock *et al.* (1949). Both methods feature progeny testing and selection within a segregating population. The selection of individuals to reproduce each segregating population is not based on their own phenotypes, but upon the merits of their cross progenies. One other distinguishing feature of both methods is that in reproducing the segregating populations, inbreeding is deliberately avoided, thus minimizing chance fixation of undesirable alleles. The major difference between the two methods is that Method 1 utilizes an inbred or single cross tester, while Method 2 features two segregating populations, each serving as the tester for the other. While these new methods offer the theoretical possibility of exploiting heterosis without the laborious task of inbreeding, their biological and practical utility must await experimental breeding tests. This presents another urgent demand on the limited available research facilities.

The foregoing discussion emphasizes the fact that theoretical contributions in quantitative genetics are far ahead of their experimental verification. In an effort to obtain a more rapid biological evaluation of these new methods and to explore further some of the conventional methods, we began pilot experiments with *Drosophila melanogaster* in 1949. Most of our studies to date have related to two quantitative traits, fecundity and egg size. While these traits possess the obvious disadvantage of being measurable only in females, they have been useful in that they differ considerably in their inheritance.

In some studies, lifetime fecundity data were taken. In order to take advantage of the rapid reproduction cycle of *Drosophila*, fecundity data usually were limited to the peak period of four to seven days of age. Techniques used in measuring fecundity were similar to those described by Gowen *et al.* (1946). The work of Gowen and our own unpublished data show fecundity to have a low heritability (5 to 15% within non-inbred stocks) and to exhibit much heterosis in crosses. Egg size was expressed as length of eggs in eye-piece micrometer units. Five eggs from each female under test were placed on a glass slide and their lengths recorded with a microscope equipped with a $4\times$ objective and a $20\times$ eye piece containing a micrometer. The early work of Warren (1924) and our own unpublished data indicated that egg size provides an excellent contrast to fecundity. In crosses it shows slight heterosis, if any, and is highly heritable with estimates in the range of 30 to 60 per cent.

In this paper we will present the results from a series of experiments pertaining to more efficient methods of exploiting heterosis. The first phase will relate to the predictive value of early testing of inbred lines, a much debated problem among corn geneticists. In turn we will consider the influence of selection within and between inbred lines and a comparison of four methods of selection for improving quantitative traits.

### EARLY TESTING OF INBRED LINES

The testing of available inbred lines in all possible single cross combinations is obviously the most accurate way of identifying superior hybrids. The feasibility of this method is immediately questioned when one realizes that for as few as 20 inbred lines there are approximately 400 combinations and for 100 inbred lines the number of combinations approaches 10,000. Therefore a practical testing program must utilize some method of screening inbred lines short of testing them in all possible combinations. In recent years many corn breeders have followed the practice of evaluating inbred lines by making topcrosses or single crosses at various stages of inbreeding ranging from open pollinated varieties to three or four generations of selfing. A recent survey of the advantages and limitations of early testing in corn was made by Sprague (1952). Nearly all the corn data support the view that testing at an early stage of inbreeding has predictive value as to the subsequent combining ability of inbred lines. The conflicting views among corn breeders seem to center around the relative degree of accuracy of this predictive value and its practical utility.

No information was available on the value of early testing of the combining ability of inbred lines of animals when our experiments were initiated. Later studies by Loh (1949) revealed that early testing of inbred lines in *Drosophila* had no value for predicting their combining ability at later stages of inbreeding. Warwick and Lewis (1954) found that growth of single cross mice could not be predicted from the early testing of the inbred lines. The first problem investigated in our laboratory related to the value of early testing for predicting the fecundity and egg size of single crosses of inbred lines (Moore, 1952). Some of the more significant results of this study are presented here.

## Foundation Stocks

Initial stocks consisted of nine wild type non-inbred stocks of *Drosophila melanogaster*. Seven of these stocks had recently been captured at various places in the United States, while the other two stocks had been maintained for several years as laboratory stocks. Fecundity and egg size data were collected on each of the nine stocks in six experiments replicated in time and involving 72 females from each stock. Inbreeding was carried out by full sib matings within each line each generation. Sufficient sublines were made each generation to avoid loss of any of the lines. As the inbreeding progressed, data were collected on the fecundity and egg size of single cross females resulting from all possible combinations of the nine lines at levels of 0, 3, 13 and 38 generations of full sibbing within the lines.

Analysis of variance for the fecundity data on the nine foundation stocks provided information which was helpful in understanding subsequent early testing results. Other than natural selection for reproductive fitness, no selection for fecundity had been practiced within any of the nine stocks. The statistical analysis revealed no significant differences among the stocks. However, a highly significant replication by stock interaction was observed. This interaction between genotype and environment would reduce the predictive value of any kind of testing unless the environmental conditions could be identified and repeated in subsequent tests. Of the total variation observed in testing these nine stocks approximately 65 per cent was due to individual differences within stocks, about 20 per cent was contributed by the genetic-environmental interaction, another 10 per cent due to replication differences and not more than 5 per cent for solely heritable differences between the stocks.

Analysis of the egg size data collected on the nine foundation stocks revealed an altogether different pattern from that shown for fecundity. In the first place, highly significant differences between the stocks were observed. Secondly, no genetic-environmental interaction was found as tested by interaction of stocks by replications. Of the total variation observed between all individual phenotypes, about 55 per cent was due to individual differences within stocks and the remaining 45 per cent were associated with differences between the stocks. This 45 per cent estimates in the broad sense the heritability of egg size found in this particular sample of stocks.

## Early Testing Results

As mentioned above, the nine stocks or lines were crossed in all possible combinations, excluding reciprocals, at 0, 3, 13 and 38 generations of inbreeding. This provided a total of 36 single crosses at each level of inbreeding for establishing the relationship of combining ability between the various intervals of inbreeding. In general, it was found that the relative level of fecundity for a single cross at any one interval had little or no value in predicting its relative position at another testing period. However, reasonable predictive values were found for egg size. The average relationship for both traits between the various test periods are shown in Table 1.

Before one attempts to relate these results to those published for corn, consideration should be given to the degree of genetic relationship existing in the two types of genetic material between individuals used in test-cross matings and those used as parents of the inbred line. Much of the early testing in corn breeding follows the procedure of self-pollinating a series of $S_0$ plants from an open pollinated variety or an $F_2$ population and outcrossing these same plants to a tester line. This establishes a 50 per cent genetic relationship between the cross progeny and first generation inbreds. In the *Drosophila* experiment described here, the individuals taken from each non-inbred stock to make the "0" generation crosses were never used for parents of the first generation inbred lines, thus no initial genetic relationship between the cross results and the inbred lines would exist as compared to the 50 per cent present in comparable corn material. If our sample of individuals used in making the "0" generation crosses were unrelated to the pair of individuals forming each inbred line, then one would not expect any relationship between these crosses and crosses made after inbreeding unless there were significant genetic differences between the foundation stocks. In the latter case, individuals sampled from any stock certainly on the average would be more alike genetically than if taken from different stocks. This in turn would lead one to expect a positive relationship between "0" generation crosses and those made after inbreeding. The correlations shown in Table 1 illustrate this point.

Since no significant differences were found for fecundity between the initial stocks, one would ex-

TABLE 1. RELATIONSHIP OF PERFORMANCE IN 36 SINGLE CROSSES AT VARIOUS GENERATIONS OF INBREEDING FOR 9 INBRED LINES

| Traits measured | Correlations by various generations of inbreeding | | | | | |
|---|---|---|---|---|---|---|
| | $0 \times 3$ | $0 \times 13$ | $0 \times 38$ | $3 \times 13$ | $3 \times 38$ | $13 \times 38$ |
| Fecundity | −0.20 | −0.04 | 0.21 | 0.22 | 0.09 | 0.25 |
| Egg size | 0.39* | 0.36* | −0.16 | 0.32 | −0.19 | 0.37* |

*Significant at the .05 level of probability.

pect the relationship to be low between $0 \times 3$, $0 \times 13$, and $0 \times 38$. Actually, two of the correlations are negative; however, none are statistically significant. Single crosses of inbred lines were somewhat more accurate in predicting subsequent performance than were the crosses of the non-inbred stocks. The three comparisons involving generations of inbreeding at 3, 13 and 38 were all positive and two of the correlations approached significance at about the 0.10 level of probability.

The predictive values shown for egg size in Table 1 are in direct contrast to the fecundity results. Here we observe significant correlations between the crosses of the non-inbred stocks and crosses at both generation 3 and 13 of inbreeding. However, it should be recalled that highly significant differences in egg size were found among the foundation stocks. This could have contributed materially to the predictive value of these early crosses. Except for the more distant comparisons ($0 \times 38$ and $3 \times 38$) egg size in single crosses had a positive and significant value in predicting egg size in these same single crosses made after further inbreeding of their parental lines.

In addition to the greater differences in the initial stocks, one other factor would contribute to higher predictive values for egg size in contrast to fecundity. The highly significant genetic-environmental interaction observed for fecundity would certainly contribute to low predictive values in a testing program. It suggests that each genotype may react differently to different environments. No such interaction was observed for egg size.

*General versus Specific Combining Ability*

These experiments were so designed to allow estimates of general and specific combining abilities of the nine lines at each level of inbreeding. General combining ability is determined largely by the additive effects of genes while specific combining ability reflects the influence of epistasis and dominance.

One of the most obvious points brought out in these analyses was that some inbred lines yielded much more uniform estimates of general and specific combining ability than other lines. On the average, the degree of inbreeding in the lines did not appear to influence the relative importance of the two types of combining ability.

If specific combining ability is largely a function of epistatic or dominant gene action and general combining ability reflects additive gene effects, then the relative importance of these two functions should vary inversely between the two quantitative traits studied. In general, such a relationship was observed. For the highly heterotic trait, fecundity, specific combining ability values were from 10 to 30 per cent of the error term while general combining ability estimates were only 5 to 15 per cent. Conversely, egg size values were 8 to 20 per cent for specific and 10 to 35 per cent for general combining ability estimates.

In summary, it can be concluded that the results from these experiments certainly add little support for the use of early testing for heterotic traits such as fecundity in *Drosophila*. In drawing this conclusion the limitations of our experiments should be kept in mind. In the first place, no significant differences were present in our initial foundation lines. Secondly, a highly significant genotype-environmental interaction was present. Thirdly, all testing periods, except one, involved in our prediction studies were separated by not less than ten generations and some by as many as 38 generations of inbreeding. Detailed analysis of the data showed that the poor average predictive values from one test period to another was not a random error involving all lines, but was largely due to two or three inbred lines changing drastically in their combining ability with all the other lines. Such lines in turn exhibited sharp changes in their general and specific combining ability estimates. If this were an isolated observation, it might wisely be credited to genetic contamination. However, this sudden change in the combining ability of an inbred line has occurred repeatedly in other experiments in our laboratory; thus indicating that an inbred line reproduced by brother-sister matings for even more than 100 generations may not be highly homozygous or genetically uniform, but is in a delicate state of genetic equilibrium between the forces of inbreeding, mutation and natural selection. While the inbreeding is a constant force toward homozygosity, natural selection acting on new mutations or recombinations of tightly linked genes could produce rapid changes in the genetic constitution of an inbred population. Durrant and Mather (1954) found evidence in a highly inbred line of *Drosophila* that suggested the occurrence of recent mutation of quantitative genes.

## SELECTION WITHIN AND BETWEEN INBRED LINES

Selection within and between inbred lines and the evaluation of the lines in hybrid combination is undoubtedly the most widely tested method of selection for hybridization of economic species of animals and plants. Sprague (1952) pointed out that this method is still the one most extensively used in corn breeding. Nearly all experimental work in the inbreeding and hybridization of farm animals, including poultry, has followed this approach. No attempt will be made to cite the abundant literature relating to this general problem. We have been interested in characterizing inbred lines in as many different ways as possible and in relating these to their general and specific combining ability for quantitative traits. One very practical criterion is the phenotype of the inbred line. In other words, how does phenotypic selection within and between inbred lines affect their general and specific combining ability?

*Selection Within Inbred Lines*

There is little experimental evidence available as to the influence of selection within an inbred line on its phenotype and on its combining ability. Under the dominance hypothesis of heterosis or even one incorporating epistasis, selection within an inbred line would be considered not only effective in minimizing the inbreeding depression but also would possibly increase the combining ability of the line. As the concept of overdominance has been revived, some workers now consider inbreeding depression to be an automatic consequence of increasing homozygosity, and probably necessary if maximum heterosis is to be obtained in crosses. The most extensive analysis of animal data on the effects of inbreeding is that of the swine results summarized by Dickerson (1952). The analysis for litter size and growth rate in 49 inbred lines from five projects with an average of nine seasons per line suggested that selection within inbred lines had been ineffective. The inbreeding depression observed under selection had been about that expected under inbreeding without selection. The report by Shultz (1953) on inbreeding and selection in poultry showed clearly that selection had moderated the inbreeding depression for such traits as fecundity and egg size.

We are studying in *Drosophila* the effects on their combining ability of selection within inbred lines. From each of a number of non-inbred foundation stocks inbred lines have been developed simultaneously by five different types of selection. Inbreeding each generation was by brother-sister matings. The kinds of selection were: (1) high fecundity, (2) low fecundity, (3) unselected (other than natural selection), (4) large egg size, and (5) small egg size. Usually, the performance of 30 females per line were measured each generation and the one showing maximum deviation in the desired direction was selected along with a full brother to reproduce the specific line. One of the foundation stocks used was comparable in some respects to improved strains and breeds of poultry or other farm animals. This particular stock had been a closed population for ten generations with selection in each generation being made on the basis of individual and family merit for both high fecundity and large egg size. Little improvement, if any, had been made in fecundity after generation five; however, egg size was still improving. By avoiding matings closer than cousins, inbreeding in this population had been kept to a minimum. The response of this population to within-line selection while inbreeding is shown in Figure 1. The reader is cautioned from attempting to compare the performance of a line between various stages of inbreeding. No control line was carried in these experiments and any differences in performance between generations is confused with possible environmental trends. The unselected inbred line should not be considered a control population for measuring environmental fluctuations since segregation and chance fixation especially during the early generations of inbreeding could have caused the observed changes in the performance of the line. In order to level out the environmental fluctuations, the points plotted at each generation in Figure 1 are three generation running averages within each line. However, the relative effects of

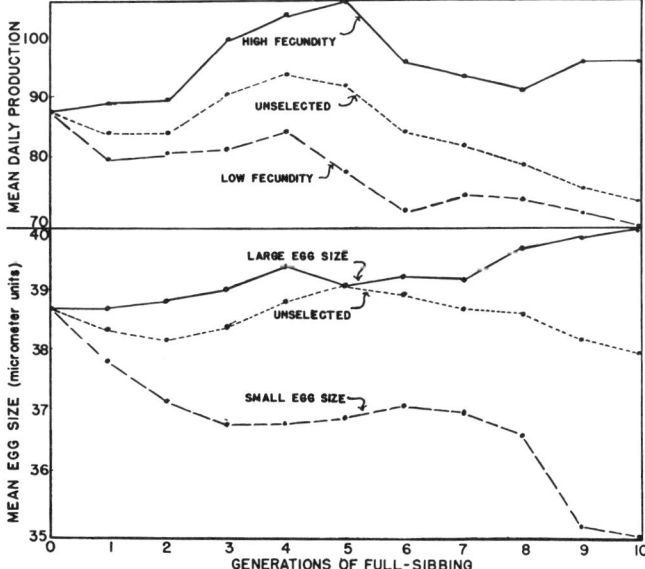

FIGURE 1. Influence of inbreeding with selection on fecundity and egg size in *D. melanogaster*.

selection can be determined accurately by comparing the three types of selection at any generation of inbreeding since these observations were made simultaneously. The effectiveness of the three types of selection for fecundity is shown in the upper graph while the lower graph shows response to egg size selection.

Inbreeding and within line selection from foundation stocks with no previous history of artificial selection showed the same general effects as illustrated in Figure 1, except that the difference between the various selection lines usually were greater, probably signifying a greater initial genetic variation. These results suggest that within line selection during inbreeding can be effective. The so-called "inbreeding depression" can, therefore, be lessened by selection pressure exerted during the inbreeding process.

*Selection between Inbred Lines*

The phenotype of an inbred line is of considerable significance in a practical breeding operation. Other things being equal one desires inbred lines with the best phenotype obtainable. Much of the inbreeding work in plants and animals has operated more or less on this principle. The bottleneck for the inbreeding and hybridization of most economic species is not the formation of inbred lines, but it is the problem of adequate testing of inbred lines in hybrid combinations. Since all inbred lines developed cannot easily be tested in crosses, the relationship of inbred phenotype to general and specific combining ability becomes of interest. If overdominance is a major influence in heterosis, as some think, one would not expect to obtain the superior hybrids by crossing only good inbred lines. Hull (1952) pointed to yield data in corn which suggested that the highest specific hybrids were found in the combination of high by low inbreds.

Our approach to the above problem is to cross a group of inbred lines of varying fecundity and egg size in all possible combinations, excluding reciprocals, and then to measure both fecundity and egg size of the resulting hybrids. Results have been summarized for four such sets of data with each replicated on three successive weeks. The number of inbred lines involved in the four sets were 8, 8, 9 and 12. These provided a total of 37 inbred lines tested in 144 hybrid combinations. Different inbred lines were used for each set of data, except in the last group of 12 lines. There were four lines used in the previous group. One way of summarizing these data is to divide the inbred lines in each set into three approximately equal classes based on the inbred's own phenotype. The three classes were designated as high, medium and low inbreds. Each set of data was summarized separately for fecundity and for egg size according to whether the parents were classified as high, medium or low inbreds and by hybrid combinations of these inbred classes.

Since the results for all four sets of data were remarkably consistent for both fecundity and egg size, only the averages are given in Table 2. In the upper part of this table one observes the magnitude of the differences which existed among the three arbitrary classifications of inbred lines. The most significant point to observe is in the lower portion of the table. Here the hybrids are summarized by the phenotype of their inbred parents. These mean values represent the average or general combining abilities of these phenotypic classes of inbreds. Thus one can conclude that the general combining ability of an inbred line for both fecundity and egg size is closely related to or predicted by its own phenotype.

The relationship of an inbred's phenotype to its specific combining ability is more significant than to its general combining ability. In the final analysis one selects a specific outstanding combination of inbred lines rather than some inbred line having high average combining ability. It is possible for an inbred line to have a high general combining ability and yet not to be a parent of any elite or superior hybrid. A practical approach to this problem is to consider the possibility of discarding or culling a certain portion of the poorer inbreds before undertaking the laborious task of testing them in hybrid combinations. Undoubtedly, this is a standard procedure to some extent in most inbreeding and hybridization programs. Some inbred lines are so poor that they automatically eliminate themselves. Others have to be discarded in choosing among lines and sublines to go into limited breeding pens and testing facilities.

In the four sets of data described above the inbred lines were crossed in all possible combinations. Therefore, it was possible to identify the best hybrids obtainable within each group. For the purpose of discussion, the best ten per cent of all hybrids within each group will be considered the elite hybrids. The major objective of any testing program is to identify these elite hybrids. Obviously, if a number of inbred lines are tested in all possible combinations, as was done in these experiments, the elite hybrids will be apparent. However, such an ex-

TABLE 2. PERFORMANCE OF INBREDS AND THEIR HYBRIDS WHEN SUMMARIZED BY INBRED PHENOTYPE
Summary of 4 sets of data involving 37 inbred lines and 144 hybrids.

| Types by inbred phenotype | Mean daily fecundity | Mean egg size (micrometer units) |
|---|---|---|
| *Inbreds* | | |
| High Inbreds | 74 | 38.3 |
| Medium Inbreds | 61 | 37.4 |
| Low Inbreds | 46 | 35.7 |
| *Hybrids* | | |
| High × High | 95 | 38.5 |
| High × Medium | 93 | 38.3 |
| High × Low | 90 | 37.7 |
| Medium × Medium | 89 | 37.9 |
| Medium × Low | 83 | 37.2 |
| Low × Low | 78 | 36.4 |

tensive testing program would drastically reduce the total number of lines which could be tested. We shall consider the possibility of eliminating certain inbred lines from the testing program on the basis of the inbred's phenotype. The distribution of the elite hybrids among the inbred parental combinations was determined for all four sets of data. The pooled frequencies are shown separately for fecundity and egg size in Table 3. The first column gives the percentage of the total hybrid matings included in each class of inbred matings. For example, the "High × High" matings represented ten per cent of the total 144 hybrid combinations or there were 15 hybrids from such matings. Columns 2 and 3 show the distribution of elite hybrids for fecundity and egg size, respectively. The most striking aspect of this table is the differential distribution of the elite fecundity hybrids as compared with the elite egg size hybrids. Note that nearly 50 per cent of the elite egg size hybrids are in the "High × High" inbred matings which included only 10 per cent of the total matings. Most of the remaining elite size hybrids are found in the "High × Medium" inbred matings. In this comparison "High" and "Medium" refer to the inbred lines with large and medium egg size. Thus from this one-third of the total hybrids better than 80 per cent of the elite hybrids are obtained.

In contrast to the egg size data, it can be observed in Table 3 that for fecundity only six per cent of the elite hybrids were found in the "High × High" fecundity inbred lines. Since ten per cent of the total hybrids were in this class, it suggests that elite fecundity hybrids might be distributed independently of parental phenotype. However, this possibility is immediately eliminated when one finds 61 per cent of the elite hybrids in the "High × Medium" parental class which contained only 23 per cent of the total hybrid combinations. The unexpected high frequency of elite hybrids in the "High × Medium" inbred class is difficult to explain on any single genetical theory of heterosis. If heterosis in fecundity were a function of overdominance one would expect the "High × Low" parental class to be the best source of elite hybrids as Hull (1952) observed in corn. On the other hand, heterosis by complete dominance or epistasis should, on the average, lead to a high frequency of elite hybrids in the "High × High" parental class. Actually there were 15 hybrids in this class and only one was classified as elite. The most plausible genetic explanation would seem to be one involving overdominance at some gene loci while other loci are occupied by alleles with incomplete dominance. There is the second alternative of explaining the results as being abnormal chance deviations but this appears rather unlikely since all four sets of data were in remarkable agreement in the distribution of elite hybrids.

Until additional data have been collected and analyzed, generalizations from our results should be made with caution. One significant point seems evident. The majority of the elite hybrids for both quantitative traits studied, fecundity and egg size, were found in the hybrid combinations of the phenotypically average to good inbred lines. Certainly poor inbred lines could have been discarded previous to testing with a minimum loss of elite hybrids.

SELECTING FOR HETEROSIS

A major objective of our laboratory has been to test new methods of selection in quantitative genetics. A practical way to test new methods is to compare their effectiveness with conventional methods. Between December, 1949 and September, 1954 we made a comparison of four methods in two experiments covering 55 generations of selection plus the initial and terminal performance tests. The four methods of selection were: (1) Conventional closed population selection based on individual and family merit; (2) Recurrent Selection for specific combinability with an inbred tester, referred to hereafter in this paper as Recurrent Cross Selection; (3) Reciprocal Recurrent Selection for specific combinability, hereafter called Reciprocal Selection; and (4) Inbreeding and hybridization. A preliminary report on these two experiments was made by Bell et al. (1953).

In comparing the effectiveness of various methods of selection it is exceedingly difficult or impossible to design experiments affording each method its optimum conditions. Theoretically, one would expect that such factors as initial genetic variation, heritability of traits involved, previous selection history of the population, size of the population, amount of selection pressure, and system of breeding would influence the outcome of such experiments. Until the relationship of these various factors are better known, one's experimental facilities and materials are major considerations in determining the scope of the experiments. In designing the experiments here reported, we endeavored to initiate all methods from the same foundation germplasm, to apply equal selection effort on each method and to study the response of quantitative traits which differed in their manifestations of heterosis.

*Experiment I.* The first experiment consisted of 16 generations of selection followed by a terminal performance test replicated on six consecutive weeks. The foundation germplasm for all methods

TABLE 3. DISTRIBUTION OF ELITE FECUNDITY AND EGG SIZE HYBRIDS AMONG THE PHENOTYPIC COMBINATIONS OF INBRED PARENTS

| Phenotypes of inbred mating | Percent of total hybrids | Percent of elite hybrids Fecundity | Egg size |
|---|---|---|---|
| High × High | 10 | 6 | 47 |
| High × Medium | 23 | 61 | 35 |
| High × Low | 26 | 17 | 12 |
| Medium × Medium | 7 | 0 | 6 |
| Medium × Low | 23 | 17 | 0 |
| Low × Low | 10 | 0 | 0 |

consisted of eight non-inbred laboratory stocks of *Drosophila melanogaster*. Three of these stocks had only recently been collected from widely separated areas, while the others had been carried as laboratory wild stocks for several years. Selections under all methods were made on the basis of a performance index combining the two quantitative traits, fecundity and egg size. A simple method of combining the expression of these two traits into a performance index was necessary to accommodate the rapid generation cycles. Previous knowledge of the variation shown by the two traits indicated that the following individual index would provide about equal selection pressure on each trait.

Performance Index = 4 days' egg production + 10 (total of 5 egg lengths)

The Closed Population Method of Selection was designed to follow as nearly as possible conventional "pure breeding" where selection is based on individual and family merit. The eight foundation stocks were combined into one segregating population or composite. In each generation of selection, nine daughters from each of 42 families were placed in individual half-pint milk bottles. Fecundity and egg size data were obtained on each of the 378 daughters. The data were summarized into individual and family indices. From the ten highest families, the 42 daughters with the highest individual indices were selected as breeders to reproduce the population. These 42 selected daughters along with their mates produced the 42 families of the next generation and the selection cycle was repeated.

In order to obtain maximum fecundity from *Drosophila* females, it is necessary to have the stimulating presence of a male. Furthermore, once a female has mated she will lay eggs fertilized by this male for an indefinite period. Therefore, the sires of the next generation of the closed population were actually taken at random from all 42 current families, except that matings closer than first cousins were avoided. As a consequence, inbreeding was held to a minimum.

The main feature of Recurrent Cross Selection is that individuals from a segregating population are top-crossed on members of an inbred line. In this experiment, the segregating population was the initial stock of the Closed Population Method. However, in the recurrent method, selection within the segregating population was based on combining or crossing ability with an inbred tester line which had been full-sibbed for 22 generations. In each generation 42 males were each mated to four tester line females to produce topcross progeny. Three days later each male was removed and mated with unrelated females of his own generation from the segregating population for reproducing the population. Each of the 42 males was evaluated for combining ability with the tester by the fecundity and egg size record of nine of his topcross daughters.

The seven males with the highest combining ability were chosen to sire the next generation of the segregating population. This next generation would already have been produced in the matings of the 42 males with their own line females. Selection was accomplished by discarding the progenies from the 35 less desirable males. From each of the seven selected progenies, six males are topcrossed on tester line females to initiate the second cycle of selection.

For Reciprocal Selection, the eight foundation stocks were combined by groups of four into two unrelated segregating populations. This provided segregating populations on both sides of the reciprocal cross and thus opportunity for complementary selection. However, in fairness to the other methods, the progeny testing for combining or crossing ability did not exceed the 378 daughters from 42 cross matings. Therefore, for one generation 42 males from one segregating population would be test-crossed to females of the complementary line, while the next generation would consist of males from the second population being test-crossed to females of the first segregating population. The procedure under Reciprocal Selection was identical to that described for Recurrent Cross Selection except that each segregating population would serve as the tester line in alternate generations.

Inbreeding and hybridization as a method of selection differs greatly in application from the other three methods and actually required much less effort as utilized in our studies. From each of the eight foundation stocks, two inbred lines were developed by continuous full-sibbing. At the end of the selection part of the experiment these inbred lines were combined into single crosses for comparison with the other methods in the terminal performance tests.

*Experiment II.* In repeating the comparison of these four methods of selection, additional technical assistance made possible several changes in experimental design. The experiment extended over 39 generations of selection with a terminal performance test replicated in seven consecutive weeks. Performance data were recorded on ten daughters from each of 40 families under each method. Selection within each method was based solely on fecundity over a three-day period between five and seven days of age. Egg size was measured, but only for the purpose of studying drift in an unselected trait. The only major change in selection techniques involved Reciprocal Selection where 20 sires from each line were test-crossed each generation. In order to keep inbreeding at a minimum, four sires out of the 20 tested in each line were selected to provide five sons each for the next generation of testing. The four sires were selected in each case on the basis of the fecundity record of their ten test-cross daughters. In this second experiment it was possible to obtain fecundity and egg size data on the segregating lines in both Recurrent Cross and Reciprocal Selection.

The foundation stock for the second experiment consisted of the Closed Population from the first experiment, an unselected non-inbred laboratory stock and two inbred lines, one being homozygous for the third chromosome gene, ebony. The wild type inbred line served as the tester line for Recurrent Cross Selection and thus was not incorporated in the segregating population of this method. One inbred line and a non-inbred population went into each of the segregating populations of Reciprocal Selection. Naturally, all four foundation stocks were combined to produce the Closed Population. In addition to the inbred lines of Experiment I, new inbred lines were developed from these foundation stocks.

In both experiments the temperature was controlled at $25° \pm 1°C$, but humidity and seasonal changes in daylight periods were uncontrolled except for the later part of the second experiment.

In setting up the generation cycles for these experiments it would have been highly desirable to have measured the performance under all methods simultaneously. This would have provided a direct comparison of the methods on a generation by generation basis, but it was not possible with the existing laboratory staff without drastically limiting the population size under each method. The alternative of collecting performance data for the different methods on consecutive weeks was taken. The inbreeding and hybridization method did not require the continuous selection effort of the other methods. Therefore, both experiments were conducted on generation cycles of three weeks with performance being measured by a different method on successive weeks.

In both experiments a control was maintained at each generation, in an effort to correct week to week, as well as seasonal, fluctuations in the performance of the various experimental populations. An inbred control was used in the first experiment and a single cross of two inbred lines in the second experiment. Each week's performance of the experimental population was corrected by the amount of that week's fluctuation in the control.

### Experimental Results

Since a detailed report of these results is to be published elsewhere, only the pertinent points will be presented here. It became apparent early in both experiments that the control populations were not correcting for weekly environmental fluctuations accurately enough to make a generation by generation comparison of the various methods. The error associated with this control population could have been nothing more than a sampling error since seldom was performance measured on more than 30 control line females during any one week. The genetic-environmental interaction shown earlier in this paper also could have contributed to this error. This interaction signifies that weekly changes or fluctuations in the environment might affect the experimental population and the control population in a different manner. This would provide a correction factor too large or too small in relation to the environmental effect on the experimental population. Since all methods were measured concurrently in the replicated terminal performance tests, their relative performances were unaffected by corrections for weekly environmental fluctuations and provide a dependable measure of the relative effects of the selection methods for the whole period.

Two methods of analysis were used to provide a comparison of the selection methods at various periods in the selection experiments. A regression analysis of performance over several generations allowed weekly fluctuations to be largely randomized among the various selection methods. A second approach was to pool the performance record of each method of selection by intervals of three or four generations. Since performance was measured for three of the methods on consecutive weeks, this pooling by several generations made possible a reasonable comparison of the methods during various periods in the experiments. Such a comparison for the first experiment is shown in Figure 2. In the upper graph the response in fecundity, the highly heterotic trait, can be observed. The high initial performance of the Recurrent Cross population somewhat confounds the comparison of it with

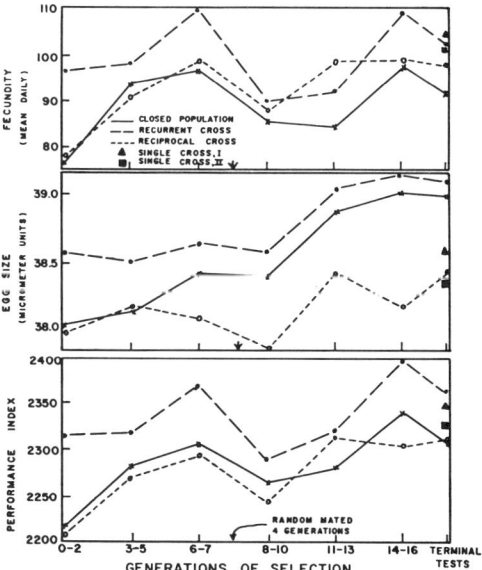

FIGURE 2. Comparison of three methods of selection over 16 generations, plus a terminal test including single cross hybrids, Experiment I.

the other methods. Superior combining ability of the inbred tester probably accounts for the high initial performance. The rate of improvement under Recurrent Cross Selection was least of all, but its initial advantage was never overtaken. A good comparison of Closed Population Selection and Reciprocal Selection is possible since both were performing at the same initial level. The significant point of this comparison is the more rapid early response in the Closed Population followed by the Reciprocal Cross overtaking and surpassing it in the later stages of the experiment.

In both experiments it was observed that the Closed Population apparently reached its highest level of fecundity early in the experiments. In the first experiment this point was at the fifth generation of selection. Thus for comparative purposes, the first experiment can be divided into an early period where improvement was being realized in the Closed Population and into a later period where this same population had apparently levelled into a plateau for improvement in fecundity. During the first period the Closed Population increased at about twice the rate of the Reciprocal Cross (6.3 eggs per generation versus 3.7). The Reciprocal Cross continued to improve beyond generation 5 and eventually surpassed the Closed Population. Over the last 11 generations of the experiment Reciprocal Selection showed 0.7 egg increase per generation as compared to 0.2 for Closed Population.

The drop in fecundity under all three methods between generations 7 and 8 followed a summer vacation when all stocks were random mated for four generations. A genetic regression in fecundity could have occurred while selection was relaxed during this period of random mating. Fecundity under all methods apparently regressed about the same amount. This would hardly be expected since the methods differed so much in their mating systems. A second and more probable reason for the apparent regression is a genetic improvement in the inbred control. Such improvement, whether by mutation, natural selection, or genetic contamination, would not necessarily be distinguished from unidentified changes in environmental conditions and would bring about an unjustified negative correction to all methods. Whatever the force acting during this period, it appeared to affect all the experimental populations in much the same manner. Therefore, the relative comparison of the various methods is valid, even though the total improvement for the 16 generations of selection under each method may actually have been more than indicated in Figure 2. In the right margin of this graph is shown the relative performance in the terminal tests of the various methods including two single crosses. A statistical analysis of these terminal tests will be presented after considering the response of the other traits in this experiment.

The center graph of Figure 2 shows the response in egg size, the highly heritable trait, under the various methods of selection. It should be kept in mind that the selection index stressed both egg size and egg production. The Recurrent Cross again was much superior in initial performance. Much of this was due to the large egg size of the inbred tester line which was not included in the other methods. Any comparison including the Recurrent Cross should be made with this reservation in mind. The Closed Population and the Reciprocal Cross were again initiated at the same level of performance and thus provide a critical test of their relative effectiveness for improving a highly heritable quantitative trait. The early points on this graph are slightly misleading in indicating that the Reciprocal Cross might be improving at a faster rate. Actually, over the first five generations the Closed Population was increasing in egg size by 0.10 units per generation as compared with 0.05 units for the Reciprocal Cross. Egg size continued to respond in the Closed Population at a rate roughly four times the improvement observed under Reciprocal Selection. The rate of increase per generation over all 16 generations was 0.08 units in the Closed Population and 0.02 units in the Reciprocal Cross.

The relative response in the performance index under the different methods of selection is shown in the lower graph of Figure 2. Since this index was an additive combination of fecundity and egg size, its response is conditioned by the observed response in the two component traits. For example, the Recurrent Cross excelled initially and terminally for both fecundity and egg size; likewise, it excelled for the performance index. Early in the experiment the Closed Population was superior to the Reciprocal Cross in both fecundity and egg size which in turn was reflected in a superior performance index for the Closed Population. However, late in the experiment the Reciprocal Cross excelled in fecundity while the Closed Population excelled in egg size; therefore, neither group established a consistent superiority in the performance index.

The most accccurate comparison of the four methods of selection in Experiment I was made in the terminal performance tests. Performance on all methods was measured concurrently in six replications run on consecutive weeks. Analyses of variance for both traits and the index are shown in Table 4. Inbreeding and hybridization were represented in these analyses by the single cross with the highest performance index. Since the number of individuals providing complete data varied between replications, each analysis of variance was made on the means for each method in each replication. An appropriate error mean square for testing significance was calculated from the pooled sum of squares. The highly significant differences between methods may have been anticipated from

TABLE 4. ANALYSIS OF THE TERMINAL PERFORMANCE TEST INVOLVING FOUR METHODS OF SELECTION COMPARED IN EXPERIMENT I

| Source of variation | Degree of freedom | Fecundity Mean square | Egg size Mean square | Performance index Mean square |
|---|---|---|---|---|
| Replications | 5 | 74.03* | 0.79* | 4356.3* |
| Methods of selection | 3 | 211.98* | 0.68* | 4445.0* |
| Replications × methods | 15 | 54.04* | 0.16* | 1858.1* |
| Within | 1835 | 7.53 | 0.03 | 174.4 |

*Significant at the .01 level of probability.

the results shown in Figure 2 and simply confirms the fact that the different methods of selection had resulted in different levels of performance. The replications × methods interaction for fecundity signifies a possible genetic-environmental interaction as mentioned earlier in this paper. The interaction mean square for egg size, while relatively small, was highly significant. Previous experiments involving other populations had not suggested an interaction between genotype and environment for egg size. These interactions could contribute to the error found in correcting performance in experimental populations with a control population.

A summary of the terminal test for Experiment I is given in Table 5. The mean fecundity, egg size and performance index for each method is given along with standard errors. This table shows that after 16 generations of selection all methods gave fecundity significantly superior to that of the Closed Population Method. In turn both single crosses and the Recurrent Cross were significantly higher than the Reciprocal Cross. There is one point to be stressed in reference to fecundity, the heterotic trait. It is that single crosses of inbred lines were found which were equal or superior in fecundity to that obtained after 16 generations of Reciprocal Selection or Recurrent Cross Selection.

The data in Table 5 show that for egg size the Closed Population was excelled only by the Recurrent Cross and this difference, although statistically significant, is only a fraction of the superiority shown by the Recurrent Cross at the beginning of the experiment. The Reciprocal Cross and

TABLE 5. RELATIVE PERFORMANCE IN VARIOUS SELECTED TRAITS AFTER 16 GENERATIONS OF SELECTION UNDER FOUR DIFFERENT METHODS
Summary of a terminal test involving 6 replications in Experiment I.

| Methods of selection | n | Mean daily fecundity | Mean egg size (micrometer units) | Performance index-mean |
|---|---|---|---|---|
| Closed population | 582 | 90.8±0.8 | 38.93±.04 | 2309.2±3.8 |
| Reciprocal cross | 573 | 97.7±0.7 | 38.42±.04 | 2314.5±3.6 |
| Recurrent cross | 544 | 102.1±0.9 | 39.16±.04 | 2366.9±4.1 |
| Single cross I | 161 | 104.3±1.1 | 38.58±.05 | 2346.0±5.0 |
| Single cross II | 164 | 101.3±1.1 | 38.38±.05 | 2324.1±5.9 |

both single crosses were significantly below the Closed Population in this highly heritable trait. With reference to the performance index, the superiority of the Recurrent Cross is clearly shown in Table 5. The single crosses ranked second with the Closed Population and the Reciprocal Cross being significantly inferior to the best single cross.

The results in Experiment I clearly demonstrated the superiority of the Closed Population Method for improving the highly heritable trait, egg size. This confirmed the theoretical expectations given by Bell et al. (1952), and since Reciprocal Selection and Recurrent Cross Selection were designed especially for improving heterotic traits, it was decided to limit selection to fecundity in the second experiment. A more critical differentiation of the methods of selection was thereby anticipated. This second experiment extended over 39 generations of selection with a terminal test of seven replications. Since the various methods were not studied simultaneously but were measured on consecutive weeks as in Experiment I, the generations were pooled for comparing the methods at various stages in the experiment. Such a comparison is presented in the upper graph of Figure 3.

The first point to be observed in Figure 3 is the low initial performance of the Recurrent Cross in contrast to the results of the first experiment. This was obtained by design rather than accident. Before initiating the experiment, the segregating population was crossed to a series of tester inbred lines in an effort to find such a poor specific combination as shown in the initial period for the Recurrent Cross. This was done to have all breeding methods starting at the same level; however, it is to be admitted that this procedure may have placed Recurrent Cross Selection under an unrealistic handicap. It has been observed in both corn and poultry data as well as in our Drosophila data that considerable heterosis is frequently realized by topcrossing. Since the initial Recurrent Cross is similar to a topcross, this initial superiority could be considered an inherent advantage of this method.

The Closed Population and the Reciprocal Cross are shown in Figure 3 to have the same initial level of fecundity. The early superiority of the Closed Population Method as shown in the first experiment is even more striking in this second experiment. The actual peak in the Closed Population occurred at generation 7. The rates of improvement as shown by the regression lines in the first graph in the lower portion of Figure 3 indicate that the Closed Population Method was increasing at roughly twice the rate of Reciprocal Selection. This corresponds very closely to the relative rates observed in the early phase of Experiment I. After reaching a ceiling at generation 7, it appears that in a few generations the Closed Population actually declined in fecundity while the other methods showed continued improvement and surpassed the

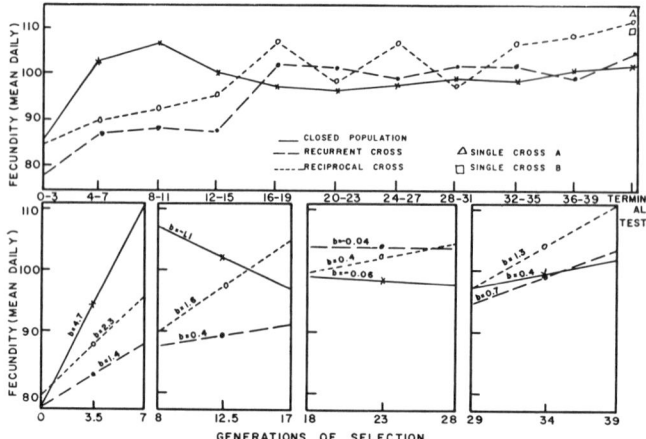

FIGURE 3. Comparison of three methods of selection, plus a terminal test including single cross hybrids (upper graph). Regression of rate of oviposition on generations of selection (lower graph). Experiment II.

plateaued population by about the fifteenth generation. The decline in the Closed Population apparently leveled off and a possible slight improvement was occurring during the last ten generations of the experiment. The Reciprocal Cross apparently continued to make gains throughout the experiment. However, this improvement certainly was not realized in a gradual manner, but apparently occurred through a somewhat rhythmical pattern of peaks and valleys in performance. The Recurrent Cross was more uniform in its response, but never reached the maximum level of performance shown by the Reciprocal Cross.

The analysis of the terminal test of the second experiment is presented in Table 6. The highly significant difference for replications is an indication of the weekly environmental fluctuations influencing fecundity. The highly significant mean square due to methods shows that under the conditions of this experiment, these methods of selection differed in their effectiveness for improving fecundity in *Drosophila*. While the interaction of replications and methods was relatively a small source of variation, it confirms the earlier observations as to genetic-environmental interactions for fecundity.

The average fecundity in the terminal test under each of the methods of selection may be seen in Table 7. The highest level of performance was made by one of the single crosses, but its superiority over the Reciprocal Cross was not significant, which in turn was not significantly superior to the second single cross. All the other paired comparisons are significantly different with the Closed Population being the poorest of the lot.

TABLE 6. ANALYSIS OF VARIANCE FOR A TERMINAL PERFORMANCE TEST INVOLVING 4 METHODS OF SELECTION COMPARED IN 7 REPLICATIONS

Test made following 39 generations of selection for high fecundity.

| Source of variation | d.f. | Mean square |
|---|---|---|
| Replications | 6 | 227.35* |
| Methods of selection | 3 | 194.66* |
| Replications × methods | 16 | 14.76* |
| Error | 2782 | 5.48 |

*Significant at the .01 level of probability.

TABLE 7. RELATIVE PERFORMANCE AFTER 39 GENERATIONS OF SELECTION UNDER 4 METHODS OF SELECTION FOR HIGH FECUNDITY

Summary of a terminal test involving 7 replications in time—Experiment II.

| Methods of selection | n | Mean daily egg production |
|---|---|---|
| Closed population | 902 | 101.3 ± .73 |
| Reciprocal | 773 | 111.5 ± .64 |
| Recurrent to tester | 912 | 105.2 ± .67 |
| Single cross I | 221 | 112.4 ± 1.15 |
| Single cross II | 256 | 110.0 ± 1.06 |

## Discussion

In general, the results from these two experiments agree with the theoretical expectations. The observed superiority of the Closed Population method for improving the highly heritable trait, egg size, was expected since the design of this method makes it especially effective in exploiting additive genetic variation. The inefficient manner in which Reciprocal Selection improved egg size suggests that the most effective method to improve highly heritable traits under Reciprocal Selection would be to practice individual selection within the segregating lines.

Thus, all the selection pressure for combining ability in the cross population could be directed toward heterotic traits.

The initial superiority of the Closed Population Method for improving the heterotic trait, fecundity, was somewhat surprising; even though Comstock, *et al.* (1949) had pointed out that initial response under Reciprocal Selection might be slow if the more favorable alleles at overdominant loci were near their equilibrium frequencies in both segregating populations. Since little artificial selection had been practiced in the foundation stocks, as well as the fact that Reciprocal Selection was initiated with newly formed segregating populations, we felt that most alleles would not be near their equilibrium frequencies. Results from both experiments showed that Reciprocal Selection eventually surpassed the Closed Population and was performing at a level of about 10 per cent higher in the terminal tests. The evidence suggests that the Closed Population plateaued for fecundity early in both experiments. A detailed analysis of this plateaued population will be reported in a separate study.

The comparison of Reciprocal Selection and Recurrent Cross Selection with single crosses of inbred lines possibly provided the widest discrepancy between the theoretical expectations and the observed results. Much less effort was expended on the inbreeding and hybridization method than on any of the other methods. Yet in both experiments single crosses were found which were equal or superior to the experimental populations of the other methods. It seems reasonable to assume that sufficient inbred lines could not be developed and tested in all hybrid combinations ever to achieve the maximum heterosis potential theoretically possible from foundation populations such as those used in our experiments. This leads to the obvious conclusion that neither Reciprocal Selection nor Recurrent Cross Selection had utilized the maximum heterosis obtainable from these populations. Among the possible reasons which could be given to account for these results, two seem to be worth mentioning. In the first place, it should be recognized that continued selection within both experiments might eventually have brought about the theoretically expected superiority of the new methods. Since selection at each generation was limited to one sex, this point could be especially significant in the first experiment where selection was limited to 16 generations. This reason is somewhat discounted by the fact that in the second experiment, even after 39 generations of selection, single crosses were still superior. We favor an alternative reason for explaining the observed results. This alternative explanation is based on the concept that both Reciprocal Selection and Recurrent Cross Selection are basically selection toward eventual homozygosity within the segregating populations. If this were not true, these methods would not achieve maximum heterosis at overdominance loci in theoretical studies. It is realized that much improvement is possible under these new methods of selection before one would expect to approach homozygosity for specific combining ability genes. Nevertheless, maximum heterosis or the limit to improvement would not be reached until the segregating populations become homozygous for those alleles which combine best in the cross. Selection *without inbreeding* has never been demonstrated to be an effective means of achieving homozygosity for genes influencing quantitative traits. Both Reciprocal Selection and Recurrent Cross Selection as originally presented and as used in our experiments avoided inbreeding within the segregating lines or populations. Overdominance for fecundity *per se* probably would not prevent these lines from becoming homozygous, as it would under the Closed Population Method. However, if heterozygosity within the segregating population was favored for reproductive fitness, natural selection acting during the reproduction of these segregating lines would oppose the force of artificial selection. Therefore, instead of reaching homozygosity, those genes involved would stabilize at equilibrium frequencies determined by the relative magnitude of the opposing forces. Incomplete analyses of genetic and phenotypic variation in the terminal populations of these experiments suggest that considerable genetic variation in fecundity was present in the segregating lines of Reciprocal Selection and Recurrent Cross Selection.

The possibility that maximum heterosis might not be obtained under Reciprocal Selection or Recurrent Cross Selection as applied in our experiments was recognized when these studies were initiated and a possible alternative or modification was suggested (Bell *et al.*, 1952). This modification calls for either intermittent or mild inbreeding within the segregating populations of Reciprocal Selection. Inbreeding would be the major force toward homozygosity while selection primarily would serve to identify superior combining genotypes. Experiments are planned to compare this new method with Reciprocal Selection as originally designed, the conventional closed population method, and inbreeding with hybridization.

*Practical Applications*

In transferring results from experiments such as those reported here to other species certain reservations are in order. It should be pointed out that many of the concepts in quantitative genetics which need biological testing or verification are products of theoretical studies and were not necessarily designed for any particular species of animals or plants. Such concepts and techniques as heritability, inbreeding coefficient and Reciprocal Selection are expected to work equally well in laboratory animals as in poultry or farm animals. Therefore, reservations concerning the transfer of results from one species to another should primarily concern the genetic conditions of the two populations. Genetic conditions refer to such variables as heritability of traits involved, previous selection history, amounts and kinds of non-additive gene effects and initial genetic variation. These are the same variables which should be considered in transferring the results from one population to expected results in another population of the same species. Naturally, such genetic peculiarities of a species as the small number of chromosomes and the lack of genetic recombination in males of *Drosophila melanogaster* should always be considered. Linkage would be expected to play a larger role in affecting expected results in this species than in a species with a large number of chromosomes, such as most domesticated animals. In order to obtain information relating to some of the variables discussed above, we have experiments under way with the flour beetle, *Tribolium castaneum*, which has ten pairs of chromosomes. The same methods of selection as reported on in this paper are being compared for their effectiveness in improving pupa size, a trait influenced to some extent by heterosis. Data are being collected in the same experiments on fecundity, fertility and hatchability.

## SUMMARY

A series of experiments relating to current problems in quantitative genetics were described. Two quantitative traits in *Drosophila melanogaster*, fecundity and egg size, were studied. Fecundity is a highly heterotic trait with low heritability, while egg size shows little or no heterosis in crosses and has relatively high heritability.

### Early Testing

Early testing of combining ability for fecundity was found to have little value in predicting subsequent combining ability for nine inbred lines crossed in all possible combinations. Factors contributing to this low predictive value included a consistent genetic-environmental interaction for fecundity, no significant differences among the initial non-inbred stocks and the long time intervals of ten generations or more between test periods.

Early testing results for egg size had more prediction value than those for fecundity. Except for the most distant comparisons, egg size in crosses had a positive and significant value in predicting egg size of these same crosses made after further inbreeding of their parental lines. In contrast to fecundity, highly significant differences in egg size existed among the initial non-inbred stocks and no genetic-environmental interaction was revealed for egg size.

Estimates of general and specific combining ability showed that the relative importance of these two functions varies inversely for the two quantitative traits studied. Specific combining ability was on the average more important than general combining ability in influencing fecundity. However, general combining ability assumed greater significance for egg size.

### Selection Within and Between Inbred Lines

Selection within lines during inbreeding was shown to be effective for both fecundity and egg size in *Drosophila*. These results consistently showed that the so-called "inbreeding depression" is conditioned by the selection pressure applied during the inbreeding process.

It was observed that the average or general combining ability of an inbred line for both fecundity and egg size is closely related to the phenotype of the inbred line. The relationship is higher, as expected, for the highly heritable trait.

Furthermore, the egg size of the parental inbred lines accurately predicted the egg size of specific hybrids. Therefore, testing for superior or elite specific hybrids in a highly heritable trait such as egg size should be limited to combinations between the best inbred lines.

The superior or elite hybrids with reference to the heterotic trait, fecundity, were usually not found in the combinations between phenotypically good inbred lines. Some special advantage seemed to exist in the combinations of "High" by "Medium" inbred lines.

The most plausible genetic explanation of the heterosis observed in these fecundity data would seem to be one involving overdominance at some gene loci while other loci are occupied by alleles with incomplete dominance.

### Selection for Heterosis

Two new methods of selection designed especially to obtain maximum heterosis were compared with conventional methods in two experiments. The methods were: (1) Individual and family selection within a closed population; (2) Reciprocal Selection for specific combining ability; (3) Recurrent Selection for specific combining ability with an inbred tester; and (4) Inbreeding and hybridization. The response in the two quantitative traits, fecundity and egg size, was measured in the two experiments over 16 and 39 generations of selection, respectively. Results from these experiments suggested that the conventional method of individual and family selection is superior to the other methods tested for improving quantitative traits determined largely by additive genes. Also, individual and family selection is initially more effective than the other methods in improving a highly heterotic trait in newly formed populations. However, in the long run, both Reciprocal Selection and Recurrent Cross Selection are superior to the Closed Population Method in obtaining maximum performance in highly heterotic traits.

Maximum heterosis is not realized under either Reciprocal or Recurrent Cross Selection, since single crosses of inbred lines derived from the same foundation stocks performed at an equal or higher level than did the cross populations of these new methods designed to take advantage of heterosis.

It is postulated that selection without inbreeding will not bring about the requisite homozygosity in the Reciprocal or Recurrent segregating lines to yield the maximum heterosis in the cross populations.

### ACKNOWLEDGMENT

The authors would like to acknowledge the valuable technical assistance contributed at various times during these studies by Margaret Eller, Joan Finehout, Susan Fisher, Brenda Hamblin, Doris Handley, Emelia Kirk and Betty Larkin.

### REFERENCES

BELL, A. E., MOORE, C. H., BOHREN, B. B., and WARREN, D. C., 1952, Systems of breeding designed to utilize heterosis in the domestic fowl. Poult. Sci. *31:* 11-22.

BELL, A. E., MOORE, C. H., and WARREN, D. C., 1953, A biological evaluation with Drosophila melanogaster of four methods of selection for the improvement of quantitative characteristics. Proc. 9th International Genetics Congress, Caryologia 6. Suppl., 851-853.

COMSTOCK, R. E., ROBINSON, H. F., and HARVEY, P. H., 1949, A breeding procedure designed to make maximum use of both general and specific combining ability. Agron. J. *41:* 360-367.

CROW, J. F., 1948, Alternative hypotheses of hybrid vigor. Genetics *33:* 477-487.

DICKERSON, GORDON E., 1952, Inbred lines for heterosis tests. In: Heterosis, pp. 330-351. Ames, Iowa State College Press.

DURRANT, ALAN, and MATHER, KENNETH, 1954, Heritable variation in a long inbred line of Drosophila. Genetica *27:* 97-119.

FALCONER, D. S., 1953, Selection for large and small size in mice. J. Genet. *51:* 470-501.

GOWEN, JOHN W., and JOHNSON, LESLIE E., 1946, On the mechanism of heterosis. I. Metabolic capacity of different races of Drosophila melanogaster for egg production. Amer. Nat. *80:* 149-179.

HULL, F. H., 1945, Recurrent selection for specific combining ability in corn. J. Amer. Soc. Agron. *37:* 134-145.
1946, Overdominance and corn breeding where hybrid seed is not feasible. J. Amer. Soc. Agron. *38:* 1100-1103.
1952, Recurrent selection and overdominance. In: Heterosis, pp. 451-573. Ames, Iowa State College Press.

KYLE, W. H., and CHAPMAN, A. B., 1953, Experimental check of the effectiveness of selection for a quantitative character. Genetics *38:* 421-443.

LERNER, I. M., 1950, Population Genetics and Animal Improvement. 342 pp. Cambridge University Press.

LOH, S. Y., 1949, Early testing as a means of evaluating $F_1$ heterosis between inbred lines of Drosophila melanogaster. Ph.D. Thesis. Iowa State College Library, Ames, Iowa.

LUSH, JAY L., 1945, Animal Breeding Plans. 3rd ed. 443 pp. Ames, Iowa State College Press.

MOORE, C. H., 1952, Performance of inbred lines of Drosophila melanogaster in single crosses as predicted by early testing. Ph.D. Thesis, Purdue University Library, Lafayette, Indiana.

ROBERTSON, F. W., and REEVE, E. C. R., 1952, Studies in quantitative inheritance. I. The effects of selection of wing and thorax length in Drosophila melanogaster. J. Genet. *50:* 414-448.

SCOSSIROLI, R. E., 1954, Effectiveness of artificial selection under irrigation of plateaued populations of Drosophila melanogaster. I. U. B. S. Symposium on Genetics of Population Structure, Series B, No. 15, Pavia, Italy, August 20-23, 1953.

SHULTZ, FRED T., 1953, Concurrent inbreeding and selection in the domestic fowl. Heredity 7: 1-21.

SPRAGUE, G. F., 1952, Early testing and recurrent selection. In: Heterosis, pp. 400-417. Ames, Iowa State College Press.

WARREN, D. C., 1924, Inheritance of egg size in Drosophila melanogaster. Genetics *9:* 41-69.

WARWICK, E. J., and LEWIS, W. L., 1954, Growth and reproductive rates of mice. J. Hered. *45:* 35-38.

[*Editor's Note:* The Discussion has been omitted.]

# SELECTION AND MAINTENANCE OF
# GENETIC VARIATION

# Editor's Comments
# on Papers 21 Through 24

**21  WRIGHT**
*The Roles of Mutation, Inbreeding, Crossbreeding and Selection in Evolution*

**22  ROBERTSON**
*Selection in Animals: Synthesis*

**23  BULMER**
Excerpts from *The Effect of Selection on Genetic Variability*

**24  LANDE**
*The Maintenance of Genetic Variability by Mutation in a Polygenic Character with Linked Loci*

 

The theory and experiments discussed in the previous sections have dealt with the rates and limits of response to directional selection, essentially with how the variation present in the population is, or can be, utilized. In this and the following section we turn to a consideration of how the variability is produced and maintained by natural selection. Obviously, with long, continued directional selection, variability would be lost. The fact that natural populations do exhibit genetic variation is clear evidence that there are forces maintaining it, notably selective forces in opposing directions acting at the same or different times, and new mutations. The papers in this section discuss, among other things, these alternatives, although none of them is solely concerned with the topic. The general title, "Selection and Maintenance of Genetic Variation," is thus no more than a guide. The papers in this section deal with theory; those on experiments follow.

Wright first proposed his multiple peak theories of adaptation—what was to become known as the shifting balance theory—in his paper "Evolution in Mendelian Populations" (Wright, 1931). His summary of his views in a shorter paper at the Sixth International Congress of Genetics is reproduced here as Paper 21. Because the 1932 paper is free of the mathematical detail on population genetics and emphasizes rather more the multiple gene combination and multiple peak ideas, it is many ways more appropriate than the 1931 paper for inclusion in this collection on quantitative genetics. Nevertheless, the paper contains no formulation familiar in the study of quantitative traits, such as are found in his 1921 papers on "Systems of Mating," primarily because of the mathematical problems of dealing with multiple interactions among loci.

Other important papers by Wright—on systems of mating (Wright 1921) and a review of quantitative inheritance (Wright, 1952)—are included in Part I, and some biographical information is given in the commentary. The theory outlined in Paper 21 is developed further in many succeeding papers, and particularly in Volumes 2 and 3 of his treatise (Wright, 1969, 1977) and later publications (Wright, 1982). He has stood by it for over fifty years; although an essential item is the role of drift, in Wright's model drift moves the population from one adaptive peak to the region of a new one and is not the source of evolution *per se,* as in Kimura's (1968) model of substitutions of neutral mutations.

In the second paper in this section, A. Robertson (Paper 22) discusses how genetic variation is maintained for quantitative traits and particularly why it is found that for traits, such as *Drosophila* bristle number, that are presumably not closely tied to fitness, most genetic variation is additive, while for traits such as reproductive performance there is little additive genetic variance. This analysis appears in the context of selection because he is attempting to explain why selection responses differ from one trait to another; among the papers he reviews is Paper 17. Robertson refers also to the results he obtained with Clayton on mutational variance (Paper 25) and to that of Haldane on selection values (Paper 11).

The analysis in Paper 22 is solely verbal and is summarized in its graphs. Traits closely related to fitness (e.g., egg number) are under strong natural selection, so most additive variance is lost; that remaining is largely due to negative correlations with other traits closely related to fitness (e.g., viability), but dominance variance remains either because of overdominance or segregation of rare recessives. Variation in traits peripheral to fitness may be maintained by stabilizing selection on the trait itself if extremes are unfit, or because of pleiotropy of the genes on the trait and on components of fitness, such as egg number, with variation maintained by heterozygote superiority on the fitness-related trait. In either case, the effects of the genes on the trait may be additive or nearly so.

Further discussions of these problems abound in the literature; for an up-to-date review, see Falconer (1981, Ch. 20) or Wright (1977). Some of the more important contributions were by Lerner, which are discussed fully in his book *Genetic Homeostasis* (Lerner, 1954), but aspects are reviewed by Lerner and Dempster (Paper 16). A basic problem with these analyses is providing direct evidence of either heterozygote superiority (other than for sickle-cell hemoglobin) or stabilizing selection. One of the few clear cases of the latter is described in Paper 26.

The final two papers included in this section, by Bulmer and by Lande, are more mathematical and deal with specific topics, each assuming many loci with genes of additive effect and infinitely large populations, such that the distributions of genotypes and phenotypes are assumed to be multivariate normal. Both illustrate how, under these multilocus models, a substantial part of the variation present at the single-locus level may not be expressed in the overall genotype because of negative correlations (negative linkage disequilibrium) among the genes. Although such a model was postulated by Mather (Paper 15), the actual pairwise correlations are very small and unlikely to be detected as a + — + — series of effects on a chromosome.

Paper 23 deals explicitly with the case of the so-called infinitesimal model in which there are assumed to be infinitely many genes of small effect, an assumption made implicity in many of the papers in Section II. Using this model, Bulmer shows that when selection reduces the phenotypic variance among those individuals surviving to have progeny, whether by reason of stabilizing or directional selection, all the consequent reduction in genotype variance among these individuals and therefore among families of their progeny is due to linkage disequilibrium. He then goes on to predict how much the variance will be reduced by continuous selection, as a balance is eventually reached between the reduction of variance due to selection and the increase due to recombination. In Paper 23 Bulmer deals solely with the case of unlinked loci, but in subsequent papers he extended the results to more general models. These are reviewed in his book *The Mathematical Theory of Quantitative Genetics* (Bulmer, 1980). Unfortunately, due to lack of space, a mathematical appendix to the paper, "The Regression Between Relatives Under the Infinitesimal Model," has been omitted.

The effect of selection on reducing variation in the population had been considered by Dickerson and Hazel (Paper 8) and more formally by Cochran (1951) at the solely statistical level. Felsenstein (1965) pointed out that truncation selection led to negative linkage disequilibrium among pairs of loci. Bulmer's important contribution was to combine these two approaches to obtain recurrence formulae that could be used in practice. Thus for truncation selection of known intensity it is possible to predict how much genetic variance, heritability, and responses will be affected for given initial heritability (see Paper 23, Table 1); it can be seen that with free recombination most of the drop in variance occurs immediately. For stabilizing selection, the analysis shows how much variation remains hidden as disequilibrium.

Bulmer's analysis takes no account of the increase in variance coming from mutation, and gene effects are assumed to be infinitesimally small. Lande (Paper 24) generalized these results (although without reference to Bulmer) for the case of stabilizing selection. He showed how recurrence formulae could be obtained for the variance-covariance matrix of effects of alleles on the same or different gametes; for example, a negative covariance of a pair of effects implies negative linkage disequilibrium of the "+" alleles at that pair of loci. Solution of these recurrence formulae enabled Lande to predict the genetic variance maintained as a function of the mutation rate per generation and the strength of the stabilizing selection. Note, however, that his solutions depend on the sizes of the gene effects produced by the mutation, a quantity that cannot be estimated. Lande (1977) subsequently gave a simplified version of the dynamic equations and equilibrium solutions for the variance components, with an extension to nonrandom mating.

The content of this volume illustrates that the majority of studies on selection of quantitative traits have been undertaken by workers interested in applications of artificial selection to plant and animal breeding. There has, however, been an increasing interest in recent years in population, ecological, and evolutionary studies of quantitative traits, perhaps in part consequent on the disappointing lack of understanding derived from studies of electrophoretic variants on starch gels. Lande's work has been an important simulus to further mathematical analysis because he dealt with a problem relating to natural selection and showed that interesting problems in quantitative genetics could, by appropriate formulation, be tackled at the whole genome level, not just by statements about loci one at a time.

Finally, let us consider what other papers might have been included in this theoretical section on production and maintenance of variation. Some of Lerner's work has already been mentioned. On the role of stabilizing selection, the optimum model of Wright (1935) has clearly had a major influence, as also did Robertson's (1956) demonstration that stabilizing selection led to fixation rather than heterozygosity. The role of mutation in maintaining variation at multiple allelic loci was nicely analyzed by Latter (1970). In the following section, we turn to more experimental approaches.

## REFERENCES

Bulmer, M. G., 1980, *The Mathematical Theory of Quantitative Genetics,* Oxford University Press, Oxford.

Cochran, W. G., 1951, Improvement by Means of Selection, in *Proceedings of the Second Berkeley Symposium on Mathematical Statistics and Probability,* J. Neyman, ed., University of California Press, Berkeley, pp. 449-470.

Falconer, D. S., 1981, *Introduction to Quantitative Genetics,* 2nd ed., Longmans, London.

Felsenstein, J., 1965, The Effect of Linkage on Directional Selection, *Genetics* **52:**349-363.

Kimura, M., 1968, Evolutionary Rate at the Molecular Level, *Nature* **217:**624-626.

Lande, R., 1977, The Influence of the Mating System on the Maintenance of Genetic Variability in Polygenic Characters, *Genetics* **86:**485-498.

Latter, B. D. H., 1970, Selection in Finite Populations With Multiple Alleles. II Centripetal Selection, Mutation and Isoallelic Variation, *Genetics* **66:**165-186.

Lerner, I. M., 1954, *Genetic Homeostasis,* Oliver and Boyd, Edinburgh.

Robertson, A., 1956, The Effect of Selection Against Extreme Deviants Based on Deviation or on Homozygosis, *J. Genet.* **54:**236-248.

Wright, S., 1921, Systems of Mating, *Genetics* **6:**111-178.

Wright, S., 1931, Evolution in Mendelian Populations, *Genetics* **16:**97-159.

Wright, S., 1935, The Analysis of Variance and the Correlations Between Relatives with Respect to Deviations from an Optimum, *J. Genet.* **30:**243-256.

Wright, S., 1952, The Genetics of Quantitative Variability, in *Quantitative Inheritance,* E. C. R. Reeve and C. H. Waddington, eds., Her Majesty's Stationery Office, London, pp. 5-41.

Wright, S., 1969, *Evolution and the Genetics of Populations. Vol. 2. The Theory of Gene Frequencies,* University of Chicago Press, Chicago.

Wright, S., 1977, *Evolution and the Genetics of Populations. Vol. 3. Experimental Results and Evolutionary Deductions,* University of Chicago Press, Chicago.

Wright, S., 1982, The Shifting Balance Theory and Macroevolution, *Ann. Rev. Genet.* **16:**1-19.

# 21

## THE ROLES OF MUTATION, INBREEDING, CROSSBREEDING AND SELECTION IN EVOLUTION

### Sewall Wright

The enormous importance of biparental reproduction as a factor in evolution was brought out a good many years ago by East. The observed properties of gene mutation—fortuitous in origin, infrequent in occurrence and deleterious when not negligible in effect—seem about as unfavorable as possible for an evolutionary process. Under biparental reproduction, however, a limited number of mutations which are not too injurious to be carried by the species furnish an almost infinite field of possible variations through which the species may work its way under natural selection.

Estimates of the total number of genes in the cells of higher organisms range from 1000 up. Some 400 loci have been reported as having mutated in Drosophila during a laboratory experience which is certainly very limited compared with the history of the species in nature. Presumably, allelomorphs of all type genes are present at all times in any reasonably numerous species. Judging from the frequency of multiple allelomorphs in those organisms which have been studied most, it is reasonably certain that many different allelomorphs of each gene are in existence at all times. With 10 allelomorphs in each of 1000 loci, the number of possible combinations is $10^{1000}$ which is a very large number. It has been estimated that the total number of electrons and protons in the whole visible universe is much less than $10^{100}$.

However, not all of this field is easily available in an interbreeding population. Suppose that each type gene is manifested in 99 percent of the individuals, and that most of the remaining 1 percent have the most favorable of the other allelomorphs, which in general means one with only a slight differential effect. The average individual will show the effects of 1 percent of the 1000, or 10 deviations from the type, and since this average has a standard deviation of $\sqrt{10}$ only a small proportion will exhibit more than 20 deviations from type where 1000 are possible. The population is thus confined to an infinitesimal portion of the field of possible gene combinations, yet this portion includes some $10^{40}$ homozygous combinations, on the above extremely conservative basis, enough so that there is no reasonable chance that any two individuals have exactly the same genetic constitution in a species of millions of millions of individuals persisting over millions of generations. There is no difficulty in accounting for the probable genetic uniqueness of each individual human being or other organism which is the product of biparental reproduction.

If the entire field of possible gene combinations be graded with respect to adaptive value under a particular set of conditions, what would be its nature? Figure 1 shows the combinations in the cases of 2 to 5 paired allelomorphs. In the last case, each of the 32 homozygous combinations is at one remove from 5 others, at two removes from 10, etc. It would require 5 dimensions to represent these relations symmetrically; a sixth dimension is needed to represent level of adaptive value. The 32 combina-

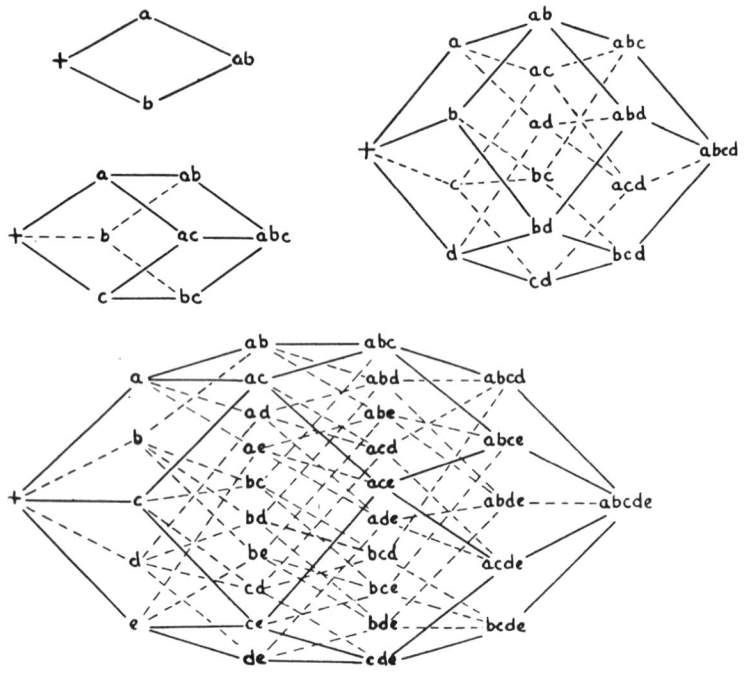

FIGURE 1.—The combinations of from 2 to 5 paired allelomorphs.

tions here compare with $10^{1000}$ in a species with 1000 loci each represented by 10 allelomorphs, and the 5 dimensions required for adequate representation compare with 9000. The two dimensions of figure 2 are a very inadequate representation of such a field. The contour lines are intended to represent the scale of adaptive value.

One possibility is that a particular combination gives maximum adaptation and that the adaptiveness of the other combinations falls off more or less regularly according to the number of removes. A species whose individuals are clustered about some combination other than the highest would move up the steepest gradient toward the peak, having reached which it would remain unchanged except for the rare occurrence of new favorable mutations.

But even in the two factor case (figure 1) it is possible that there may be two peaks, and the chance that this may be the case greatly increases with each additional locus. With something like $10^{1000}$ possibilities (figure 2) it may be taken as certain that there will be an enormous number of

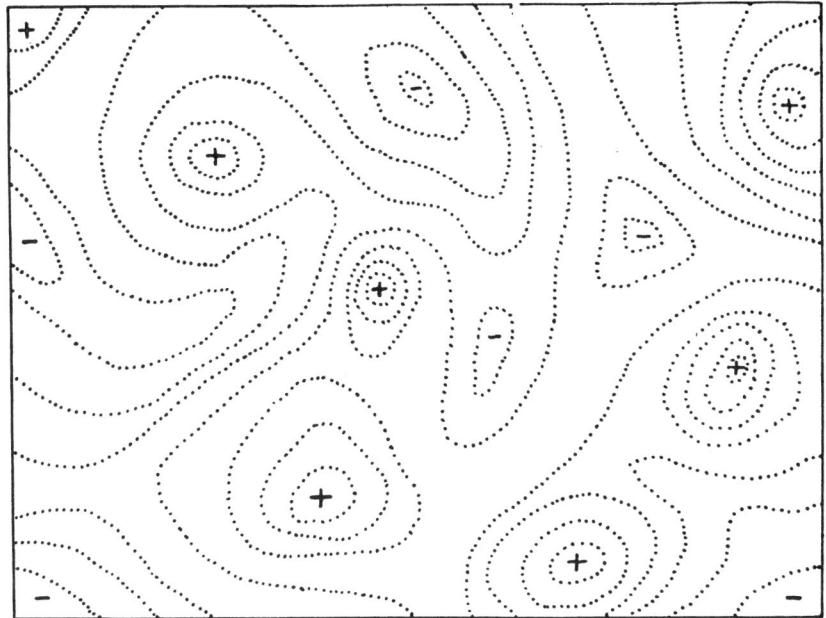

FIGURE 2.—Diagrammatic representation of the field of gene combinations in two dimensions instead of many thousands. Dotted lines represent contours with respect to adaptiveness.

widely separated harmonious combinations. The chance that a random combination is as adaptive as those characteristic of the species may be as low as $10^{-100}$ and still leave room for $10^{800}$ separate peaks, each surrounded by $10^{100}$ more or less similar combinations. In a rugged field of this character, selection will easily carry the species to the nearest peak, but there may be innumerable other peaks which are higher but which are separated by "valleys." The problem of evolution as I see it is that of a mechanism by which the species may continually find its way from lower to higher peaks in such a field. In order that this may occur, there must be some trial and error mechanism on a grand scale by which the species may explore the region surrounding the small portion of the field which it occupies. To evolve, the species must not be under strict control of natural selection. Is there such a trial and error mechanism?

At this point let us consider briefly the situation with respect to a single locus. In each graph in figure 3 the abscissas represent a scale of gene frequency, 0 percent of the type genes to the left, 100 percent to the right. The elementary evolutionary process is, of course, change of gene frequency, a

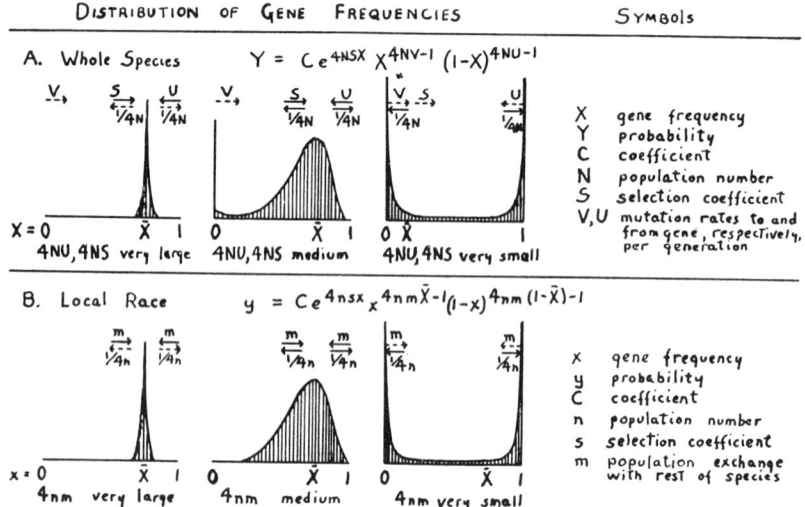

FIGURE 3.—Random variability of a gene frequency under various specified conditions.

practically continuous process. Owing to the symmetry of the Mendelian mechanism, any gene frequency tends to remain constant in the absence of disturbing factors. If the type gene mutates at a certain rate, its frequency tends to move to the left, but at a continually decreasing rate. The type gene would ultimately be lost from the population if there were no opposing factor. But the type gene is in general favored by selection. Under selection, its frequency tends to move to the right. The rate is greatest at some point near the middle of the range. At a certain gene frequency the opposing pressures are equal and opposite, and at this point there is consequently equilibrium. There are other mechanisms of equilibrium among evolutionary factors which need not be discussed here. Note that we have here a theory of the stability of species in spite of continuing mutation pressure, a continuing field of variability so extensive that no two individuals are ever genetically the same, and continuing selection.

If the population is not indefinitely large, another factor must be taken into account: the effects of accidents of sampling among those that survive and become parents in each generation and among the germ cells of these, in other words, the effects of inbreeding. Gene frequency in a given generation is in general a little different one way or the other from that in the preceding, merely by chance. In time, ge e frequency may wander a long way from the position of equilibrium, although the farther it wanders the greater the pressure toward return. The result is a frequency distribution within which gene frequency moves at random. There is considerable spread even with very slight inbreeding and the form of distribution becomes U-shaped with close inbreeding. The rate of movement of gene frequency is very slow in the former case but is rapid in the latter (among unfixed genes). In this case, however, the tendency toward complete fixation of genes, practically irrespective of selection, leads in the end to extinction.

In a local race, subject to a small amount of crossbreeding with the rest of the species (figure 3, lower half), the tendency toward random fixation is balanced by immigration pressure instead of by mutation and selection. In a small sufficiently isolated group all gene frequencies can drift irregularly back and forth about their mean values at a rapid rate, in terms of geologic time, without reaching fixation and giving the effects of close inbreeding. The resultant differentiation of races is of course increased by any local differences in the conditions of selection.

Let us return to the field of gene combinations (figure 4). In an indefinitely large but freely interbreeding species living under constant conditions, each gene will reach ultimately a certain equilibrium. The species will occupy a certain field of variation about a peak in our diagram (heavy broken contour in upper left of each figure). The field occupied remains constant although no two individuals are ever identical. Under the above conditions further evolution can occur only by the appearance of wholly new (instead of recurrent) mutations, and ones which happen to be favorable from the first. Such mutations would change the character of the field itself, increasing the elevation of the peak occupied by the species. Evolutionary progress through this mechanism is excessively slow since the chance of occurrence of such mutations is very small and, after occurrence, the time required for attainment of sufficient frequency to be subject to selection to an appreciable extent is enormous.

The general rate of mutation may conceivably increase for some reason. For example, certain authors have suggested an increased incidence of cosmic rays in this connection. The effect (figure 4A) will be as a rule a spreading of the field occupied by the species until a new equilibrium is reached. There will be an average lowering of the adaptive level of the species. On the other hand, there will be a speeding up of the process discussed above, elevation of the peak itself through appearance of novel favorable mutations. Another possibility of evolutionary advance is that the spreading of the field occupied may go so far as to include another and

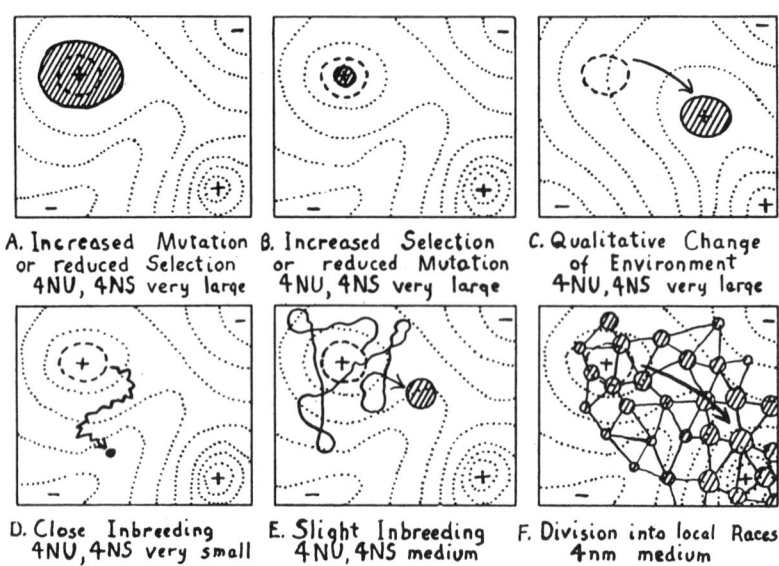

FIGURE 4.—Field of gene combinations occupied by a population within the general field of possible combinations. Type of history under specified conditions indicated by relation to initial field (heavy broken contour) and arrow.

higher peak, in which case the species will move over and occupy the region about this. These mechanisms do not appear adequate to explain evolution to an important extent.

The effects of reduced mutation rate (figure 4B) are of course the opposite: a rise in average level, but reduced variability, less chance of novel favorable mutation, and less chance of capture of a neighboring peak.

The effect of increased severity of selection (also 4B) is, of course, to increase the average level of adaptation until a new equilibrium is reached. But again this is at the expense of the field of variation of the species and

362   S. Wright

reduces the chance of capture of another adaptive peak. The only basis for continuing advance is the appearance of novel favorable mutations which are relatively rapidly utilized in this case. But at best the rate is extremely slow even in terms of geologic time, judging from the observed rates of mutation.

Relaxation of selection has of course the opposite effects and thus effects somewhat like those of increased mutation rate (figure 4A).

The environment, living and non-living, of any species is actually in continual change. In terms of our diagram this means that certain of the high places are gradually being depressed and certain of the low places are becoming higher (figure 4C). A species occupying a small field under influence of severe selection is likely to be left in a pit and become extinct, the victim of extreme specialization to conditions which have ceased, but if under sufficiently moderate selection to occupy a wide field, it will merely be kept continually on the move. Here we undoubtedly have an important evolutionary process and one which has been generally recognized. It consists largely of change without advance in adaptation. The mechanism is, however, one which shuffles the species about in the general field. Since the species will be shuffled out of low peaks more easily than high ones, it should gradually find its way to the higher general regions of the field as a whole.

Figure 4D illustrates the effect of reduction in size of population below a certain relation to the rate of mutation and severity of selection. There is fixation of one or another allelomorph in nearly every locus, largely irrespective of the direction favored by selection. The species moves down from its peak in an erratic fashion and comes to occupy a much smaller field. In other words there is the deterioration and homogeneity of a closely inbred population. After equilibrium has been reached in variability, movement becomes excessively slow, and, such as there is, is nonadaptive. The end can only be extinction. Extreme inbreeding is not a factor which is likely to give evolutionary advance.

With an intermediate relation between size of population and mutation rate, gene frequencies drift at random without reaching the complete fixation of close inbreeding (figure 4E). The species moves down from the extreme peak but continually wanders in the vicinity. There is some chance that it may encounter a gradient leading to another peak and shift its allegiance to this. Since it will escape relatively easily from low peaks as compared with high ones, there is here a trial and error mechanism by which in time the species may work its way to the highest peaks in the general field. The rate of progress, however, is extremely slow since change of gene frequency is of the order of the reciprocal of the effective population size and this reciprocal must be of the order of the mutation rate in order to meet the conditions for this case.

Finally (figure 4F), let us consider the case of a large species which is subdivided into many small local races, each breeding largely within itself but occasionally crossbreeding. The field of gene combinations occupied by each of these local races shifts continually in a nonadaptive fashion (except in so far as there are local differences in the conditions of selection). The rate of movement may be enormously greater than in the preceding case since the condition for such movement is that the reciprocal of the population number be of the order of the proportion of crossbreeding instead of the mutation rate. With many local races, each spreading over a considerable field and moving relatively rapidly in the more general field about the controlling peak, the chances are good that one at least will come under the influence of another peak. If a higher peak, this race will expand in numbers and by crossbreeding with the others will pull the whole species toward the new position. The average adaptiveness of the species thus advances under intergroup selection, an enormously more effective process than intragroup selection. The conclusion is that subdivision of a species into local races provides the most effective mechanism for trial and error in the field of gene combinations.

It need scarcely be pointed out that with such a mechanism complete isolation of a portion of a species should result relatively rapidly in specific differentiation, and one that is not necessarily adaptive. The effective intergroup competition leading to adaptive advance may be between species rather than races. Such isolation is doubtless usually geographic in character at the outset but may be clinched by the development of hybrid sterility. The usual difference of the chromosome complements of related species puts the importance of chromosome aberration as an evolutionary process beyond question, but, as I see it, this importance is not in the character differences which they bring (slight in balanced types), but rather in leading to the sterility of hybrids and thus making permanent the isolation of two groups.

How far do the observations of actual species and their subdivisions conform to this picture? This is naturally too large a subject for more than a few suggestions.

That evolution involves nonadaptive differentiation to a large extent at the subspecies and even the species level is indicated by the kinds of differences by which such groups are actually distinguished by systematists. It is only at the subfamily and family levels that clear-cut adaptive differences become the rule (ROBSON, JACOT). The principal evolutionary mechanism in the origin of species must thus be an essentially nonadaptive one.

That natural species often are subdivided into numerous local races is indicated by many studies. The case of the human species is most familiar. Aside from the familiar racial differences recent studies indicate a distribution of frequencies relative to an apparently nonadaptive series of allelomorphs, that determining blood groups, of just the sort discussed above. I scarcely need to labor the point that changes in the average of mankind in the historic period have come about more by expansion of some types and decrease and absorption of others than by uniform evolutionary advance. During the recent period, no doubt, the phases of intergroup competition and crossbreeding have tended to overbalance the process of local differentiation, but it is probable that in the hundreds of thousands of years of prehistory, human evolution was determined by a balance between these factors.

Subdivision into numerous local races whose differences are largely nonadaptive has been recorded in other organisms wherever a sufficiently detailed study has been made. Among the land snails of the Hawaiian Islands, GULICK (sixty years ago) found that each mountain valley, often each grove of trees, had its own characteristic type, differing from others in "nonutilitarian" respects. GULICK attributed this differentiation to inbreeding. More recently CRAMPTON has found a similar situation in the land snails of Tahiti and has followed over a period of years evolutionary changes which seem to be of the type here discussed. I may also refer to the studies of fishes by DAVID STARR JORDAN, garter snakes by RUTHVEN, bird lice by KELLOGG, deer mice by OSGOOD, and gall wasps by KINSEY as others which indicate the role of local isolation as a differentiating factor. Many other cases are discussed by OSBORN and especially by RENSCH in recent summaries. Many of these authors insist on the nonadaptive character of most of the differences among local races. Others attribute all differences to the environment, but this seems to be more an expression of faith than a view based on tangible evidence.

An even more minute local differentiation has been revealed when the methods of statistical analysis have been applied. SCHMIDT demonstrated the existence of persistent mean differences at each collecting station in certain species of marine fish of the fjords of Denmark, and these differences were not related in any close way to the environment. That the differences were in part genetic was demonstrated in the laboratory. DAVID THOMPSON has found a correlation between water distance and degree of differentiation within certain fresh water species of fish of the streams of Illinois. SUMNER's extensive studies of subspecies of Peromyscus (deer mice) reveal genetic differentiations, often apparently nonadaptive, among local populations and demonstrate the genetic heterogeneity of each such group.

The modern breeds of livestock have come from selection among the products of local inbreeding and of crossbreeding between these, followed by renewed inbreeding, rather than from mass selection of species. The recent studies of the geographical distribution of particular genes in livestock and cultivated plants by SEREBROVSKY, PHILIPTSCHENKO and others are especially instructive with respect to the composition of such species.

The paleontologists present a picture which has been interpreted by some as irreconcilable with the Mendelian mechanism, but this seems to be due more to a failure to appreciate statistical consequences of this mechanism than to anything in the data. The horse has been the standard example of an orthogenetic evolutionary sequence preserved for us with an abundance of material. Yet MATHEW's interpretation as one in which evolution has proceeded by extensive differentiation of local races, intergroup selection, and crossbreeding is as close as possible to that required under the Mendelian theory.

Summing up: I have attempted to form a judgment as to the conditions for evolution based on the statistical consequences of Mendelian heredity. The most general conclusion is that evolution depends on a certain balance among its factors. There must be gene mutation, but an excessive rate gives an array of freaks, not evolution; there must be selection, but too severe a process destroys the field of variability, and thus the basis for further advance; prevalence of local inbreeding within a species has extremely important evolutionary consequences, but too close inbreeding leads merely to extinction. A certain amount of crossbreeding is favorable but not too much. In this dependence on balance the species is like a living organism. At all levels of organization life depends on the maintenance of a certain balance among its factors.

More specifically, under biparental reproduction a very low rate of mutation balanced by moderate selection is enough to maintain a practically infinite field of possible gene combinations within the species. The field actually occupied is relatively small though sufficiently extensive that no two individuals have the same genetic constitution. The course of evolution through the general field is not controlled by direction of mutation and not directly by selection, except as conditions change, but by a trial and error mechanism consisting of a largely nonadaptive differentiation of local races (due to inbreeding balanced by occasional crossbreeding) and a determination of long time trend by intergroup selection. The splitting of species depends on the effects of more complete isolation, often made permanent by the accumulation of chromosome aberrations, usually of the balanced type. Studies of natural species indicate that the conditions for such an evolutionary process are often present.

### LITERATURE CITED

CRAMPTON, H. E., 1925 Contemporaneous organic differentiation in the species of Partula living in Moorea, Society Islands. Amer. Nat. 59:5-35.
EAST, E. M., 1918 The role of reproduction in evolution. Amer. Nat. 52:273-289.
GULICK, J. T., 1905 Evolution, racial and habitudinal. Pub. Carnegie Instn. 25:1-269.
JACOT, A. P., 1932 The status of the species and the genus. Amer. Nat. 66:346-364.
JORDAN, D. S., 1908 The law of geminate species. Amer. Nat. 42:73-80.
KELLOGG, V. L., 1908 Darwinism, today. 403 pp. New York: Henry Holt and Co.
KINSEY, A. C., 1930 The gall wasp genus Cynips. Indiana Univ. Studies. 84-86:1-577.
MATHEW, W. D., 1926 The evolution of the horse. A record and its interpretation. Quart. Rev. Biol. 1:139-185.
OSBORN, H. F., 1927 The origin of species. V. Speciation and mutation. Amer. Nat. 49:193-239.
OSGOOD, W. H., 1909 Revision of the mice of the genus Peromyscus. North American Fauna 28:1-285.
PHILIPTSCHENKO, J., 1927 Variabilität and Variation. 101 pp. Berlin.
RENSCH, B., 1929 Das Prinzip geographischer Rassenkreise und das Problem der Artbildung. 206 pp. Berlin: Gebrüder Borntraeger.
ROBSON, G. C., 1928 The species problem. 283 pp. Edinburgh and London: Oliver and Boyd.
RUTHVEN, A. G., 1908 Variation and genetic relationships of the garter snakes. U. S. Nat. Mus. Bull. 61:1-301.
SCHMIDT, J., 1917 Statistical investigations with *Zoarces viviparus* L. J. Genet. 7:105-118.
SEREBROVSKY, A. S., 1929 Beitrag zur geographischen Genetic des Haushahns in Sowjet-Russland. Arch. f. Geflügelkunde, Jahrgang 3:161-169.
SUMNER, F. B., 1932 Genetic, distributional, and evolutionary studies of the subspecies of deer mice (Peromyscus). Bibl. genet. 9:1-106.
THOMPSON, D. H., 1931 Variation in fishes as a function of distance. Trans. Ill. State Acad. of Sci. 23:276-281.
WRIGHT, S., 1931 Evolution in Mendelian populations. Genetics 16:97-159.

# SELECTION IN ANIMALS: SYNTHESIS

## ALAN ROBERTSON

Institute of Animal Genetics, Edinburgh, Scotland

I am perhaps fortunate as the synthesizer of the four papers presented today in that they form a fairly homogeneous group. They deal with selection results using characters of quite different kinds —body size, egg size, and egg production in *Drosophila*; body size, lactation performance and litter size in mice, and finally over-all economic performance in poultry, itself made up of many different traits. I should like to deal briefly with the problem of why we find different situations when we analyze the genetic variation in different quantitative characters.

It is natural to expect that little or no additive genetic variance will be present in characters closely connected with fitness. As Fisher's fundamental theorem of natural selection states, the rate of change of fitness of a population is proportional to the additive genetic variance remaining in it. The population which has had opportunity to reach equilibrium with its present environment should thus have little or no additive genetic variance in fitness, although there may be differences in genotypes as far as their over-all fitness is concerned. To my mind, the surprising fact is that for many characters of trivial importance to the organism, it is impossible to escape the conclusion that in populations that have not been under artificial selection, the hereditary variance appears to be almost entirely of the additive kind. In some experiments on abdominal bristles in *Drosophila* done by myself and my colleagues, the additive genetic variance appears to make up over 80 per cent of the total genetic variance. The same appears to be true for other bristle systems in *Drosophila* and for such characters as fat content of milk in cattle. However, it is not possible to cite many cases because of the difficulty of measuring the total genetic variance in characters, except where it is possible either to make duplicate measurements of a character or to measure the variability of a genetically homogeneous but heterozygous group. From the latter type of measurement we can suggest that probably at least half of the genetic variance in body size in *Drosophila* is simply additive. At the other extreme we have the evidence of Wallace's results with *Drosophila* that the genetic variation in larval survival is only to a small extent additive.

Now, if almost all the genetic variation in a population is additive, we cannot escape the conclusion that it must be due to genes which act additively, in the sense that the average value of the heterozygote is close to the mean of those for the two homozygotes. Of course, recessive genes can cause some additive variance but this is usually a small proportion of the total unless the frequency of all recessives is high, which is rather improbable. Historically, this is rather an odd situation. In the early development of quantitative genetic theory, additive action was often assumed because of the simple parallel between the heterozygosis in a population and the genetic variance. This has been the source of some criticism by classical geneticists who pointed out that for the genes with which they were used to working, a recessive-dominant relationship was the rule and that intermediate heterozygotes are infrequent. (I might point out in parenthesis here, that a change of meaning in the usage of the word "dominant" from the classical one has gradually taken place. Its present use, generally to describe a mutant gene, indicates that the heterozygote is distinguishable from wild type organisms, and does not imply that the homozygous mutant is not distinguishable from the heterozygous form, as the classical usage would. Thus, a mutant acting additively would on the present usage of the word be called "dominant"). However, we now find that this assumption of additive action originally made for convenience in algebraic treatment is in fact true for many characters with which we have to deal.

Why should this be so? I suggest that we have to look for the answer in the evolution of dominance, whether we accept the model proposed by Fisher or that proposed by Wright. On these hypotheses there will be a selection for genes which will reduce the expression of a deleterious gene when it is in the heterozygous state, as almost all rare genes are. But this selection will only operate to reduce the effect on reproductive fitness in the heterozygote. We can represent the situation diagrammatically in Figure 1.

This is a very simple schematic diagram of an organism in as far as its evolutionary properties are concerned. At the left-hand side we have the primary gene reactions which, through an exceedingly complex interrelationship of chemical and physical processes, produce in the end the adult phenotype. But, in the sense of purpose, the end is the reproductive fitness of the organism. Now, the evolution of dominance will only alter the results of different gene substitutions in so far as they affect the reproductive fitness of the organism. Characters which are in the direct line of development from primary gene action to reproductive fitness will also show to some extent this phenomenon of dominance. One might expect, perhaps, that the later they occur in the chain of development reactions, the more dominance they will show and those early in the chain will show the basic action of the genes more closely. But there are

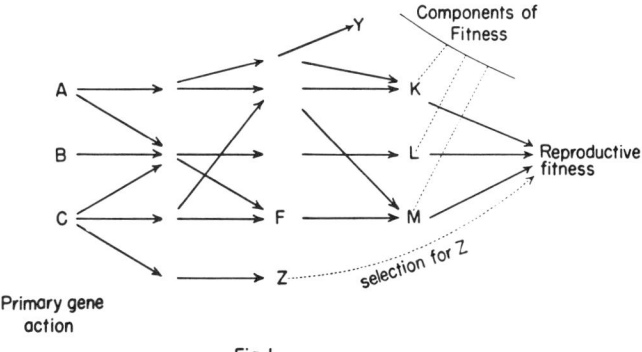

FIGURE 1. The evolution of dominance in different characters.

many characters, Y, Z for instance, perhaps once of importance to the species and now only remaining by evolutionary accident, in which minor changes of the phenotype can have no effect on reproductive fitness. We might call these "trivial" or "peripheral" characters in the sense that they are outside the main pathway from primary gene action to reproductive fitness. Now there is no reason why the evolution of dominance as far as fitness is concerned, perhaps by affecting reactions later in the process than the point at which the development of a peripheral character branches off, should affect the dominance relationships in this character. It is tempting to suggest that the higher proportion of genetic variation in such characters which we find to be additive relates to the basic patterns of gene action unmodified by natural selection.

Before leaving Figure 1, it can be used to discuss one further aspect of natural selection. I have taken it as axiomatic that in a population at equilibrium there will be no additive genetic variance in reproductive fitness. In the major components of fitness such as viability and fertility (K, L, and M) there may be some utilizable genetic variation so that we can obtain some response to selection. But this must be at the expense of some other component of fitness because by definition we cannot increase over-all fitness. We may therefore expect a negative additive genetic correlation between the main components of fitness, as Dickerson has shown in his paper. However, I am indebted to Dr. Forbes Robertson for pointing out that this does not imply an over-all negative genetic correlation, as one might expect for instance that heterozygotes would be superior to homozygotes for all the major fitness components. The over-all value will then depend on the relative magnitudes of the two effects but that determined by the usual method of measuring genetic correlations which utilises only the additive genetic variance would be expected to be negative.

If we take a "peripheral" character (Z) and submit it to intense selection we are altering its position in relation to fitness so that it has now become a major fitness component. We shall eventually reach a new equilibrium position in which there is no further utilizable genetic variance left, in which the genetic situation is similar to that in natural fitness components. We can then expect negative additive genetic correlations between it and the natural fitness components. It may be possible to use such stable populations for the investigations of the genetics of fitness components, which themselves are often rather difficult to work with because of measurement problems and sensitivity to environmental fluctuations. I and my colleagues have a population of *Drosophila melanogaster* which after some 35 generations of selection downwards for abdominal bristles has stabilized itself at a mean count of about five bristles in females (as compared to 38 in the original population). It would be easy to use this population for work on such programs as reciprocal recurrent selection, while it would be of little interest to do such an experiment on the base population in which almost all the genetic variance was additive.

I would now like to go into more detail about the genetic connection between different characters and fitness in what is in essence an attempt to classify characters according to the way in which the genetic variation in the population is made up. I shall be discussing the situation entirely in terms of genotypic values. Thus, to any genotype (represented by the individual genes making it up) there corresponds a series of genotypic values for different characters, each value being the mean measurement of all organisms of that genotype under the given array of environmental conditions. The population is then a collection of genotypes each of which has a series of genotypic values for different characters. I now want to deal with the average genotypic value for fitness of organisms with a given genotypic value for a metric character. I will

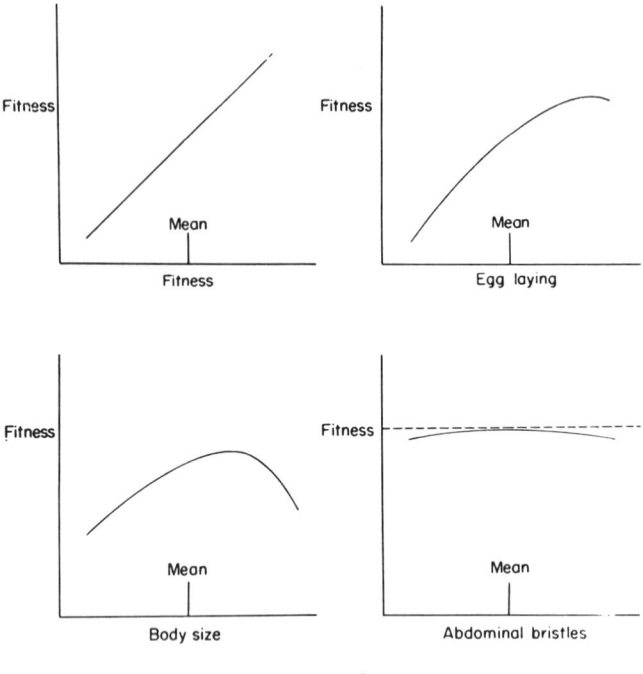

Fig 2

FIGURE 2. The classification of characters according to their relationship with fitness.

divide the metric characters into four broad classes (see Figure 2).

a. Fitness itself. We will of course, have a linear identical relationship between fitness and fitness in which some genotypes will have average values well above the mean of the population, but to which the population as a whole cannot change because these extreme values are those of heterozygotes. Thus, on the simple overdominance hypothesis the heterozygotes will be above the mean for fitness and the homozygotes on the whole below it. There will be no additive genetic variance, but inbreeding depression and its converse, heterosis.

b. A major component of fitness such as egg laying in *Drosophila* (K, L, and M). Females genetically inclined to lay more eggs than average will probably be fitter than females which lay less eggs. We might expect the curve to turn over a little at the top end. Only a small proportion of the genetic variation will be additive and there will be fairly considerable inbreeding depression.

c. Characters with slight direct effect on fitness such as perhaps body size in *Drosophila* (F). The genotypic value with optimum fitness will be slightly displaced from the mean value in the population. One would expect a proportion of additive genetic variance but also some inbreeding depression.

d. Peripheral characters with no direct effect on fitness such as abdominal bristles in *Drosophila* (Y and Z). The optimum value of the character will then coincide with the population mean, although the extreme values may be less fit than the mean. A large part of the genetic variance would then be additive and there would be no inbreeding depression.

I must emphasize that these ideas relating the genetic situation concerning a character with its position in the development of the organism are not yet fully worked out. A good many will probably turn out to be incorrect or to need major modification. However, they may represent a useful generalization to indicate the sort of genetic situations we might expect in different characters and some possible reasons why.

## THE INVESTIGATION OF CAUSAL RELATIONSHIPS BETWEEN CHARACTERS AND FITNESS

Turning now to Figure 2D, I should like to stress particularly the need for caution in interpreting such relationships when they appear in practice, as for instance in the connection between egg size and hatchability found for duck eggs by Rendel. These represent an observed relationship between two measurements but one should be very cautious in inferring any causal relationship between the two. It is possible, on the one hand, that the extremes are less fit simply because they are extremes. That is to say, the causation is at the phenotypic level. Or on the other hand they may be less fit because they are more homozygous and we know that in general homozygotes are less fit than heterozygotes. The causation is then much more fundamental in the development of the organism. Now these two models are just as acceptable as descriptions of this particular relationship but may have quite different consequences. Continued selection of an intermediate leads to genetic fixation rather than the maintenance of genetic variability, as Wright has shown. However, the maintenance of genetic variability is a premise of the second model as the heterozygotes are assumed a priori to be more fit than the homozygotes. We may draw a diagram like Figure 3 to illustrate the causal relationships between the different phenomena but perhaps other people would put in the causal arrows in a different way.

of time that response to selection will continue. These changes in gene frequency will in turn depend on the individual effects of the separate genes. The greater the effect the greater the change in gene frequency and the quicker the selection limits are reached, as Falconer has shown with selection for lactation performance. I would like to resuscitate in this connection a formula due to Haldane—derived in 1927 for quite another purpose—in which he showed that the selective advantage in quantitative selection due to a difference $\delta$ is equal to $\frac{\bar{i}\delta}{\sigma}$ where $\bar{i}$ is the selection intensity (in standard units) and $\sigma$ the standard deviation of the remaining variance (partly due to other genes, and partly due to environment). The change in frequency in one generation of an additive gene whose total effect is a is then given by

$$\Delta q = \frac{\bar{i}a}{2\sigma} pq$$

In some of my own selection experiments as well as some of those of Dr. Falconer and of Drs. Forbes

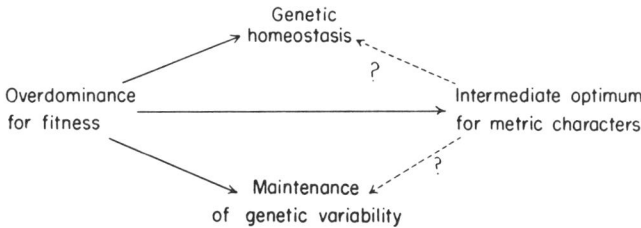

Fig. 3

FIGURE 3. Suggested causal relationships between different phenomena.

We have so far been dealing with populations in equilibrium. To what extent can we predict what will happen if the equilibrium is disturbed? The answer is that we can only do this in a very general way. We may attempt to predict the response of the population mean to selection using such parameters as the additive genetic variance but we do not know how long this response will continue because we can't predict how the parameters themselves will change under selection. The description of a population, in such terms as Kempthorne has done in this meeting, is a static one. As Kempthorne has himself said to me "It is a descriptive and not a mechanistic model." Even on quite simple mechanistic models it is easy to show that on selecting in one direction for a character the additive genetic variance may increase and of selecting the other way it will decrease. Thus, gene frequencies do not enter into the descriptive model but they do into the mechanistic model. On the change in gene frequencies in each generation will depend the length

Robertson and Reeve the selection response has lasted 20 or more generations in both directions. If we therefore, substitute possible values of $\Delta q$ of 0.02 we can obtain probable values of $\frac{a}{\sigma}$ of the order of 0.2 to 0.3. This gives us an idea of the effect of the more important genes we are dealing with in such experiments. However, such concepts do not help us greatly to predict how long response to selection will last when we are in the early stages of the program. It seems that the only thing we can do is try it and see.

It is the frequent experience of those carrying out *Drosophila* selection experiments such as those reported here by Dr. Forbes Robertson and Dr. Bell as well as by myself and my colleagues that cessation of response to selection does not mean genetic fixation. This maintenance of genetic variance perhaps occurs as a result of the negative genetic correlation between the characters selected for and fitness components. In such situations, heritability estimates are useless in predicting response to selection, although they may have been quite adequate in the early generations. Indeed it can be argued that if there is a differential viability of zygotes between fertilization and measurement the concept of additive genetic variance ceases to have any value. All one can usefully discuss are. the operational estimates obtained by the different methods. In some of our cases these fitted neatly together in agreement with other diagnoses deriving from the use of methods peculiar to *Drosophila*, in others the solutions still have to be found.

I should like, finally, to present a little evidence on the rate of production of new variation by mutation, which has continually entered into these discussions as a possible explanation for otherwise unexplained phenomena. In continuous variation we can no longer measure mutation rates at individual loci. We shall have to specify the rate of increase of variance each generation. Characters like abdominal bristles in *Drosophila* in which most of the variance is additive are the simplest to deal with in this respect. Taking two investigations by Mather and his collaborators as a source of evidence it is possible to get at least an idea of the order of magnitude of the rate of occurrence of new variance for this character. Two independent estimates gave values of about 0.01 units per generation compared to a total amount of five units in wild populations. It would thus take about 500 generations for a genetically homogeneous population to reach the present levels of variation that we find in wild populations. This is an indication that for this character at least we should not worry too much about spontaneous variation in laboratory experiments except in special circumstances when we are dealing with genetically homogeneous stocks. But from the evolutionary point of view this is a high rate of emergence of new variation and suggests that mutation may be important in maintaining variation in quantitative characters in wild populations. The figures suggest that after an inbred line has been mass mated for of the order of 30 to 40 generations, detectable variations should have accumulated in bristle characters. We are proposing to carry out such experiments with several characters as the existing knowledge on this point is very scant indeed.

# THE EFFECT OF SELECTION ON GENETIC VARIABILITY

M. G. BULMER

Department of Biomathematics, University of Oxford, England

This paper arose from the following apparently paradoxical consideration. If a metric character is subjected to stabilizing selection, which reduces the phenotypic variance in each generation by weeding out extreme deviations from the norm, one would expect the genetic variance to be reduced proportionately in each generation and ultimately to be eliminated altogether. But Crow and Kimura (1970) show that the change in the variance under selection decreases as the number of loci involved increases and must eventually tend to zero when the number of loci becomes effectively infinite. Also, most metric characters investigated in natural and domesticated populations possess much genetic variability, and are likely subject to stabilizing selection (Falconer 1960). The intuitively obvious argument that stabilizing selection will lead to the rapid elimination of genetic variability must therefore be wrong. Why?

To state the problem precisely, consider a metric character whose phenotypic value in an individual, $Y$, measured before the operation of selection can be regarded as the sum of the genotypic value, $G$, and an independent environmental deviation, $E$, so that $Y = G + E$. The total phenotypic variance, $V$, is therefore the sum of a genetic component, $V_G$, and an environmental component, $V_E$, so that

$$V = V_G + V_E. \qquad (1)$$

It is emphasized that these variances are assumed to be measured in the population *before* selection; the symbol $V'$ will be used for the total variance measured *after* the operation of selection. The genetic variance may contain an additive and a dominance component,

$$V_G = V_A + V_D; \qquad (2)$$

but it is assumed that there are no epistatic components, so that there is no interaction between the effects of different loci. For any one individual, the total genotypic value, $G$, may therefore be written as the sum of the genotypic values from the different loci affecting the character,

$$G = \sum_{i=1}^{N} g_i, \qquad (3)$$

where $g_i$ is the genotypic value at the $i$th locus and $N$ is the number of loci involved. It should be noted that $g_i$ depends on both alleles represented at the $i$th locus and may thus contain a dominance component due to the interaction between the two alleles at that locus. For the population we may write

$$V_G = \text{Var}(G) = \text{Var}(\sum_{i=1}^{N} g_i) = \sum_{i=1}^{N} \text{Var}(g_i) + \sum_{i \neq j} \text{Cov}(g_i, g_j), \quad (4)$$

where $\text{Var}(g_i)$ is the variance at the $i$th locus and $\text{Cov}(g_i, g_j)$ is the covariance between the $i$th and $j$th loci. Under random mating in the absence of selection, the covariances would, of course, be zero, so that equation (4) would take the simpler form

$$V_G = \sum_{i=1}^{N} \text{Var}(g_i). \quad (5)$$

It should be observed that the genetic variance $V_G$ refers to the variability of the total genotypic values, $G$, between individuals in the population, *not* to the variability of the $g_i$'s at different loci in the same individual. Likewise $\text{Var}(g_i)$ refers to the variability at a fixed locus, and $\text{Cov}(g_i, g_j)$ to the covariance at a fixed pair of loci, over different individuals in the population.

We shall now suppose that the number of loci involved is effectively infinite and that the environmental deviation, $E$, is normally distributed. This model was first studied by Fisher (1918) and is usually called the infinitesimal model. Under random mating in the absence of selection, it can now be shown, not only that the character is normally distributed, but also that the phenotypic values of two or more related individuals follow a multivariate normal distribution in the absence of linkage; this result also holds in the presence of linkage if the related individuals are identical twins or offspring and one or both parents, but not if they are related in any other way. These results are derived in the Appendix. Their importance is that, if the phenotypic values in a set of related individuals are distributed as multivariate normal variables, any regression of one of them on the rest must be linear and homoscedastic. The regression equation and the residual variance can be calculated from the correlations between the individuals by standard regression theory; the calculations for the three special cases considered in this paper are given in the Appendix. Note that these results hold good even when there is dominance, although the correlations between relatives and hence the regression equation depend on the degree of dominance. However, these are only limiting results. When the number of loci is small, the variability will usually be larger in the middle than at the periphery of the regression line or plane, owing to the greater amount of heterozygosis, and dominance will usually cause departures from linearity.

Let us now consider the effect of one generation of selection on the variance in the next generation. In the absence of selection, the joint regression of offspring ($Y$) on both parents ($P_1$ and $P_2$) is given by

$$Y = a + bP_1 + bP_2 + e, \tag{6}$$

where

$$b = \tfrac{1}{2}h^2$$
$$\mathrm{Var}(e) = (1 - \tfrac{1}{2}h^4)V. \tag{7}$$

These results are derived in the Appendix; $h^2$ is the heritability in the strict sense, $h^2 = V_A/V$. This regression equation holds good whatever values are attributed to the independent variables, $P_1$ and $P_2$, and is therefore unaffected if we allow selection to operate in the parental generation. If selection acts in such a way as to change the variance in the parental generation from $V$ to $V' = V + dV$, and if individuals mate at random after selection, then the variance in the offspring generation will become

$$\begin{aligned}\mathrm{Var}(Y) &= b^2\mathrm{Var}(P_1) + b^2\mathrm{Var}(P_2) + \mathrm{Var}(e) \\ &= 2b^2(V + dV) + \mathrm{Var}(e) = V + \tfrac{1}{2}h^4 dV.\end{aligned} \tag{8}$$

Thus, the variance of the offspring distribution is changed by the amount $\tfrac{1}{2}h^4 dV$; this must be a change in the genetic variance, since the environmental variance is assumed to be constant. (If selection acts only on one parent, as will be the case for a maternal character like clutch size or egg weight, the change in the variance will only be half the above amount.) Under stabilizing selection, as under most types of selection except disruptive selection, $dV$ will be negative, so that the variance will of course be reduced. But if this process is repeated in each generation, the genetic variance will decrease quite rapidly until the heritability becomes negligible. How can this be reconciled with the conclusion of Crow and Kimura (1970) that there is no change in the variance under selection when the number of loci is effectively infinite and with the fact that the heritability is quite high in many characters which are probably subject to stabilizing selection? Conversely, if $dV$ is positive, one is led by the same argument to the absurd conclusion that fresh genetic variability is continually generated by the action of disruptive selection.

The resolution of this paradox depends on the fact that the genetic variance under selection is the sum of two components, which will be called the equilibrium genetic variance and the disequilibrium contribution. The disequilibrium contribution is the component of the genetic variance due to the phenomenon usually known as "linkage disequilibrium" but which will here be called "joint disequilibrium" since its existence does not depend on linkage; the equilibrium genetic variance is the value which the genetic variance would have in the absence of disequilibrium. From equation (4), we find that the genetic variance can be expressed as

$$V_G = \sum_{i=1}^{N} \mathrm{Var}(g_i) + \sum_{i \neq j} \mathrm{Cov}(g_i, g_j). \tag{9}$$

The first term on the right-hand side of this equation is the equilibrium genetic variance, and the second term is the disequilibrium contribution.

Under random mating in the absence of selection, the disequilibrium contribution would, of course, be zero; but this will not be true under selection, since one generation of random mating is sufficient to restore Hardy-Weinberg equilibrium at each individual locus but not to restore joint equilibrium between pairs of loci (e.g., Li 1955 or Falconer 1960). In particular, stabilizing selection, which favors values near the mean, will introduce a negative correlation between pairs of loci, part of which will be retained in the next generation, so that the disequilibrium contribution will be negative and the actual genetic variance will be less than the equilibrium genetic variance. If selection were relaxed, joint equilibrium would gradually be restored, and the genetic variance would return to its equilibrium value; if none of the loci were linked, the disequilibrium contribution would be halved in each generation after selection ceases, but linkage would slow down the rate of return to equilibrium.

We shall now show that, under the infinitesimal model, the change in the genetic variance demonstrated in equation (8) is due entirely to the disequilibrium contribution. To prove this, let us consider the effect of one generation of selection on the variance among their grandchildren. In the absence of linkage and selection, the regression of a grandchild on his four grandparents is given by

$$Y = a + bG_1 + bG_2 + bG_3 + bG_4 + e, \qquad (10)$$

where

$$b = \tfrac{1}{4}h^2$$
$$\mathrm{Var}(e) = (1 - \tfrac{1}{4}h^4)V, \qquad (11)$$

as derived in the Appendix. This formulation is only valid in the absence of linkage, whereas the formulation of the parent-offspring regression in equations (6) and (7) is also valid in the presence of linkage. If we now allow selection to change the variance in the grandparental generation from $V$ to $V + dV$, then it can be shown in exactly the same way as before that the variance among the grandchildren is

$$\mathrm{Var}(Y) = V + \tfrac{1}{4}h^4 dV. \qquad (12)$$

It is essential to this argument that there be no selection in the parental generation, since it is assumed that all surviving individuals in the grandparental generation have the same chance of having grandchildren. Thus, the change in the variance has been halved when compared with its value in the generation immediately following selection (see eq. [8]). This is exactly what would be expected in the absence of linkage if the change in the variance is restricted entirely to the disequilibrium contribution and proves that there has been no permanent change in the equilibrium genetic variance. This argument is based on the assumption that there is no linkage; but there is no reason to suppose that it should not also hold in the presence of linkage, with the modification that the disequilibrium contribution is not halved but is reduced by a fraction somewhat less than one-half in each generation after selection ceases. The argument on which equation (8) is based is valid in the presence of linkage, so that the disequilibrium contribution in the offspring generation generated by selection in the previous generation is given by $\frac{1}{2}h^4 dV$ whether linkage is present or not.

It can be concluded that the change in the variance resulting from one generation of selection does not reflect a permanent change in the equilibrium genetic variance but is due entirely to the correlation between loci induced by selection and therefore disappears rapidly when selection ceases. It must be stressed, however, that this is a limiting result, which is only true exactly when the effective number of loci is infinite. The restriction of the argument to an infinite number of loci cannot be removed in any general way since, for the equilibrium variance to remain constant, there must be no change in the mean gene frequency even if the phenotypic mean changes as a result of selection, which implies an infinite number of loci. It is obvious that selection can cause a permanent change in the equilibrium genetic variance in any actual situation, however large the number of loci may be. The limiting result obtained here shows, however, that the magnitude of any permanent change must decrease as the number of loci involved increases and that when the number of loci is large it is likely to be much less than the temporary change due to the induced correlation between loci, that is, to joint disequilibrium.

Before considering the effect of several generations of selection, we must investigate the effect of the disequilibrium contribution on the genetic structure of the population. To do this, we consider the regression of an individual on one parent and on both grandparents on the same side, which in the absence of linkage is given by

$$Y = a + bP + cG_1 + cG_2, \qquad (13)$$

where

$$b = \tfrac{1}{2}h^2(2 - h^2)/(2 - h^4),$$
$$c = \tfrac{1}{4}h^2(1 - h^2)/(2 - h^4). \qquad (14)$$

The derivation is in the Appendix. If selection acts on the grandparents so as to change their variance from $V$ to $V + dV$, then the covariance of $Y$ and $P$ is given by

$$\begin{aligned}
\mathrm{Cov}(Y, P) &= \mathrm{Cov}(bP + cG_1 + cG_2, P) = b\mathrm{Var}(P) + c\mathrm{Cov}(G_1 + G_2, P) \\
&= b\mathrm{Var}(P) + \tfrac{1}{2}ch^2\mathrm{Var}(G_1 + G_2), \text{ from equations (6) and (7)}, \\
&= b(V + \tfrac{1}{2}h^4 dV) + ch^2(V + dV) = \tfrac{1}{2}(h^2 V + \tfrac{1}{2}h^4 dV).
\end{aligned} \qquad (15)$$

The additive genetic variance in the parental generation is twice the covariance between parent and offspring, which is equal to $h^2 V + \tfrac{1}{2}h^2 dV$; this is the sum of the additive genetic variance before selection and of the disequilibrium contribution. The disequilibrium contribution can therefore be regarded as a component of the additive genetic variance.

Let us now consider the effect of several generations of selection. We shall write $A$ instead of $V_A$ for the additive genetic variance, and we shall suppose that in the $i$th generation of selection the additive genetic variance is $A_i$, the phenotypic variance (measured before selection) $V_i$, the heritability $h_i^2 = A_i/V_i$, and the disequilibrium contribution $d_i$; if the additive genetic variance and the phenotypic variance before the onset of selection were $A_0$ and $V_0$, respectively, then

$$A_i = A_0 + d_i, \qquad (16)$$
$$V_i = V_0 + d_i.$$

If the effect of selection in the $i$th generation is to change the phenotypic variance from $V_i$ to $V_i + dV_i$, the disequilibrium contribution in the next generation, in the absence of linkage, is

$$d_{i+1} = \tfrac{1}{2}d_i + \tfrac{1}{2}h_i^4 dV_i, \qquad (17)$$

since only half the disequilibrium contribution present in the previous generation is preserved. This recurrence relationship allows us to calculate the disequilibrium contribution in successive generations of selection. If $dV_i$ is constant or approaches a limiting value $dV^*$, then $d_i$ will rapidly tend to a limiting value, $d^*$, which can be evaluated by putting $d_{i+1} = d_i$ in equation (17); hence,

$$d^* = h^{*4} dV^*, \qquad (18)$$

where $h^{*2}$ is the limiting value of the heritability. In many circumstances, $dV_i$ will be proportional to the phenotypic variance, $V_i$, so that we may write $dV_i = kV_i$. Substituting this value in equation (18) and solving it in terms of $V_0$ and $h_0^2$, we find that

$$d^* = V_0\{2kh_0^2 - 1 + [1 - 4kh_0^2(1 - h_0^2)]^{\frac{1}{2}}\}/2(1 - k). \qquad (19)$$

We shall now consider the significance of these results under artificial and natural selection.

### ARTIFICIAL SELECTION

Consider a mass-selection experiment in which the individuals with the highest phenotypic values are chosen to be the parents of the next generation. If the phenotype is normally distributed in the parent generation with mean $M$ and variance $V$ and if a proportion $P$ is selected then the mean in the selected group of parents is $M + IV^{\frac{1}{2}}$, where $I$, the intensity of selection, is equal to $f(x)/P$, $x$ being the standard normal deviate corresponding to $P$ and $f(x)$ the standard normal density function at $x$; for example, if $P = 0.2$, then $x = 0.8416$, $f(x) = 0.2800$, and $I = 1.400$. The response to selection among their offspring is equal to $Ih^2V^{\frac{1}{2}}$. (See Falconer 1960 for further explanation.) It can be shown similarly that the variance in the selected group of parents is $V[1 - I(I - x)]$, so that the change in the variance as a result of selection is $-I(I - x)V$, which is equal to $-0.7818V$ when $P = 0.2$. Putting $k = -0.7818$ in equation (19) and taking $h_0^2 = \tfrac{1}{2}$, we find that the disequilibrium contribution after several generations of selection of this intensity will reach a limiting value of $-0.125V_0$. Thus, if the phenotypic variance was 100 and the additive genetic variance 50 before selection began, then after several generations of selection the phenotypic variance will be reduced to 87.5 and the additive genetic variance to 37.5, so that the heritability is reduced from 0.5 to 0.428. The response to selection is reduced from 7.00 at the beginning of the experiment to 5.61, which is a reduction of 20%. The original values of the variance and heritability will, of course, be rapidly restored when selection ceases. The theoretical course of the selection experiment, calculated from the recurrence relationship contained in equation (17) is shown in table 1.

TABLE 1
THE EFFECT OF DISEQUILIBRIUM IN AN ARTIFICIAL SELECTION EXPERIMENT*

|  | GENERATION | | | | | |
|---|---|---|---|---|---|---|
|  | 0 | 1 | 2 | 3 | 4 | ∞ |
| Disequilibrium contribution ... | 0 | −9.8 | −11.9 | −12.4 | −12.5 | −12.5 |
| Phenotypic variance ......... | 100 | 90.2 | 88.1 | 87.6 | 87.5 | 87.5 |
| Heritability ................ | 0.500 | 0.446 | 0.432 | 0.429 | 0.428 | 0.428 |
| Response to selection ........ | ... | 7.00 | 5.93 | 5.68 | 5.63 | 5.61 |

* Expected values for several parameters in a directional selection experiment computed for four generations. The top 20% of individuals are selected in each generation.

The limiting values are attained very quickly, nearly all the change occurring in the first two generations. If linkage were taken into account, an even larger effect would be expected.

NATURAL SELECTION

The intensity of selection is likely to be considerably less in nature than in artificial breeding experiments. As an example of stabilizing selection, Haldane (1953) has quoted the data of Rendel (1943) on the weights of 960 ducks' eggs of which 64.5% hatched. The mean weight of the original population was 73.9 gm and that of those which hatched 73.8 gm, so that there is no evidence of any directional selection on the mean weight. On the other hand, the variance of the weights was reduced from 52.7 to 43.9, which is highly significant and indicates that eggs of intermediate weight are at a selection advantage compared with either light or heavy eggs. The change in the variance as a result of selection is $dV = -8.8$; but before applying equation (18) we must allow for the fact that egg weight is maternally determined, so that selection acts only on one parent. In this case,

$$d_{i+1} = \tfrac{1}{2}d_i + \tfrac{1}{4}h_i^4 dV_i, \tag{20}$$

so that in the limit

$$d^* = \tfrac{1}{2}h^{*4} dV^*. \tag{21}$$

Assuming that $h^{*2} = 0.6$ (this is the heritability of egg weight in poultry quoted by Falconer 1960), we conclude that $d^* = -1.6$. If there were no selection, the phenotypic variance would increase from 52.7 to 54.3, and the heritability from 0.60 to 0.61, which is a negligible difference. The important point, however, is that once the phenotypic variance has been reduced from 54.3 to 52.7, due to a reduction in the effective additive genetic variance from 33.2 to 31.6, there will be no further reduction in the variance under continuous stabilizing selection at the same intensity; furthermore, if selection ceases, the small loss of variance will be quickly recovered. Attention must again be drawn to the proviso that this analysis is only strictly valid when the number of loci is effectively infinite and in the absence of linkage.

The effect of joint disequilibrium will probably be similar under directional selection and under stabilizing selection, since both can be regarded as cases of selection for an optimal value. The most plausible model of this type of selection, first considered by Haldane (1953), is that the fitness of an individual with phenotypic value $Y$ is equal to $\exp - c(Y - T)^2$, where $T$ is the optimal value and $c$ determines the intensity of selection. If $M$ and $V$ are the phenotypic mean and variance, this seems equally suitable as a model of stabilizing selection (when $M = T$) or of directional selection (when $M \neq T$). Furthermore, if $Y$ is normally distributed, it is quite easy to show that the change in the mean and variance in the parental generation as a result of selection are, respectively,

$$dM = 2cV(T - M)/(1 + 2cV),$$
$$dV = -2cV^2/(1 + 2cV). \qquad (22)$$

The response to selection is $h^2 dM$. Since the disequilibrium contribution is determined by $dV$, which is independent of $M$ and $T$, it follows that it will be the same, for this model, under either stabilizing or directional selection.

Finally, mention must be made of disruptive selection, under which it is supposed that extreme phenotypic values (either large or small) are favored at the expense of intermediate values. This form of selection will lead to an increase in the phenotypic variability, so that the disequilibrium contribution will be positive rather than negative. Thus disruptive selection will cause a small increase in the additive genetic variance and, hence, in the heritability.

SUMMARY

If a metric character is determined by an effectively infinite number of loci, selection cannot cause any permanent change in the genetic variance but will cause a temporary change which is rapidly reversed when selection ceases. This change is due entirely to the correlation between pairs of loci which is induced by selection; the correlation is negative, leading to a reduction in the genetic variance under stabilizing or directional selection, and is positive, leading to an increase in the variance under disruptive selection. When selection ceases, the correlation rapidly disappears as joint equilibrium at pairs of loci is reestablished, and the variance returns to its original value. An expression is derived for the predicted amount of change in the genetic variance due to disequilibrium in the absence of linkage. The change is likely to be small under selection intensities found under natural conditions, but it may be appreciable under intense artificial selection. This limiting result shows that the magnitude of any permanent change in the variance due to selection must decrease as the number of loci involved increases and that, when the number of loci is large, it is likely to be much less than the temporary change due to disequilibrium.

[*Editor's Note:* The Appendix has been omitted.]

LITERATURE CITED

Cockerham, C. C. 1956. Effects of linkage on the covariances between relatives. Genetics 41:138–141.

Crow, J. F., and M. Kimura. 1970. An introduction to population genetics theory. Harper & Row, New York. 591 p.

Falconer, D. S. 1960. Introduction to quantitative genetics. Oliver & Boyd, Edinburgh and London. 365 p.

Fisher, R. A. 1918. The correlation between relatives on the supposition of Mendelian inheritance. Roy. Soc. (Edinburgh), Trans. 52:321–341.

Haldane, J. B. S. 1953. The measurement of natural selection. 9th Int. Congr. Genet., Proc., p. 480–487.

Li, C. C. 1955. Population genetics. Univ. Chicago Press, Chicago. 366 p.

Rendel, J. M. 1943. Variations in the weights of hatched and unhatched ducks' eggs. Biometrika 33:48–58.

Schnell, F. W. 1963. The covariance between relatives in the presence of linkage, p. 468–483. *In* W. D. Hanson and H. F. Robinson [ed.], Statistical genetics and plant breeding. Nat. Acad. Sci. Nat. Res. Council Pub. 982.

# 24

Copyright © 1976 by Cambridge University Press
Reprinted from *Genet. Res. (Cambridge)* **26**:221-235 (1976)

## The maintenance of genetic variability by mutation in a polygenic character with linked loci

By RUSSELL LANDE

*Museum of Comparative Zoology, Harvard University, Cambridge, Mass. 02138, U.S.A.*

### SUMMARY

It is assumed that a character under stabilizing selection is determined genetically by $n$ linked, mutable loci with additive effects and a range of many possible allelic effects at each locus. A general qualitative feature of such systems is that the genetic variance for the character is independent of the linkage map of the loci, provided linkage is not very tight. A particular detailed model shows that certain aspects of the genetic system are moulded by stabilizing selection while others are selectively neutral. With reference to experimental data on characters of *Drosophila* flies, maize, and mice, it is concluded that large amounts of genetic variation can be maintained by mutation in polygenic characters even when there is strong stabilizing selection. The properties of the model are compared with those of heterotic models with linked loci.

### 1. INTRODUCTION

It is now generally accepted that the evolution of species usually proceeds by a continuous transformation of their characters. The study of extant populations shows that these characters often display considerable heritable variation and typically are influenced by many genes of small effect (Falconer, 1960). The question of how heritable variation for such characters is maintained is still largely unresolved.

It has been argued that most polygenic or quantitative characters are under natural selection for an intermediate optimum phenotype, and such stabilizing selection has been demonstrated for many characters (Rendel, 1943; Lerner, 1954; Allard & Jain, 1962; Lack, 1966). Fisher (1930) showed theoretically that stabilizing selection depletes genetic variability in polygenic characters and this has been confirmed by other workers both theoretically (Robertson, 1956; Lewontin, 1964*b*) and experimentally (e.g. Waddington, 1960; Scharloo, 1964; Gibson & Bradley, 1974). Lewontin (1964*b*) found that with directional dominance many genes can be kept segregating by stabilizing selection, but that only a limited amount of genetic variance can be maintained (less than with a single locus). Gale & Kearsey (1968) and Kearsey & Gale (1968) gave examples where two and three linked loci can be kept segregating if they have very unequal effects; this requirement implies that even with more loci the effective number of loci must be small and that only a limited amount of genetic variance could be maintained. Such schemes cannot account for the genetic variability in typical polygenic characters which have large effective numbers of loci (Falconer, 1960).

It is usually thought that mutation cannot by itself offset the loss of genetic variability caused by stabilizing selection to maintain the levels of heritable variation observed in natural artificial populations. Selective mechanisms for maintaining genetic variability are often invoked, usually heterozygote advantage and heterogeneous environments. But even in nearly constant, uniform environments sexual populations are known to maintain considerable genetic variation for quantitative characters. An example is the Kaduna population of *Drosophila melanogaster* which has been kept in such conditions for hundreds of generations (Lopez-Fanjul & Hill, 1973). Heterozygote advantage or 'heterosis' is a commonly observed phenomenon but there are two possible explanations for it (East & Jones, 1919). The first is heterozygote superiority at many individual loci and the second is that different inbred lines are homozygous for different partially recessive deleterious genes. It is particularly important to distinguish between these two possible causes of hybrid vigour when considering the maintenance of genetic variability as the first can contribute to the amount of variability maintained while the second cannot. It is well established that many deleterious recessive and partly recessive genes exist in normal outbred individuals. Furthermore, inbred lines have fitnesses as great as wild type under some conditions (East & Jones, 1919; Dobzhansky, Holtz & Spassky, 1942; Dobzhansky & Spassky, 1944). Many species of plants preferentially inbreed to various degrees when there is the possibility of increased outbreeding. These observations seem to argue against any general heterosis at the level of single loci; and very few cases of single locus heterozygote advantage are known (Lewontin, 1974).

Because of computational difficulties, all analytic models and simulations of polygenic systems have ignored either linkage or mutation. All theoretical knowledge of the behaviour of linked genes has thus been derived from models which have only selective mechanisms for the maintenance of genetic variability. This has diverted attention from the possibility that mutation may be a powerful force maintaining genetic variation in polygenic characters and fostered the notion that heterozygote advantage at the level of single loci is necessary to explain observed amounts of genetic variation in characters of adaptive significance.

Here a model of a phenotypic character under stabilizing selection influenced by many linked, mutable genes is investigated. Because stabilizing selection acts not to preserve but to destroy genetic variation, this provides a framework for evaluating the power of mutation to maintain genetic variability in polygenic characters. This evaluation is made in light of the available experimental evidence on mutation in quantitative characters. The effect of linkage on the correlations of alleles at different loci is also derived and a comparison is made with previous models of linkage that incorporate selective mechanisms for maintaining genetic variability.

# Polygenic characters

## 2. ASSUMPTIONS

In modelling a quantitative character it is certainly more realistic, at least in large populations, to consider that at each locus there are many alleles with a range of effects rather than simply two effects, $+$ and $-$, as is usually assumed. This is evident from the molecular structure of proteins which allows a potentially large number of amino acid substitutions which would have some effect on the function of the molecule and therefore on the phenotype as well. The first assumption concerns the mutation process.

(1) At the $i$th locus, each allele mutates with probability $\mu_i$ each generation, and the distribution of mutant effects is the same for all alleles. The variance of the mutational changes at the $i$th locus is denoted as $m_i^2$. At each locus, the changes produced by mutation have a mean of zero so that there is no mutation pressure influencing the population mean.

To simplify matters it is further assumed that:

(2) The population mates randomly with respect to the character.

(3) The effective population size is large enough to ignore genetic drift.

(4) There is no genotype-environment interaction and the allelic effects are additive within and between loci.

## 3. QUALITATIVE ANALYSIS

Consider a metrical character affected by $n$ diploid loci. The effect of an allele at locus $i$ is denoted $x_i$. The covariance of allelic effects between locus $i$ and locus $j$ in the gametes in generation $t$ will be written as $C_{ij}(t)$. From the assumptions it is easily shown that mutation changes the $C_{ij}(t)$ by an amount $\delta_{ij}\mu_i m_i^2$ per generation, where $\delta_{ij} = 1$ if $i = j$ and is zero otherwise. Thus mutation increases the variances but does not alter the covariances. This is not intended as a universal description of the mutation process. In reality, there must be a limit to the range of allelic effects and hence the genetic variance at each locus. If the genetic variance at a locus were at its maximum or saturation level, mutation could not increase the genetic variance at that locus and would decrease the magnitude of the covariances between that locus and the other loci. Such extreme saturation must be rare because mutagenic agents generally increase the genetic variance of quantitative characters. There is some experimental evidence that the amount of new genetic variance produced by mutation each generation is roughly constant independent of the background level of genetic variance. The data of Scossiroli & Scossiroli (1959), Yamada & Kitagawa (1961) and Clayton & Robertson (1964) show that the genetic variance produced in quantitative characters of *Drosophila* by a given dose of X-rays is approximately the same in lines with low and high heterozygosity. No other evidence bearing on this point was found. The present description of mutation may thus be of some relevance to natural populations.

The above property of mutation will now be used to derive a qualitative feature of systems of linked mutating loci under stabilizing selection. The assumption of additive effects allows each gamete to be assigned a value of $\sum_{i=1}^{n} x_i$. Stabilizing selection, in favouring those gametes with values closest to the optimum, produces negative correlations in allelic effects at closely linked loci, so that positive and negative deviations from the optimum tend to cancel (Mather, 1941; Lewontin, 1964b). This creates a pool of hidden genetic variation which is stored in linked combinations.

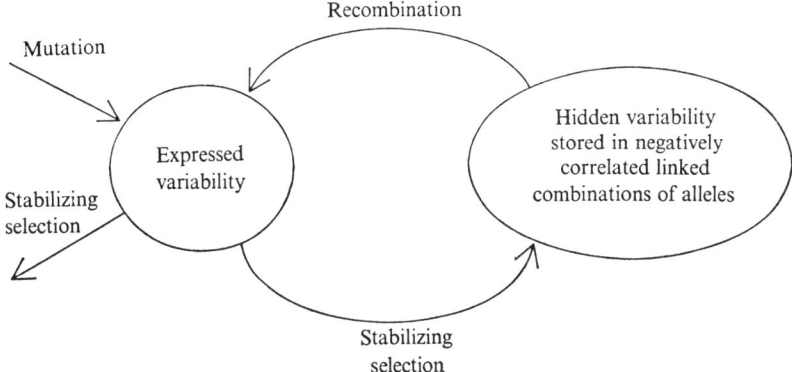

Fig. 1. Flow diagram of genetic variability. See text for explanation.

The influence of stabilizing selection, mutation and recombination on the expressed and hidden genetic variance can be summarized as follows: stabilizing selection converts expressed variance into hidden variance and depletes expressed variation. Mutation does not alter the hidden variation but contributes directly to the expressed genetic variance. Recombination decreases the correlations between loci, converting hidden variation into expressed variation. These relations are depicted in a flow diagram of genetic variability in Fig. 1. When this system has reached equilibrium, the flow out of expressed variation into hidden variation must equal the flow into expressed variation from hidden variation, regardless of the rate of recombination of the loci. The amount of expressed variability is then determined only by the input from mutation and the output on the left due to stabilizing selection. Therefore if there are no saturation effects, the linkage structure of the loci will have no influence on the amount of expressed genetic variance maintained by mutation.

## 4. A DETAILED MODEL

Kimura (1965) analyzed a one locus, multiallelic model of mutation and stabilizing selection. He used continuous time and a quadratic deviations fitness function. With the previous assumptions and the additional assumption that the changes in allelic effect produced by mutation are small, his model shows that the equilibrium distribution of allelic effects is approximately normal with the average effect at the optimum. Latter (1970) investigated a discrete generation model of a single locus with mutation and a Gaussian fitness function. Using Kimura's result, he derived a simple formula for the amount of genetic variance maintained. The approach of Latter is extended here to describe many linked loci.

A population with discrete generations is considered in which the cyclic order of events is selection, recombination, mutation and random mating. A primed letter is used to denote loci from one gamete, say the paternal gamete, the unprimed letters denote loci from the maternal gamete. A subscript $w$ indicates that selection has operated.

In zygotes formed by random mating in generation $t$, there may be a covariance between the effects of alleles at distinct loci $i$ and $j$ in the same gamete (in terms of covariances, $C_{ij}(t) = C_{i'j'}(t) \neq 0$), but there is no covariance between the effects of alleles from different gametes ($C_{ij'}(t) = C_{i'j}(t) = 0$). Selection creates covariances between the effects of alleles from different gametes, which are denoted as

$$b_{ij}(t) = (C_{ij'}(t))_w = (C_{i'j}(t))_w \neq 0.$$

If locus $i$ and locus $j$ recombine, in the new gametes $i$ will be linked to $j'$ and $i'$ will be linked to $j$, and the recombined fraction will have covariance $b_{ij}(t)$ while the non-recombined fraction will have covariance $(C_{ij}(t))_w$. Letting the recombination fraction between locus $i$ and locus $j$ be $r_{ij}$, the effect of selection, recombination and mutation on the variances and covariances is summarized as

$$C_{ij}(t+1) = (1-r_{ij})(C_{ij}(t))_w + r_{ij}b_{ij}(t) + \delta_{ij}\mu_i m_i^2. \tag{1}$$

A specific model of selection will be adopted to express the variables $(C_{ij}(t))_w$ and $b_{ij}(t)$ in terms of the variables $C_{ij}(t)$ to solve these equations.

The stabilizing selection which acts on the phenotypes is taken to be a Gaussian function so that fitness decreases with deviation, $z$, away from the optimum as

$$W(z) = \exp\left\{\frac{-z^2}{2w^2}\right\}. \tag{2}$$

Roughly $w$ gives the width or range of phenotypes around the optimum which have a high fitness. A large $w$ indicates weak stabilizing selection and a small $w$ indicates strong stabilizing selection. In keeping with the assumption of no genotype-environment interaction, the distribution of phenotypes, $z$, among individuals with a genotype value $X$, is taken to be normal with a variance of $\sigma_e^2$ which is the environmental variance.

$$p(z|X) = \frac{1}{\sqrt{(2\pi\sigma_e^2)}} \exp\left\{\frac{-(z-X)^2}{2\sigma_e^2}\right\}. \tag{3}$$

The relative fitness of individuals with genotype $X$ is

$$\tilde{W}(X) \propto \int W(z)p(z|X)\,dz \propto \exp\left\{\frac{-X^2}{2(w^2+\sigma_e^2)}\right\}. \tag{4}$$

The environmental variance, $\sigma_e^2$, acts in conjunction with the strength of selection on the phenotypes to limit the strength of selection on the genotypes; even for very strong selection on the phenotypes (small $w$) the width of the fitness function acting on the genotypes cannot be less than $\sigma_e$.

The joint distribution of allelic effects in maternal gametes is denoted as

$$f(x, t) = f(x_1, \ldots, x_n, t)$$

and in paternal gametes as $f(x', t) = f(x'_1, \ldots, x'_n, t)$. It is assumed that there is no sexual dimorphism so that $f(x, t) = f(x', t)$ for $x = x'$. With random mating the joint distribution of allelic effects in zygotes is $F(x, x', t) = f(x, t) f(x', t)$. Since the genes are additive $X = \sum_{i=1}^{n} (x_i + x'_i)$ and the distribution of alleles in zygotes after selection is

$$F_w(x, x', t) \propto f(x, t) f(x', t) \, \tilde{W}\left(\sum_{i=1}^{n} (x_i + x'_i)\right). \tag{5}$$

It will be shown later than under certain conditions the distribution of allelic effects in gametes is approximately multivariate normal and so can be written

$$f(x, t) \propto \exp\left\{-\tfrac{1}{2}(x - \bar{x}(t)) \, C^{-1}(t) \, (x - \bar{x}(t))^T\right\}, \tag{6}$$

where $C^{-1}(t)$ is the inverse of the variance–covariance matrix $C(t)$ and

$$\bar{x}(t) = (\bar{x}_1(t), \ldots, \bar{x}_n(t))$$

is the vector of mean effects of alleles at the various loci. The superscript $T$ denotes transpose.

The distribution of allelic effects in zygotes after selection is then

$$F_w(x, x', t) \propto \exp\left\{-\tfrac{1}{2}\left[(x - \bar{x}(t)) \, C^{-1}(t) \, (x - \bar{x}(t))^T + (x' - \bar{x}(t)) \, C^{-1}(t) \, (x' - \bar{x}(t))^T \right.\right.$$
$$\left.\left. + \frac{1}{w^2 + \sigma_e^2} (x\mathbf{1}x^T + 2x'\mathbf{1}x^T + x'\mathbf{1}x'^T)\right]\right\}, \tag{7}$$

where $\mathbf{1}$ is a matrix every element of which is 1. This expression has the multivariate normal form

$$F_w(x, x', t) \propto \exp\left\{-\tfrac{1}{2}(x - \bar{x}_w(t), x' - \bar{x}_w(t)) \, K^{-1}(t) \, (x - \bar{x}_w(t), x' - \bar{x}_w(t))^T\right\}, \tag{8}$$

where $K^{-1}(t)$ is the inverse of the variance–covariance matrix of allelic effects within and between gametes in the selected zygotes and $\bar{x}_w(t)$ is the vector of mean effects after selection. Because of the symmetry between maternal and paternal gametes the matrix $K(t)$ must be of the form

$$K(t) = \begin{pmatrix} C_w(t) & b(t) \\ b(t) & C_w(t) \end{pmatrix}, \tag{9}$$

where $C_w(t)$ and $b(t)$ are symmetric. Separating linear and quadratic terms in (7) and (8) and ignoring constant factors it is found that

$$K^{-1}(t) = \begin{pmatrix} C_w(t) & b(t) \\ b(t) & C_w(t) \end{pmatrix}^{-1} = \begin{pmatrix} C^{-1}(t) + \dfrac{1}{w^2 + \sigma_e^2} & \dfrac{1}{w^2 + \sigma_e^2} \\ \dfrac{1}{w^2 + \sigma_e^2} & C^{-1}(t) + \dfrac{1}{w^2 + \sigma_e^2} \end{pmatrix} \tag{10}$$

and

$$(\bar{x}_w(t), \bar{x}_w(t)) \, K^{-1}(t) = (\bar{x}(t), \bar{x}(t)) \begin{pmatrix} C^{-1}(t) & 0 \\ 0 & C^{-1}(t) \end{pmatrix}. \quad (11)$$

The dynamics of the mean effects will be analyzed first. Because recombination and mutation do not change the mean effects, $\bar{x}_w(t) = \bar{x}(t+1)$. Equation (11) is then

$$(\bar{x}(t+1), \bar{x}(t+1)) \, K^{-1}(t) \begin{pmatrix} C(t) & 0 \\ 0 & C(t) \end{pmatrix} = (\bar{x}(t), \bar{x}(t)). \quad (12)$$

Using (10) this reduces to two identical matrix equations, which yield the relations

$$\bar{x}_i(t+1) + \frac{2R_i(t)}{w^2 + \sigma_e^2} \sum_{i=1}^{n} \bar{x}_i(t+1) = \bar{x}_i(t), \quad (13)$$

where $2R_i(t) = 2\sum_{j=1}^{n} C_{ij}(t)$ is the expressed genetic variance attributable to locus $i$. Since the mean phenotype in generation $t$ is equal to the mean genotype by assumption 4,

$$\bar{z}(t) = 2 \sum_{i=1}^{n} \bar{x}_i(t).$$

Denoting the total expressed genetic variability as

$$\sigma_g^2(t) = 2 \sum_{i=1}^{n} R_i(t)$$

equations (13) may be summed and rearranged to give

$$\Delta \bar{z}(t) = - \left( \frac{\sigma_g^2(t)}{w^2 + \sigma_e^2 + \sigma_g^2(t)} \right) \bar{z}(t). \quad (14)$$

The mean phenotype converges to the optimum at a rate which depends on the expressed genetic variance but not on the hidden genetic variance. The change in the mean effect of alleles at locus $i$ can be obtained from (13) and (14).

$$\Delta \bar{x}_i(t) = - \left( \frac{R_i(t)}{w^2 + \sigma_e^2 + \sigma_g^2(t)} \right) \bar{z}(t). \quad (15)$$

The direction of change in the mean effect of alleles at any locus is determined only by whether the mean phenotype, $\bar{z}(t)$, is above or below the optimum.

We now proceed to the analysis of the genetic variation. Multiplying both sides of equation (10) by $K(t)$ in the form of (9) gives two matrix equations for $C_w(t)$ and $b(t)$ which can be solved to give

$$C_w(t) = \tfrac{1}{2} \left( C^{-1}(t) + \frac{2\mathbf{1}}{w^2 + \sigma_e^2} \right)^{-1} + \tfrac{1}{2} C(t), \quad (16a)$$

$$b(t) = C_w(t) - C(t). \quad (16b)$$

Upon performing the matrix inversion (16a) becomes

$$C_w(t) = C(t) - \frac{C(t) \, \mathbf{1} \, C(t)}{w^2 + \sigma_e^2 + \sigma_g^2(t)}. \quad (17)$$

The $ij$th term in this equation is

$$(C_{ij}(t))_w = C_{ij}(t) - \frac{R_i(t)\,R_j(t)}{w^2 + \sigma_e^2 + \sigma_g^2(t)}. \tag{18}$$

(16$b$) and (18) may be used to convert the basic equations (1) into recurrence relations

$$\Delta C_{ij}(t) = -\frac{R_i(t)\,R_j(t)}{w^2 + \sigma_e^2 + \sigma_g^2(t)} - r_{ij} C_{ij}(t) + \delta_{ij}\mu_i m_i^2. \tag{19}$$

Since the primary interest is in the maintenance of genetic variability, the equilibrium condition, $\Delta C_{ij}(t) = 0$, will be examined. A hat, $\hat{\phantom{a}}$, signifies equilibrium value. The diagonal equations in (19) with $i = j$ are then $\hat{R}_i^2 = \mu_i m_i^2 (w^2 + \sigma_e^2 + \hat{\sigma}_g^2)$. The only admissible solution to this equation is

$$\hat{R}_i = +\sqrt{(\mu_i m_i^2 (w^2 + \sigma_e^2 + \hat{\sigma}_g^2))} \tag{20}$$

because $2R_i(t)$ is the amount of variance expressed by locus $i$ which must be positive. Using (20), the off-diagonal equations in (19) have the solution

$$\hat{C}_{ij} = -\frac{\sqrt{(\mu_i m_i^2 \mu_j m_j^2)}}{r_{ij}} \quad \text{for} \quad i \neq j. \tag{21a}$$

From the definition of $R_i(t)$

$$\hat{C}_{ii} = \sqrt{(\mu_i m_i^2)} \left( \sqrt{(w^2 + \sigma_e^2 + \hat{\sigma}_g^2)} + \sum_{\substack{j=1 \\ j \neq i}}^{n} \frac{\sqrt{(\mu_j m_j^2)}}{r_{ij}} \right). \tag{21b}$$

The total expressed genetic variation is obtained from $\hat{\sigma}_g^2 = 2 \sum_{i=1}^{n} \hat{R}_i$.

$$\hat{\sigma}_g^2 = 2\left(\sum_{i=1}^{n} \sqrt{(\mu_i m_i^2)}\right) \sqrt{\left(w^2 + \sigma_e^2 + \left(\sum_{i=1}^{n} \sqrt{(\mu_i m_i^2)}\right)^2\right)} + 2\left(\sum_{i=1}^{n} \sqrt{(\mu_i m_i^2)}\right)^2, \tag{21c}$$

Equations (21) are the unique equilibrium solution to the dynamic equations (19). The dynamic system (19) does not seem to be of the type that would give limit cycle behaviour. A computer study of two and three locus cases confirmed that this equilibrium solution is always globally stable.

It is evident from (21$b$) that this solution cannot be valid when linkage is very tight ($r_{ij} \to 0$) because in actuality there must be some upper limit to the $\hat{C}_{ii}$ as explained in the qualitative analysis; this model is intended to apply only for those ranges of parameters where there are no saturation effects. However, linkage could be a cause of saturation only for extremely low recombination rates because $\sqrt{(\mu_i m_i^2)} \ll \sqrt{(w^2 + \sigma_e^2 + \hat{\sigma}_g^2)}$.

The equilibrium correlations can be written in terms of the solution (21) as

$$\hat{\rho}_{ij} = \frac{\hat{C}_{ij}}{\sqrt{\left(\left(\hat{R}_i + \sum_{\substack{j=1 \\ j \neq i}}^{n} |\hat{C}_{ij}|\right)\left(\hat{R}_j + \sum_{\substack{i=1 \\ i \neq j}}^{n} |\hat{C}_{ij}|\right)\right)}}. \tag{22}$$

We now return to the question of when the equilibrium distribution of allelic effects in the gametes, $f(x, t)$, is approximately multivariate normal. One way of ascertaining this is to start with a multivariate normal distribution for $f(x,t)$ and then to inquire under what conditions one generation of selection, recombination and mutation will yield an approximately multivariate normal distribution. Each of these will be dealt with in turn.

(i) From equation (7) the distribution of effects in gametes after selection on zygotes is multivariate normal because it is a marginal distribution of $F_w(x, x', t)$ which is multivariate normal.

(ii) Recombination produces a gametic distribution which is a sum of multivariate normal distributions with equal sets of variances, but unequal sets of covariances and correlations (see equation (1)). Such a sum will be approximately multivariate normal if the component distributions have approximately equal correlations,

$$|(C_{ij}(t))_w - b_{ij}(t)|/\sqrt{((C_{ii}(t))_w (C_{jj}(t))_w)} \ll 1 \quad \text{for} \quad i \neq j$$

or, by application of (16b)

$$\sqrt{\left(\frac{C_{ii}(t) C_{jj}(t)}{(C_{ii}(t))_w (C_{jj}(t))_w}\right)} |\rho_{ij}(t)| \ll 1 \quad \text{for} \quad i \neq j.$$

Near equilibrium, the first factor is close to one because from (1), $\hat{C}_{ii} = (\hat{C}_{ii})_w + \mu_i m_i^2$ and the solution (21b) shows that $\mu_i m_i^2 \ll \hat{C}_{ii}$. The conditions arising from recombination are thus

$$|\hat{\rho}_{ij}| \ll 1 \quad \text{for} \quad i \neq j. \tag{23}$$

(iii) Kimura (1965) demonstrated for a one locus model that, if the changes produced by mutation are small, the equilibrium distribution of allelic effects is approximately normal. It is necessary to know precisely how small the mutational changes must be. If $m_i^2 \ll (C_{ii}(t))_w$ then the effects of the new mutants will be almost normally distributed with variance $m_i^2 + (C_{ii}(t))_w$ and the new mutants will have nearly the same variance as the non-mutants. Since mutation does not alter the covariances, these conditions ensure that the various component distributions, and hence their sum, will remain approximately multivariate normal. As in (ii) it is noted that near equilibrium $(C_{ii}(t))_w \simeq \hat{C}_{ii}$ and these conditions become $m_i^2 \ll \hat{C}_{ii}$, or using the solution (21b)

$$m_i \ll \sqrt{(\mu_i)} \left( \sqrt{(w^2 + \sigma_e^2 + \hat{\sigma}_g^2)} + \sum_{\substack{j=1 \\ j \neq i}}^{n} \frac{\sqrt{(\mu_j m_j^2)}}{r_{ij}} \right). \tag{24}$$

For a single locus this reduces to $m_i \ll \sqrt{(\mu_i(w^2 + \sigma_e^2))}$ so the standard deviation of mutational changes must be much smaller, possibly by orders of magnitude, than the width of the selection function acting on the genotypes. For more loci these conditions become less restrictive, with linkage permitting larger mutational changes.

## 5. DISCUSSION

The equilibrium state of this genetic system has a somewhat peculiar type of stability. The equilibrium phenotype distribution is normal with mean at the optimum. Equation (14) is equivalent to the well-known result that the response of the population mean is equal to the heritability times the selection pressure on the mean (Falconer, 1960). The variance–covariance structure of allelic effects at the different loci in the gametes converges to a unique globally stable equilibrium independent of the mean effects at the individual loci. The peculiarity is that, even though the variance–covariance structure and the grand mean effect of alleles at all loci (the mean genotype or phenotype) both have a stable equilibrium, the mean effects at the individual loci do not (equation (15)). The mean effect of alleles at any given locus then depends largely on historical conditions and chance events. Random genetic drift should be particularly important as there are many patterns of change in the mean effects which are selectively neutral because they do not alter the phenotype distribution. With $n$ loci there are $n$ degrees of freedom for changing the mean effects. The single restriction on these changes, that the mean genotype converges to the optimum, leaves $n-1$ degrees of freedom for selectively neutral changes to occur. Therefore most random changes in the mean effects of alleles at the individual loci are selectively neutral. As a consequence, considerable genetic differentiation between populations could take place by random genetic drift without any phenotypic divergence. Such a process may be, in part, responsible for the genetic differences observed between sibling species (Lewontin, 1974).

We now consider the equilibrium structure of the genetic variation and how it is maintained. In the case of one locus, the equilibrium solution to the detailed model (21) is identical to that of Latter (1970) when infinite population size is used in his formulas. Latter has shown his expressions, which employ a Gaussian fitness function, to be in agreement with those of Kimura (1965) when selection is weak, as implicitly assumed by Kimura in his use of the quadratic deviations fitness function. With weak selection the Gaussian fitness function becomes equivalent to a quadratic deviations fitness function. Kimura (1965) applied his one locus calculation to a polygenic character, assuming that the loci were uncorrelated. In general, with weak selection, the equilibrium amount of expressed genetic variability agrees with Kimura's formula. Even though he assumed there were no correlations between the loci, his calculation is correct because the equilibrium expressed genetic variance is not a function of the linkage relation of the loci.

Both the qualitative analysis and the solution to the detailed model demonstrate that the amount of expressed variability maintained by mutation does not depend on the arrangement of the loci in the genome. Thus the mean fitness of individuals in the population is also independent of the structure of the linkage map. This behaviour contrasts sharply with heterotic models where the mean fitness is strongly dependent on the linkage arrangement of the loci (Lewontin, 1964a, 1971; Franklin & Lewontin, 1970).

Unlike the rate of recombination between the loci, the number of recombining

# Polygenic characters

loci has a strong influence on the amount of expressed genetic variability maintained This can be seen most easily by considering a character with an effective number, $n_E$, of equally mutable loci ($\mu_i m_i^2 = \mu_j m_j^2$). Now the total rate of production of new genetic variation per generation by mutation is

$$\sigma_m^2 = 2 \sum_{i=1}^{n} \mu_i m_i^2.$$

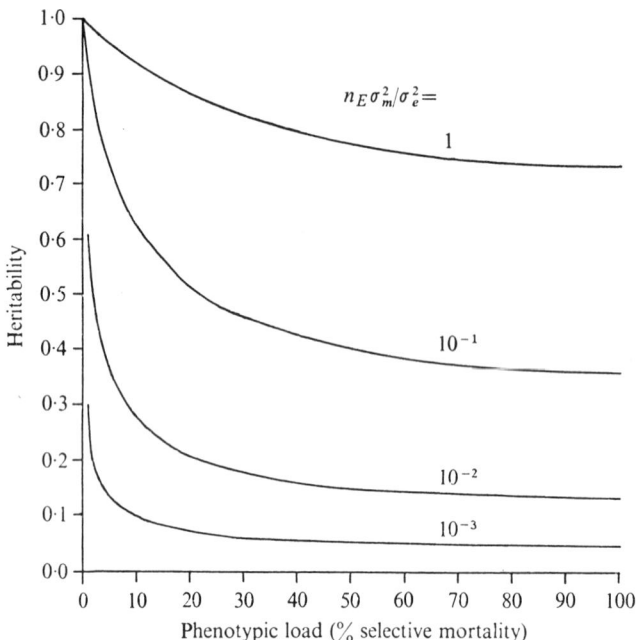

Fig. 2. The heritability maintained by mutation as a function of the strength of stabilizing selection. The heritability is calculated from equation (25) as $h^2 = \hat{\sigma}_g^2/(\sigma_e^2 + \hat{\sigma}_g^2)$. The per cent selective mortality is $100(1 - \sqrt{(w^2/(w^2 + \sigma_e^2 + \hat{\sigma}_g^2))})$ which is calculated from the fitness function in equation (2) acting on a normal phenotype distribution with variance $\sigma_e^2 + \hat{\sigma}_g^2$ and mean at the optimum. The numbers above each line on the graph are values of the parameter $n_E \sigma_m^2/\sigma_e^2$.

This is a measurable parameter, whereas the mutation rates, $\mu_i$, and the magnitude of mutational changes, $m_i$, at the individual loci are not. With $n_E$ equally mutable loci, $\mu_i m_i^2 = \sigma_m^2/(2n_E)$ and the expressed genetic variance can be rewritten from (21c) as

$$\hat{\sigma}_g^2 = \sqrt{\left(2n_E \sigma_m^2 \left(w^2 + \sigma_e^2 + \frac{n_E \sigma_m^2}{2}\right)\right)} + n_E \sigma_m^2. \tag{25}$$

Thus for a fixed total rate of production of genetic variance by mutation, $\sigma_m^2$, and environmental variance, $\sigma_e^2$, the higher the effective number of loci the more expressed genetic variability is maintained.

Fig. 2 shows the heritability which can be maintained under various amounts of selective mortality due to stabilizing selection for different values of the parameter $n_E \sigma_m^2/\sigma_e^2$. It can be seen that when selection is weak, a small increase in the strength of selection can produce a major decrease in the heritability, but when selection is strong any further increase in the strength of selection has little influence on the heritability. This occurs because there is a limit to the strength of stabilizing selection that can be exerted on the genotypes which is determined by the environmental variance of the character (equation (4)). The power of mutation to maintain genetic variation against the force of stabilizing selection is best evaluated from some specific examples.

The first measurements of the production of genetic variation by mutation were performed on *Drosophila melanogaster*. Paxman (1957) estimated $\sigma_m^2$ for abdominal and sternopleural bristles and also summarized previous measurements by other workers. Estimates of $\sigma_m^2/\sigma_e^2$ are given in Table 1. It seems reasonable to use $1 \times 10^{-3}$ as an estimate of $\sigma_m^2/\sigma_e^2$ for both of these characters. The effective number of loci influencing abdominal bristles in *D. melanogaster* has been estimated with low accuracy as about 100 (Falconer, 1960). A rough estimate of the composite parameter $n_E \sigma_m^2/\sigma_e^2$ for this character is therefore about $1 \times 10^{-1}$. An estimate of 16 effective loci influencing sternopleural bristles was made by Barnes & Kearsey (1970). A rough estimate of $n_E \sigma_m^2/\sigma_e^2$ for sternopleural bristles is thus about $1 \cdot 6 \times 10^{-2}$.

Table 1. *Rates of production of genetic variance for quantitative characters of* D. melanogaster

| | $\sigma_m^2$ | | | | |
| Character | Durrant & Mather | Clayton & Robertson | Paxman | $\sigma_e^2$ | Estimated $\sigma_m^2/\sigma_e^2$ |
|---|---|---|---|---|---|
| Abdominal bristles | 0·006 | 0·005 | 0·0035 | 5·0 | $1 \times 10^{-3}$ |
| Sternopleural bristles | 0·002 | — | 0·001 | 2·2 | $1 \times 10^{-3}$ |

$\sigma_e^2$ for sternopleural bristles is from Lopez-Fanjul & Hill (1973) and was calculated as $(1-h^2)$ times the phenotypic variance for the Kaduna population. $\sigma_e^2$ for abdominal bristles is from Clayton, Morris & Robertson (1957) and is the average variance of males and females from inbred lines of the Kaduna population. Values of $\sigma_m^2$ were summarized by Paxman (1957), whose own values have been multiplied by 5/2 because they applied only to chromosome II.

Russell, Sprague & Penny (1963) measured mutation rates for nine quantitative characters of maize in lines that had been selfed at least ten generations. They detected an average of 0·064 mutations per gamete per character. A conservative estimate of $\sigma_m^2/\sigma_e^2$ for these characters can be obtained as $2/\sigma_e^2$ times the rate of mutation detected per gamete times the minimum detectable variance produced by a mutation. The factor of 2 accounts for diploidy. The authors measured 100 offspring from each plant to determine the significance of differences in the line means by the $t$ test and found that almost all the detected differences were significant at the 1 % level. With additive effects, and taking $\sigma_e^2$ as the variance within lines, the minimum detected effect is $2m > t_{0 \cdot 01} \sqrt{(2\sigma_e^2/100)}$ so the minimum variance produced by a detectable mutation is $m^2 > 3 \cdot 3 \times 10^{-2} \sigma_e^2$. A conservative estimate of $\sigma_m^2/\sigma_e^2$ averaged over the nine characters is thus $2(0 \cdot 064)(3 \cdot 3 \times 10^{-2}) = 4 \times 10^{-3}$. With as few as 10 effective loci, most of these characters would have $n_E \sigma_m^2/\sigma_e^2$ in the range of $10^{-2}$ to $10^{-1}$ or larger.

Many investigators have measured the divergence between the mean values of quantitative characters in two or more inbred lines derived from a single long-inbred line. Grewal (1962) and Hoi-Sen (1972) have reviewed several such studies. These authors estimated the average rate of increase of variance between lines for 27 and 31 mouse skeletal characters respectively as $3 \times 10^{-3}\sigma_e^2$ per generation. To convert this into units of $\sigma_m^2/\sigma_e^2$ two correction factors must be taken into account. First, because divergence between lines is due to homozygous mutant alleles, the measured effect is $2m$ rather than $m$ (if there is no dominance) which inflates the variance by a factor of 4. Second, as pointed out by Grewal (1962) this technique detects only a haploid complement of mutations which deflates the variance by a factor of 2. Therefore the average value of $\sigma_m^2/\sigma_e^2$ for these characters is $\frac{2}{4}(3 \times 10^{-3}) = 1 \cdot 5 \times 10^{-3}$. With as few as 10 effective loci, most of these characters would have values of $n_E \sigma_m^2/\sigma_e^2$ near $10^{-2}$.

These examples indicate that for quantitative characters, $n_E \sigma_m^2/\sigma_e^2$ may typically be near $10^{-2}$ or $10^{-1}$. It can be seen from Fig. 2 that when $n_E \sigma_m^2/\sigma_e^2$ is in this range, high heritabilities can be maintained even when stabilizing selection is very strong. It is concluded that mutation can be a potent force maintaining genetic variation in polygenic characters under stabilizing selection.

Up to this point only the genetic variance which is expressed in the phenotype has been discussed. It is also of some interest to examine the amount of hidden genetic variance which is stored in the negative covariances between effects of alleles at different loci in the gametes. From (21a) it can be observed that the covariances, and thus the hidden genetic variance, do not depend on the strength of selection, but are purely a function of the internal properties of the genetic system: the mutability and linkage map of the loci. Large amounts of hidden genetic variation can be accumulated by closely linked loci.

The equilibrium structure of the correlations between alleles at different loci in this model of stabilizing selection differs greatly from that in heterotic models. Consider two loci, $i$ and $j$. From (22) it can be seen that if other loci are moved closer to $i$ and $j$ on the linkage map the magnitude of correlation between $i$ and $j$ will decrease. This behaviour is the opposite of the imbedding effect found in the heterotic systems (Lewontin, 1964a). Furthermore, increasing the number of loci greatly decreases the average magnitude of correlation between the loci, again opposite to the result for the heterotic systems (Franklin & Lewontin, 1970). This can be deduced from the fact that the correlation matrix, $\hat{P}$, must be non-negative definite (Cramér, 1951, pp. 295–6) and in particular

$$(1, ..., 1)\hat{P}(1, ..., 1)^T = n(n-1)\overline{\hat{\rho}_{ij}} + n \geq 0,$$

where $\overline{\hat{\rho}_{ij}}$ is the average correlation between loci. As all of the correlations are negative, this relation may be expressed in terms of the average magnitude of correlation,

$$|\overline{\hat{\rho}_{ij}}| \leq \frac{1}{n-1}. \tag{26}$$

Thus a polygenic character under stabilizing selection must have a small average magnitude of correlation between loci, regardless of the linkage arrangement of the loci. This result is quite robust and does not depend on the assumptions used in this model, but only on all of the correlations being negative, which is a general feature of stabilizing selection (Mather, 1941; Lewontin, 1964b).

Thoday & Gibson (1972) have detected natural and artificial stabilizing selection on sternopleural bristle number in *Drosophila melanogaster* by measuring a negative correlation between the effects of chromosomes II and III. Similar measurements could also be made on parts of chromosomes. Measuring such correlations can yield information about the type of selection operating on the character, but to construe this as a test of the model would be erroneous, since if the correlations are all negative it is impossible for condition (26) to be violated. However, the predictions concerning the amount of genetic variation maintained, and the independence of the covariances and the strength of selection, could be tested by artificial selection experiments. With *Drosophila* there are various experimental approaches for modifying recombination rates to determine whether or not they influence the amount of genetic variation maintained.

Finally it should be mentioned that models with assortative mating and multivariate fitness functions (which allow for pleiotropic gene action) could be investigated by methods similar to those used here.

I thank Jerry Coyne, Marcus Feldman, Donal Hickey, Richard Lewontin, Alan Robertson and Montgomery Slatkin for encouragement and criticism. An anonymous reviewer helped improve the generality of (26) and made other useful suggestions. This work was partially supported by an Atomic Energy Commission grant to R. Lewontin and a National Science Foundation grant to M. Feldman.

## REFERENCES

ALLARD, R. W. & JAIN, S. K. (1962). Population studies in predominantly self-pollinated species. II. Analysis of quantitative genetic changes in a bulk-hybrid population of barley. *Evolution* **16**, 90–101.

BARNES, B. W. & KEARSEY, M. J. (1970). Variation for metrical characters in *Drosophila* populations. I. Genetic analysis. *Heredity* **25**, 1–10.

CLAYTON, G. A., MORRIS, J. A. & ROBERTSON, A. (1957). An experimental check on quantitative genetical theory. I. Short-term responses to selection. *Journal of Genetics* **55**, 131–151.

CLAYTON, G. A. & ROBERTSON, A. (1964). The effects of X-rays on quantitative characters. *Genetical Research* **5**, 410–422.

CRAMÉR, H. (1951). *Mathematical Methods of Statistics*. Princeton: Princeton University Press.

DOBZHANSKY, TH., HOLTZ, A. M. & SPASSKY, B. (1942). Genetics of natural populations. VIII. Concealed variability in the second and fourth chromosomes of *Drosophila pseudoobscura* and its bearing on the problem of heterosis. *Genetics* **27**, 463–490.

DOBZHANSKY, TH. & SPASSKY, B. (1944). Genetics of natural populations. XI. Manifestation of genetic variants in *Drosophila pseudoobscura* in different environments. *Genetics* **29**, 270–290.

EAST, E. M. & JONES, D. F. (1919). *Inbreeding and Outbreeding: Their Genetic and Sociological Significance*. Philadelphia: Lippincot.

FALCONER, D. S. (1960). *Introduction to Quantitative Genetics*. New York: Ronald Press.

FISHER, R. A. (1930). *The Genetical Theory of Natural Selection*. Oxford: Clarendon Press.

Franklin, I. & Lewontin, R. C. (1970). Is the gene the unit of selection? *Genetics* **65**, 707–734.
Gale, J. S. & Kearsey, M. J. (1968). Stable equilibria in the absence of dominance. *Heredity* **23**, 553–561.
Gibson, J. B. & Bradley, B. P. (1974). Stabilizing selection in constant and fluctuating environments. *Heredity* **33**, 293–302.
Grewal, M. S. (1962). The rate of divergence of sublines in the C57BL strain of mice. *Genetical Research* **3**, 226–237.
Hoi-Sen, Y. (1972). Is sub-line differentiation a continuing process in inbred strains of mice? *Genetical Research* **19**, 53–59.
Kearsey, M. J. & Gale, J. S. (1968). Stabilizing selection in the absence of dominance: an additional note. *Heredity* **23**, 617–620.
Kimura, M. (1965). A stochastic model concerning the maintenance of genetic variability in quantitative characters. *Proceedings of the National Academy of Science U.S.A.* **54**, 731–736.
Lack, D. (1966). *Population Studies of Birds*. Oxford: Clarendon Press.
Latter, B. H. D. (1970). Selection in finite populations with multiple alleles. II. Centrepetal selection, mutation and isoallelic variation. *Genetics* **66**, 165–186.
Lerner, I. M. (1954). *Genetic Homeostasis*. Edinburgh: Oliver and Boyd.
Lewontin, R. C. (1964a). The interaction of selection and linkage. I. General considerations. Heterotic models. *Genetics* **49**, 49–67.
Lewontin, R. C. (1964b). The interaction of selection and linkage. II. Optimum models. *Genetics* **50**, 757–782.
Lewontin, R. C. (1971). The effect of genetic linkage on the mean fitness of a population. *Proceedings of the National Academy of Science U.S.A.* **68**, 984–986.
Lewontin, R. C. (1974). *The Genetic Basis of Evolutionary Change*. New York: Columbia University Press.
Lopez-Fanjul, C. & Hill, W. G. (1973). Genetic differences between populations of *D. melanogaster* for a quantitative trait. *Genetical Research* **22**, 51–68.
Mather, K. (1941). Variation and selection of polygenic characters. *Journal of Genetics* **41**, 159–193.
Paxman, G. J. (1957). A study of spontaneous mutation in *Drosophila melanogaster*. *Genetica* **29**, 39–57.
Rendel, J. M. (1943). Variations in the weights of hatched and unhatched ducks' eggs. *Biometrika* **33**, 48–58.
Robertson, A. (1956). The effect of selection against extreme deviants based on deviation or homozygosis. *Journal of Genetics* **54**, 236–248.
Russell, W. A., Sprague, G. F. & Penny, H. L. (1963). Mutations affecting quantitative characters in long-time inbred lines of maize. *Crop Science* **3**, 175–178.
Scharloo, W. (1964). The effect of stabilizing and disruptive selection on the expression of *cubitus interruptus* in *Drosophila*. *Genetics* **50**, 553–562.
Scossiroli, R. E. & Scossiroli, S. (1959). On the relative role of mutation and recombination in responses to selection for polygenic traits in irradiated populations of *D. melanogaster*. *International Journal of Radiation Biology* **1**, 61–69.
Thoday, J. M. & Gibson, J. B. (1972). A simple test for stabilizing and disruptive selection. *Egyptian Journal of Genetics and Cytology* **1**, 47–50.
Waddington, C. H. (1960). Experiments on canalizing selection. *Genetical Research* **1**, 140–150.
Yamada, Y. & Kitagawa, O. (1961). Doubling doses for polygenic mutations in *Drosophila melanogaster*. *Japanese Journal of Genetics* **36**, 76–83.

## ERRATUM

Page 222, line 6 should read: "... observed in natural and artificial populations."

# NATURE OF QUANTITATIVE
# GENETIC VARIATION

# Editor's Comments
# on Papers 25 Through 28

**25  CLAYTON and ROBERTSON**
   *Mutation and Quantitative Variation*

**26  LINNEY, BARNES, and KEARSEY**
   *Variation for Metrical Characters in* Drosophila *Populations.
   III. The Nature of Selection*

**27  "STUDENT"**
   *A Calculation of the Minimum Number of Genes in Winter's
   Selection Experiment*

**28  THODAY**
   *Location of Polygenes*

   In the experiments included so far in this volume, responses to directional selection have been investigated for a range of traits in a range of organisms. In this section we consider experimental evidence on the sources and maintenance of genetic variance and on the number and effects of the genes contributing to this variability and to response. Because the topic is wide and the number of papers included is small, we can only sample the field.
   Direct evidence on the amount of new variation produced in each generation by mutation can be obtained in a number of different ways. One is to accumulate mutations on chromosomes balanced against a marker stock in which crossing over is not permitted, a feasible plan in *Drosophila,* and then to estimate the variance in effect on some trait among them. This method was used by Durrant and Mather (1954) for bristle number and subsequently by Mukai and coworkers (e.g., Mukai et al., 1972) for the viability component of fitness. Another method is to start from a highly inbred stock and allow variation to accumulate, generally using selection either immediately or after a period of generations to estimate the amount of variability. The same procedures can be used for natural mutations and for those induced by, for example, x-radiation, and the possible role of induced mutagenesis in increasing response to artificial selection can be investigated.

*Editor's Comments on Papers 25 Through 28*

Paper 25 by Clayton and A. Robertson has been chosen for reprinting because it was one of the earliest to use selection to estimate mutational variance. It considers the value of induced mutagenesis, and includes calculations putting the results into the standard quantitative genetic framework. Their final estimate, a variance per generation for bristle number in *Drosophila melanogaster* from new natural mutations of 0.001 times the existing variance in wild populations, has generally been supported by more recent studies (e.g. Clayton and Robertson, 1964, and see Paper 24 for a review). As Clayton and Robertson point out, this level of mutation would be sufficient to maintain variability in large populations without invoking selection, although it does not explain why particular mean levels of the character are maintained. Their conclusion (p. 156) that "the effect of mutation may in general be regarded as negligible in selection experiments in laboratory animals" is more contentious, however. There is both experimental (Frankham, 1980) and theoretical (Hill, 1982) evidence that mutations have contributed substantially to response in long-term experiments, particularly with large populations (e.g; Enfield, 1980; Yoo, 1980). The importance of mutations in the long run was also illustrated in the early experiment of selecting from an inbred line by Mather and Wigan (1942), which is discussed by Clayton and Robertson (Paper 25), and the early generations of which are described by Mather (Paper 15), and, by inference, by Dudley in the Illinois corn experiment (Paper 14).

The role of stabilizing selection was reviewed by Robertson (Paper 22), who argued that it may not be important in maintaining variability and that variation in metric traits peripheral to fitness could be due to selection acting on correlated fitness traits. Nevertheless, stabilizing selection has featured widely in models for theoretical analyses on maintenance of variance (see, for example, Papers 23 and 24), even though evidence for it is largely inferred from the constancy of population mean rather than from direct experimental studies, The classic examples are of birth weight in man and of clutch size in birds (Lack, 1966), but other examples are few. A nice study was, however, undertaken by Linney, Barnes, and Kearsey (1971) of Birmingham University; it deals with bristle number in *Drosophila* and therefore ties in well with several of the other papers in this volume, including that preceding. It is reprinted as Paper 26.

Linney et al. confirmed a previous study (Kearsey and Barnes, 1970) in showing that in crowded cultures the variability in bristle score is very much less than in uncrowded cultures. They also found that the variance among families reared in a standard environment was much less when the parents were reared at high density than at low density (1.16 versus 1.61, Table 1), although the variance within families was the same. The regression of progeny on parent was higher when the parent was reared at low density. They argue that the change in parental variance was attributable to a reduction in genetic variance among those reared in crowded cultures, and from a second experiment with inbred populations, where a reduction in variance again occurs, argue that the effects observed cannot be due to selection for heterozygotes. The results are put together to show fitness as a function of chaeta number (Figure 1). There are, nevertheless, some problems of interpreting such experiments, as Kearsey and Barnes (1970, p. 17) point out; bristle number is not expressed in the larva, so because they are studying survival to adults, "selection acts at a pre-adult stage and hence cannot be for chaeta number as such, but presumably for some pleiotropic effect of the same loci".

In Papers 27 and 28 we turn to description of the genetics of quantitative traits in terms of the individual genes. The paper by "Student" is of historical interest because it was of one of the first in which an attempt was made to estimate the number of genes influencing a trait, developing the arguments in a very clear way; it is also of interest because it uses data from the first twenty eight generations of the Illinois corn experiment (Winter, 1929), for which later results were presented by Dudley in Paper 14. "Student" refers to this as "Winter's selection experiment."

"Student" was, as every statistician knows, a pseudonym for W. S. Gossett of Guiness Breweries, Dublin, and this study (Paper 27) was his only notable contribution to genetics, apart from the remarks he published earlier on the same topic and refers to in this paper.

Basically, "Student's" method is to compute the genetic standard deviation from the phenotypic standard deviation and regression of response to selection differential (i.e., realized heritability) and to compare this to the range estimated from the divergence between high and low lines. The method is similar to that used by Wright (1952, Paper 9 of Part 1). "Student" is careful to point out that he is making a minimum estimate of numbers of genes on the assumption of additive genes of equal effect and initial frequency 0.5. He concludes: "Nevertheless, though I do not feel that the above calculation can be altogether absolved from the charge of 'playing with figures,' I think that it does really afford some evidence that the oil percentage of Winter's maize was conditioned by the presence, or absence, of a number of genes, at least of the order of 20–40, possibly of 200–400, and not at all likely to be of the order 5–10." (p. 81). This result was important, albeit indirect, evidence of the polygenic inheritance of quantitative traits.

Examination of Figure 1 of Paper 14 shows that the divergence of 8 percent oil found by generation 28 and used by "Student" had reached about 18 percent by generation 76, with considerable asymmetry of response, presumably explainable at least in part by simple scale effects. Dudley repeated "Student's" calculations, using rather different values for the additive variance, and found that the minimum number of genes would have to be roughly doubled. Such a calculation should not be taken too seriously, because it assumes that all the variants utilized were initially present in the population. In view of the almost linear response of the high line (or somewhat less than linear on a logarithmic scale), it seems likely that much of the recent response derived from mutations after the experiment started.

The approach used in Paper 27 is based solely on population means and variances and can never identify the effects of individual genes or their location on the chromosomes. Another approach is to use chromosome substitution to identify on which chromosome the genes influencing the trait in question are located, a technique that is feasible in *Drosophila melanogaster* and that has been widely used (see, for example, Paper 17). It is further possible, by the use of marker stocks, to estimate the total effects of segments of chromosomes (e.g., Breese and Mather, 1957). With even more labor, it is possible in principle to estimate the position and effect of individual genes for the metric trait and thus to describe the genetics of the character completely.

In the final paper in this volume, J. M. Thoday (Paper 28) of the University of Cambridge outlines how this identification can proceed for *Drosophila melanogaster*. In essence, a high-scoring and a low-scoring line are taken, one of which has marker genes incorporated along its length. A series of crosses are made, chromosomes are extracted, and the "effects" of each are calculated by scoring large numbers of individuals with the same constitution, a straightforward trick in *Drosophila*. If there is a single gene influencing the trait between two markers, then recombinant chromosomes carrying one, but not both, of the markers will segregate into two groups in terms of their effect on the trait.

In practice, the method is very laborious, and the greater the labor, the greater the possibility of identifying more genes of smaller effect. Another disadvantage is that the method relies heavily on one line being solely + + +... the other − −−... for the relevant genes, as McMillan and Robertson (1974) point out in their discussion of some of the statistical problems. Nevertheless, Thoday's method has appeal in taking gene identification to its natural endpoint short of actual DNA sequencing.

In his short paper, Thoday mentions that his group found pairs of closely linked genes on chromosomes II and III and could then explain some of the responses to selection found in previous experiments. For more information on the results obtained using the method, see, for example, Spickett (1963) and the review by Thoday (1977).

It is notable, however, that direct analysis of individual gene effects such as that of Thoday can be carried out only in *Drosophila melanogaster*, which has well-marked chromosomes and for which various tricks can be applied. In other species it may be possible to identify the effects of individual regions that are associated with markers, providing enough markers are available. All these techniques require substantial inputs of labor, and this is likely to remain the case. Perhaps in the not too distant future we will be able to say that this cow gives more milk than that because of base or gene substitutions at such and such a series of positions on the DNA or chromosome. Meanwhile, those concerned with improving the performance of populations still have to deal in quantitative genetic terms.

## REFERENCES

Breese, E. L., and K. Mather, 1957, The Organisation of Polygenic Activity Within a Chromosome in *Drosophila*. I. Hair Characters, *Heredity* **11:**373-395.

Clayton, G. A., and A. Robertson, 1964, The Effects of X-rays on Quantitative Characters, *Genet. Res. (Cambridge)* **5:**410-422.

Durrant, A., and K. Mather, 1954, Heritable Variation in a Long Inbred Line of *Drosophila, Genetica* **27:**97-119.

Enfield, F. D., 1980, Long-Term Effects of Selection; the Limits to Response, in *Selection Experiments in Laboratory and Domestic Animals*, A. Robertson, ed., Commonwealth Agricultural Bureaux, Slough, pp. 69-86.

Frankham, R., 1980, Origin of Genetic Variation in Selection Lines, in *Selection Experiments in Laboratory and Domestic Animals*, A. Robertson, ed., Commonwealth Agricultural Bureaux, Slough, pp. 56-68.

Hill, W. G., 1982, Rates of Change in Quantitative Traits from Fixation of New Mutations, *Natl. Acad. Sci. (U.S.A.) Proc.* **79:**142-145.

Kearsey, M. J., and B. W. Barnes, 1970, Variation for Metrical Characters in *Drosophila* Populations. II. Natural Selection, *Heredity* **25:**11-21.

Lack, D., 1966, *Population Studies of Birds*, Clarendon Press, Oxford.

Mather, K., and L. G. Wigan, 1942, The Selection of Invisible Mutations, *R. Soc. (London) Proc.* **B131:**50-64.

McMillan, I. and A. Robertson, 1974, The Power of Methods for the Detection of Major Genes Affecting Quantitative Characters, *Heredity* **32:**349-356.

Mukai, T., S. I. Chigusa, L. E. Mettler, and J. F. Crow, 1972, Mutation Rate and Dominance of Genes Affecting Viability in *Drosophila melanogaster, Genetics* **72:**335-355.

Spickett, S. G., 1963, Genetic and Developmental Studies of a Quantitative Character, *Nature* **199**:870–873.

Thoday, J. M., 1977, Effects of Specific Genes, in *Proceedings of the International Conference on Quantitative Genetics,* E. Pollak, O. Kempthorne, and T. B. Bailey, Jr., eds., Iowa State University Press, Ames, pp. 141–159.

Winter, F. L., 1929, The Mean and Variability as Affected by Continuous Selection for Composition in Corn, *J. Agric. Res.* **39**:451–475.

Wright, S., 1952, The Genetics of Quantitative Variability, in *Quantitative Inheritance,* E. C. R. Reeve and C. H. Waddington, eds., Her Majesty's Stationery Office, London, pp. 5–41.

Yoo, B. H., 1980, Long-Term Selection for a Quantitative Character in Large Replicate Populations of *Drosophila melanogaster, Genet. Res. (Cambridge)* **35**:1–17.

# 25

Copyright © 1955 by The University of Chicago
Reprinted from Am. Nat. **89**:151-158 (1955)

## MUTATION AND QUANTITATIVE VARIATION

### G. CLAYTON AND ALAN ROBERTSON

Institute of Animal Genetics, Edinburgh

Almost all our knowledge of the origin of new variation by mutation comes from work on individual genes with lethal or visible effects. The extent to which new variation arises in continuously variable characters has received little attention. Gustafsson (1953) found that irradiation of barley produced variants of practical value though most of these may be simple Mendelian segregants. More recently, Scossiroli (1953) and Buzzati-Traverso (1953) have shown that new variation induced by radiation can be utilized by either natural or artificial selection to give striking changes in a population. The former was using two lines selected by Mather for the number of sterno-pleural chaetae until there was no further response. By irradiation and selection, he was able to change the mean of the up line from its previous plateau of 26 bristles to 44 bristles in 17 generations of radiation and selection. He was however unable to change the down line. It seemed possible to us that in the up line there had remained some unfixable genetic variation which had been released by the radiation perhaps by the production of chiasmata in unusual regions. Stimulated by this work, we decided to investigate on a small scale the effects of radiation in a strain in which genetic variation had been reduced to a low level by inbreeding.

In our other work on abdominal chaetae (Clayton, Morris and Robertson, in press) we have used as base population a stock which shortly after captivity in the wild has been kept in a population cage with average numbers around 5000. Mr. B. K. Sen, working on egg production in Drosophila, made several inbred lines from this by continued full-sib mating and one of these was chosen for the radiation work because, of the surviving lines, it had the least variance in the count of abdominal chaetae. It had then been inbred for 28 generations. Four selection lines were started in two groups. The two irradiated lines were given 1800 r of X-rays as adults each generation and the two control lines were not treated. Within each group, one line was selected upwards for the total number of chaetae on the fourth and fifth abdominal segments and the other downwards. The selection procedure was the same in all cases—25 ♂ and 25 ♀ were measured and the extreme 10 of each sex chosen as parents of the next generation. Culture was in half-pint bottles. The selection was continued for 17 generations.

The current effect of the radiation on lethals was not checked but at the end of the experiments, a sample of third chromosomes was made homozygous, by the usual technique, for each of the four lines. The numbers of chromosomes lethal and non-lethal when homozygous are given in table 1.

There is no doubt about the accumulation of lethals in the irradiated stocks. In a population of this small size, the equilibrium between mutation

## TABLE 1

|  |  | Lethal | Non-lethal |
|---|---|---|---|
| Control | High (HC) | 2 | 32 |
|  | Low (LC) | 0 | 32 |
| Irradiated | High (HR) | 58 | 9 |
|  | Low (LR) | 29 | 9 |

and selection is reached fairly quickly. If the heterozygotes are not at a disadvantage, Wright (1937) has shown that lethal recessives reach an equilibrium frequency of $\mu\sqrt{2\pi N}$ where $\mu$ is the mutation rate per generation and N the effective population size. If we take the actual size to be the effective size in our case (although the latter may in fact be less), the equilibrium frequency is about $11\mu$ and should therefore have been reached during the experiment. The observed frequencies of lethal chromosomes are 3 per cent and 83 per cent in the control and irradiated lines respectively. Taking these as the equilibrium figures and allowing for chromosomes carrying more than one lethal, the mutation rate per generation works out at 0.3 per cent and 16 per cent, which is not unreasonable.

The effects of the selection are given in table 2 in which, to smooth out differences between generations due to the small numbers measured, the

## TABLE 2

| Generation | HC | LC | HR | LR |
|---|---|---|---|---|
| 1-5 | 28.93 | 28.88 | 29.65 | 29.26 |
| 6-10 | 29.52 | 29.29 | 31.20 | 29.18 |
| 11-17 | 28.99 | 28.93 | 32.26 | 28.69 |
| 17A | 29.02 | 28.72 | 31.64 | 28.30 |

generations are given in three groups 1-5, 6-10 and 11-17. The mean count of the two sexes has been taken. It had seemed possible in the early stages of the work that the radiation might be having a direct effect on the mean. This could arise because of the smaller number of eggs hatching in the radiation lines due to dominant lethals. For the last generation, 17A, there was therefore neither selection nor radiation and 50 flies of each sex were counted instead of 25.

The control lines change remarkably little during the course of the experiment. HC is higher than LC in the last two entries, the differences being $0.06 \pm 0.36$ and $0.30 \pm 0.54$ respectively. The radiated lines do show a slight drop of 0.50 bristles in 17A from the previous entry in agreement with the idea that the lower degree of crowding in these lines may have increased the mean count. Crowding is known to depress the count (Rasmusson, 1952) but it is a little surprising that our control conditions, which we had considered optimum, should have such an effect. LR is below the average of the two controls in gen. 17A ($0.67 \pm 0.39$) and the value of the differences in the generations 11-17 of the experiment suggests that this difference may be real. HR does show a definite increase above the controls (2.77 ± 0.29) in gen. 17A. Our impression during the experiment was that this occurred in the early generations and the final figure was not as large as we expected earlier. Perhaps this is due to an isolated event rather than an accumulation of small effects.

The average variance of the two sexes calculated within generations is given in table 3. The two control lines have been averaged because of their similar behavior. The figures for each line are based on 48 d.f. per generation except for 17A where they are based on 98.

TABLE 3

| Generation | Controls | HR | LR |
|---|---|---|---|
| 1–5 | 5.03 | 6.39 | 5.72 |
| 6–10 | 4.60 | 7.55 | 5.44 |
| 11–17 | 4.43 | 8.91 | 4.62 |
| 17A | 3.87 | 6.07 | 6.79 |

The trends are not as clear as they are for the means. The controls decline slightly but the effect is not significant. HR is more variable than the controls as might be expected from the response to selection. However, had all this extra variance been utilizable by selection, the response should have been greater. Our base outbred population has a mean variance of 12.2 and under this intensity of selection would respond at 1.5 chaetae per generation. LR is only slightly more variable than the controls except for generation 17A where the behavior of the two irradiated lines is surprising.

We had hoped to carry on selection for sterno-pleural chaetae simultaneously with this work but some of the lines were lost in the early generations. However, at generation 10 of LR, selection for sterno-pleurals was carried out for 5 generations in both directions under similar conditions to that described above for abdominal chaetae. There was no detectable divergence between the lines.

In experiments of this sort, it is essential to guard against the possibility of contamination. A response to selection may be due to a single contaminant. When this work was started no suitable inbred stocks were available in this laboratory though they have now been obtained. It was not realized until too late that our line was in fact marked. Examination of salivary chromosomes from LR at the end of the experiment, kindly made by Mr. G. R. Knight, showed that the line was homozygous for In(3R)K known to be present at low frequency in the base population. Unfortunately HR, the line with the greatest response, had by then been discarded. It may be noted that no new inversions were found in the 10 larvae examined in LR.

## DISCUSSION

In mutation work on lethal or visible genes, the mutation rate can be simply stated as the proportion of chromosomes in each generation which gives rise to a change of a certain type. But for quantitative characters, we have no such absolute measurement and no way of bringing the results for the new variation on to the same plane of reference as those for "good" genes. Presumably our measurement of the effect of radiation in producing new continuous variation would be in terms of the increase in variance for a given dosage. One is still faced with the problem of comparing the new variation arising in different characters. A possible standard of comparison might be the genetic variance of the character usually observed in wild populations. This is perhaps vague, particularly for domestic animals, but it does have the merit of giving a figure which is also of evolutionary interest, i.e. the problem of how long it would take for a genetically invariant population to acquire as much variation as is usually found in the wild, solely by the accumulation of mutations.

In the character we used, the few observations on wild populations of *Drosophila melanogaster* suggest fortunately that the greater part of the genetic variance is additive and can therefore be measured from the response to selection. The latter can then be expressed as the product of the heritability of the character concerned and the selection differential. The heritability is equal to $\dfrac{\sigma^2 g}{\sigma^2 p}$ the ratio of the additive genetic variance $\sigma^2 g$, to the phenotypic variance $\sigma^2 p$ and, on the assumption of normality, the selection differential can be expressed as $\bar{i}\sigma p$, where $\bar{i}$, depending on the degree of selection, can be obtained from tables (Fisher and Yates 1938, table 20). We then have for the average response in each generation

$$\Delta G = \frac{\sigma^2 g}{\sigma^2 p} \cdot \bar{i}\sigma p; = \bar{i}\,\frac{\sigma^2 g}{\sigma p}$$

For the two control lines, the last two entries show HC higher than LC by 0.06 ± 0.36 and 0.30 ± 0.64. We may combine these to give a mean value of 0.12 ± 0.32. As an upper limit for this divergence, we can take the mean plus twice the standard error, i.e. 0.76 bristles, so that the response in each line would be 0.38. This would occur in about 14 generations giving an average response each generation of not more than 0.027. In the controls, $\bar{i} = 0.94$ (corresponding to 10 selected out of 25) and $p = 2.1$ (table 3). We have then

$$0.027 = 0.94 \times \frac{\sigma^2 g}{2.1}; \text{ i.e. } \sigma^2 g = 0.060$$

This is an average value over the 14 generations. The line had been continuously mated full-sib before the start of this experiment and assuming that equilibrium had been reached between the loss of variance due to the mating system (at a rate of 19.1 per cent per generation) and new variation arising per generation by mutation we would have for $\sigma^2 g(o)$ the genetic variance present at the start where

$$0.191\,\sigma^2 g(o) = \sigma^2 gm$$

where $\sigma^2\text{gm}$ is the amount of new variation produced each generation. Thus $\sigma^2\text{g(o)} = 5.2\sigma^2\text{gm}$. During the experiment, the relative loss of variability will be $\frac{1}{2N}$, where N is the number of parents, in this case 20. Thus at generation $K: \sigma^2\text{g}(K) = 0.975\sigma^2\text{g}(K-1) + \sigma^2\text{gm}$. This gives for the fourteenth generation $\sigma^2\text{g} = 15\sigma^2\text{gm}$ giving an average over the experiment of $10\sigma^2\text{gm}$. The upper limit to $\sigma^2\text{gm}$ is thus 0.006 units per generation. In the few wild populations that have been examined, the genetic variance has been about 5 units. The rate of spontaneous production of new variance in each generation is thus probably less than about .001 of that present in wild populations.

Two experiments bearing on the spontaneous origin of variation in this character have been carried out by Mather and Wigan (1942) and by Durrant and Mather (1954) both using the same Oregon stock, full-sib mated for 78 generations by 1942 and for over 300 generations by 1954. Mass selection for abdominal and sterno-pleural chaetae was done in the first experiment. The intensity of selection was higher than in our work (the extreme 1 in 20, corresponding to an $\bar{i}$ of 1.87) and the mating system involved slight inbreeding which would cause a loss of variance of about 5 per cent per generation. One might expect that the genetic variance would gradually increase until it stabilized at around 20 $\sigma^2\text{gm}$. Progress was slow but definite in the first phase of the abdominal selection but increased in the later stages. The results in the last phase can be put into our framework as follows. The rate of divergence from generations 34-53 was roughly 0.2 bristles per generation, giving a rate of change in each line of 0.1. $\sigma p$ is given in table 3 of the paper as about 2.5. We have then

$$0.1 = 1.87 \times \frac{\sigma^2 g}{2.5}; \sigma^2 g = 0.13$$

If we assume this to be equal to 20 $\sigma^2\text{gm}$, we obtain a value for the latter of 0.007 units per generation, of the same magnitude as our upper limit.

In the second experiment, a sample of 10 second chromosomes was taken from the inbred line and all possible zygotic combinations were formed. Significant differences between chromosomes were found in the bristle counts of zygotes containing them and the component of variance between chromosomes was 0.056. The rate of production of new variance between chromosomes is thus, by the argument presented earlier a fifth of this, 0.011, as the line had been mated full-sib. The new variance between diploid individuals is a half of this giving a value of $\sigma^2\text{gm}$ for the second chromosome alone of 0.006 units per generation, again of the same order of magnitude.

Selection of sterno-pleurals was only done for 21 generations, producing in that time a divergence of one chaeta. This gives a value of 0.020 for the average value of $\sigma^2\text{g}$ and of 0.002 units per generation for $\sigma^2\text{gm}$. The chromosome assay gives 0.002 units for $\sigma^2\text{gm}$ for the second chromosome only. In our outbred population, the genetic variance in sterno-pleurals is 1.7 units.

The experiments of Mather and his co-workers both suggest that the rate of production of new variance by mutation each generation is of the order of .002 to .001 of that present in wild populations both for abdominal and sterno-pleural chaetae. If rates of this magnitude are found for other characters and in other animals, the effect of mutation may in general be regarded as negligible in selection experiments in laboratory animals and, even more so, in domestic animals where such experiments rarely include even as many as five generations of selection. But from the evolutionary point of view, this rate must be considered as high. Many recent investigations have revealed a surprising amount of variability latent in wild populations which has in turn led to speculation as to the mechanisms preserving this variability in the population. It would seem from these results that, for characters such as numbers of sterno-pleural or abdominal chaetae, which are probably in themselves selectively neutral, the existing variation could well be maintained by the equilibrium between inbreeding and mutation, without the necessity of invoking mechanisms involving selection. In the absence of selection, equilibrium between mutation and inbreeding will be reached when the genetic variation in the population is equal to $2N\sigma^2 gm$ where N is Wright's effective population size. The variation found in our populations would thus only require that N should be of the order of hundreds. The only estimates of N for Drosophila, that of Wright, Dobzhansky and Hovanitz (1942) in *D. pseudoobscura*, were of the order of thousands or tens of thousands.

If we take the HR line, as that which showed the greatest response under selection, we find in generation 17A a deviation from the control means of 2.77 chaetae. Putting $\sigma p = 2.8$ in this case (table 2) we have $2.77/17 = 0.94 \times \sigma^2 g/2.8$ i.e. $\sigma^2 g = 0.49$. Taking this average value over all generations to be the actual value at the eighth generation, we get for the rate of accumulation of new variation a value of 0.06 units per generation or about 10 times the value calculated as the higher limit for spontaneous mutation. Thus at this rate it would need about 80 generations with a dose of 1800 r a generation to produce variance equal to that in wild populations.

However the observed increase in variance at the 6-10th generations was about 3 units compared with the calculation of the part utilizable by selection as 0.49. Perhaps the accumulation of lethals and steriles in the population may mask the utilization in selection of the new variation because many of the chromosomes will be unfixable, unless a crossover occurs at the right place. We propose to irradiate the random-bred base population and observe the effect on its utilizable variation. If this is important, it may be that other breeding programs, involving measures to reduce the frequency of recessive lethals, would be more successful than simple selection in making use of the new variation.

Apart from the experiments of Scossiroli and Buzzati-Traverso, referred to earlier, there are few other records of selection after irradiation. Lewis

(1949) reports work carried out by Harrison rather similar in design to our own. Males from an inbred line were given 4000 r each generation and selection was for abdominal chaeta number. Responses were obtained in the down lines but only in one of the up lines.

We are indebted to Dr. I. M. Lerner for drawing our attention to two papers by Russian workers who selected for sterno-pleural chaetae after irradiation. Serebrovsky (1935) started with a "Florida" stock of *Drosophila melanogaster*, which, judging by the variability at the start of the experiment, must have been inbred, though no mention is made of this. He gave 2000 r to males, 1000 r to females and used a mixture of family and individual selection with full-sib mating. After 8 generations the divergence between the control lines was greater than that between the irradiated lines. Using the four selected lines as starting material, he then changed to individual selection with much lower inbreeding. In the succeeding 10 generations, he obtained a divergence of 6 chaetae between the control lines and of only 2 between the irradiated lines. It is surprising that such variation should have remained in the control lines after 8 generations of full-sib matings.

Rokizky (1936), a pupil of Serebrovsky, then extended this work in more detail. In the first series he used the same Florida stock as Serebrovsky and a similar breeding and selection system as in the latter's first experiment. He only selected upwards with three lines, control, irradiated (3000 r in ♂) and temperature shocked (35–36°C in larvae). Selection for 12 generations produced little change in any of the lines. In his second series, starting this time from a Caucasus stock, he selected upwards 40 lines, 20 controls and 20 irradiated. Full-sib mating was carried out and selection was based on individual score for 25 generations. He again gave males 3000 r each generation. The results are summarized in table 4.

TABLE 4

| Generation | Radiated | | Control | |
|---|---|---|---|---|
| | Mean | S.D. | Mean | S.D. |
| 1–5 | 20.01 | 2.11 | 19.80 | 1.96 |
| 20–25 | 20.61 | 2.63 | 19.96 | 2.17 |

The inbreeding system practiced severely reduces possible responses. The slight increase in the controls is not inconsistent with Mather's results. The radiated lines show a more definite increase and, as do our irradiated lines, an increase in variability. There was considerable heterogeneity of behavior amongst the irradiated lines. One gained 3.5 bristles with an average standard deviation in the later generations of 2.31 and another gained 1.4 bristles with an average increase in standard deviation to 3.26.

Our own results would suggest that the production of new variation in abdominal bristles by mutation is slow. Under irradiation, new variation can be detected but that utilizable by direct selection is small. The general tenor of the reports of similar experiments by other workers is in agreement, with the exception of those of Scossiroli (1953). In the latter case, the disagreement may be due to the fact that we started from an inbred line and he used a line which had ceased to respond to selection. Some of our results are not completely explainable and we regard this only as a pilot experiment leading to more detailed work with suitably marked stocks on both spontaneous and induced variation.

## SUMMARY

Selection for abdominal chaetae has been carried out in an inbred line of *D. melanogaster*, both with and without irradiation of 1800 r of X-rays each generation. The response in the control stocks in 17 generations was not significant. The irradiated lines responded to selection but slowly compared with wild populations.

This is discussed in relation to the results of other workers. Two papers by Mather and co-workers are found to give consistent estimates of the rate of spontaneous production of new variance in abdominal chaetae of the order of 0.01 units each generation, which is not inconsistent with our results. The variance found in several wild populations is about 5 units. The evolutionary aspect of these results is discussed.

## LITERATURE CITED

Buzzati-Traverso, A. A., 1953, On the role of mutation rate in evolution. Proc. IX Intern. Cong. Genetics, Bellagio.
Clayton, G., J. A. Morris, and A. Robertson, 1953, Selection for abdominal chaetae in a large population of *Drosophila melanogaster*. Proc. Symp. Genetics Pop. Struct., Pavia 7-15.
Durrant, A., and K. Mather, 1954, Heritable variation in a long inbred line of Drosophila. Genetica 27: 97-119.
Fisher, R. A., and F. Yates, 1938, Statistical Tables. Oliver and Boyd, Edinburgh.
Gustafsson, A., 1953, New genes and chromosomes in agricultural plants. Proc. IX Intern. Cong. Genetics, Bellagio.
Lewis, D., 1949, 40th Report of the John Innes Horticultural Institution, 30 pp.
Mather, K., and L. G. Wigan, 1942, The selection of invisible mutations. Proc. Roy. Soc., B, 131: 50-64.
Rasmusson, M., 1952, Variation in bristle number in *D. melanogaster*. Acta Zool. 33: 278-306.
Rokizky, P., 1956, Experimental analysis of the problems of selection by X-ray irradiation. Uspehi Zootehničeskih Nauk 2: 161-202.
Scossiroli, R. E., 1953a, Effectiveness of artificial selection under irradiation of plateaued populations of *D. melanogaster*. Proc. Symp. Genetics of Pop. Struct., Pavia 42-66.
  1953b, Artificial selection of a quantitative trait in *D. melanogaster* under increased mutation rate. Proc. IX Intern. Cong. Genetics, Bellagio.
Serebrovsky, R. E., 1935, Acceleration of the rate of selection of quantitative characters in *D. melanogaster* by the action of X-rays. Zoologičeskii Zurnal 14: 465-480.
Wright, S., 1937, The distribution of gene frequencies in populations. Proc. Nat. Acad. Sci. 23: 307-320.
Wright, S., Th. Dobzhansky, and W. Hovanitz, 1942, Genetics of natural populations VII. Genetics 27: 363-394.

Copyright © 1971 by The Genetical Society of Great Britain
Reprinted from *Heredity* **27**:163-174 (1971)

# VARIATION FOR METRICAL CHARACTERS IN *DROSOPHILA* POPULATIONS

## III. THE NATURE OF SELECTION

R. LINNEY, B. W. BARNES and M. J. KEARSEY
*Department of Genetics, University of Birmingham*

Received 5.ix.70

### 1. INTRODUCTION

AN earlier paper (Kearsey and Barnes, 1970) reported the results of experiments investigating the relationship between sternopleural chaeta number and fitness in a cage population of *Drosophila melanogaster*. The population studied was derived from a cross between two lines, produced by selection for high and for low number of chaetae from our "Texas" population (Barnes and Kearsey, 1970), and subsequently allowed to evolve under cage conditions.

It was shown that the phenotypic variance of adults captured in the cage, the survivors of intense larval competition, was approximately one-quarter that of their contemporaries raised at very low density. Furthermore, this decrease in phenotypic variance was shown to be due, almost exclusively, to a decrease in genetic variance. That is, there has been selective elimination of certain genotypes at the pre-adult stage. This selective elimination was related to chaeta number. It was then possible to estimate the relative fitness of different phenotypes by comparing their relative frequencies at the two larval densities.

By this means fitness was shown to be greatest for phenotypes with a value close to that of the $F_1$ between the parental selection lines and to decline markedly with deviations in both directions from this optimum, reaching a fitness of zero within the phenotypic range. Such a reduced fitness of extreme phenotypes might be due to association between genes controlling chaeta number and subvital genes fixed during the establishment of the parental selection lines. That is, the selection lines may contain a number of recessive subvital genes linked to genes for sternopleural chaeta number and this association might still be present at the time of the experiment. Extreme phenotypes will tend to be homozygous for such subvitals, whereas intermediate phenotypes will be heterozygous and hence have higher fitness. However, the detailed relationship between phenotype and fitness was not consistent with the relationship generally found in such cases. Although linked subvitals might in part be the cause, it was argued that the results are more compatible with stabilising selection.

The relationship between chaeta number and fecundity, on the other hand, was entirely consistent with a system of dispersed subvital genes. However, females extreme for chaeta number do not reach adulthood under cage conditions and as a consequence the variation in fecundity had no impact on fitness. The major component of fitness in this case appears to be egg to adult survival.

The experiments to be described here were designed to test the validity of the two principal conclusions drawn previously. Firstly, that selection is concerned directly with genes controlling chaeta number as opposed to linked deleterious genes (Experiment 1). Secondly, given that the first conclusion is correct, that selection is based on metric deviation rather than heterozygote advantage at the loci controlling chaeta number (Experiment 2). Selection on the basis of metric deviation implies that the fitness of an individual is solely a function of its phenotypic deviation from an intermediate optimum and does not depend directly on the number of loci at which it is heterozygous.

## 2. Experiment 1

If the decreased fitness of extreme phenotypes is due to linked subvital genes, then the relationship between chaeta number and fitness found in the derived population should not apply in the base population. The first experiment was designed to assess selection in the base population.

The " Texas " population has been maintained for 6 years in this laboratory and the mean and variance have not changed noticeably in that time. Given the hypothesis that the genes controlling chaetae are neutral, such genes will be in linkage equilibrium with respect to any subvital genes since the population is large (about 3500). Hence the hypothesis is disproved if a reduction in the *genetic* variance for chaeta number can be demonstrated, as a result of crowding. In the previous experiment (Kearsey and Barnes, *loc. cit.*) the decline in *phenotypic* variance was so great as to be explicable only in terms of a reduction in *genetic* variance. The *phenotypic* variance of " Texas ", on the other hand, is small even at low larval densities and declines little on crowding. This leads to difficulties in interpretation. While it is easy to demonstrate a significant decline in phenotypic variance (if sufficient flies are scored) the difference, being small, could be due to one, or a combination, of the following factors operating at high density:

1. A reduction in the effect of individual gene substitutions.
2. A reduction of the environmental variance.
3. The selective elimination of extreme phenotypes.

The first two points involve genotype environment interaction while (3) alone involves selection. The regression techniques used previously are insufficiently sensitive to discriminate between these possibilities in the present case. However, the effects of genotype environment interaction may be excluded by progeny testing phenotypes in the same environment. This was achieved by crossing a random sample of the males surviving at both high and low density to virgin females of an inbred tester line. Their genetic variance for chaeta number was then assessed from the performance of their progeny raised under uniform environmental conditions. Selection at high density will result in a reduced variance between families of high-density males compared to that of their contemporaries raised at low density. Furthermore, fitnesses can be estimated from the relative frequencies of different family means derived from parents at high and low densities. The advantage of this approach is that it excludes genotype environment interaction affecting the variances and also obviates the need to correct the phenotypes of flies raised at high density for the direct environmental depression of chaeta number produced by food deprivation.

(a) *Method*

The wild population used in these studies, " Texas ", originated from 30 inseminated females caught at Austin in Texas in October 1965, and subsequently maintained in a population cage (Barnes and Kearsey, 1970). One thousand eggs were collected from the cage population and incubated in tubes containing standard oatmeal medium at a density of 50 eggs per tube. After eclosion, a random sample of 250 males was collected and scored (low-density sample), and at the same time a sample of 250 males was collected from the cage and scored (high-density sample). Two days later, all the males surviving in the two samples were mated individually to four virgin females from the inbred line Oregon, and the progeny raised in tubes. All the families were incubated in a single randomised block at 25° C. Five female progeny were scored from all successful matings 11 days later.

The size of the experiment was determined from a pilot study involving only 30 males from each density carried out by our colleague Mr A. Birley.

(b) *Results and discussion*

The mean and variance of the male parents used in the progeny tests and of their female offspring are shown in table 1. The mean chaeta number

TABLE 1

*Summary of data from progeny tests on males from low- and from high-density conditions. Five female progeny scored from crosses between sample males and Oregon females*

|  | Density | |
| --- | --- | --- |
|  | High | Low |
| Number of ♂♂ sampled | 133 | 187 |
| Mean of ♂♂ sampled | 15·8045 | 16·9393 |
| Variance of ♂♂ sampled | 2·6736 | 3·2693 |
| Mean of ♀♀ offspring | 19·5895 | 19·5359 |
| Variance of offspring family means | 1·1602 | 1·6093 |
| Average variance within families | 2·7975 | 2·7390 |

of flies raised at high densities is significantly less ($P < 0.001$) than for flies raised at low density, as was found previously (Kearsey and Barnes, *loc. cit.*). The variance, although reduced at the high density, is not significantly less than the low-density variance ($0.10 > P > 0.05$).

Turning now to the offspring data, we find that the means do not differ significantly ($0.3 > P > 0.2$). However, the variance of family means of progeny obtained from male parents raised at high density is very significantly reduced ($P \simeq 0.025$). A conventional one-tail variance ratio test was used here as, from our previous evidence, we expect the variance to be greatest at low density. Furthermore, this decrease in the variation *between* families must indicate a reduction in the genetic variance at high density as the *within* family variances at the two densities are homogeneous.

This reduction of the genetic variance can be shown more clearly by the regression analyses given in table 2. The variation between family means has been partitioned into the following items.

1. The variation due to the regression of progeny onto male parent.
2. The variation resulting from departures from linearity of this regression.
3. The variation between families derived from male parents of the same phenotype.

The remainder mean square (2) is not significant for either treatment. Thus there is no evidence of non-additive genetic variation. Items 2 and 3 have therefore been combined to give the pooled residual items of table 2. Furthermore, there is no significant difference between the residual items from high and low density. The regression items are highly significant at

TABLE 2

*Regression analysis (based on family means)*

| | Density | | | | | |
|---|---|---|---|---|---|---|
| | High | | | Low | | |
| Item | d.f. | M.S. | P | d.f. | M.S. | P |
| 1. Regression on ♂ parent | 1 | 14·0938 | <0·0001 | 1 | 77·5796 | <0·0001 |
| 2. Remainder | 7 | 1·3380 | >0·20 | 8 | 1·1645 | N.S. |
| 3. Between families within ♂ phenotypes | 124 | 1·0458 | <0·001 | 177 | 1·2002 | <0·001 |
| Pooled residual | 131 | 1·0615 | — | 185 | 1·1986 | — |
| Within families | 532 | 0·5595 | — | 748 | 0·5478 | — |

Regression of family mean on to ♂ parent $\hat{b}_H = 0.19984 \pm 0.0548$ $\hat{b}_L = 0.35718 \pm 0.0444$

both densities, indicating the presence of additive genetic variation. The estimated slopes of these regressions $\hat{b}_H$ and $\hat{b}_L$ are given in table 2. The difference between the slopes is highly significant (P $\simeq$ 0·025), $\hat{b}_H$ being approximately half $\hat{b}_L$.

Thus the regression analysis has clearly shown that the decline in genetic variance amongst the males raised in the cage can be explained solely by a reduction in the additive genetic variance. The regression slopes, $\hat{b}_H$ and $\hat{b}_L$, are equivalent to half the narrow heritability, $h^2$, of the two male samples. In the absence of non-additive variance we can estimate the magnitude of the additive genetic variance, $V_A$, and the environmental variance, $V_E$. These estimates are as follows.

| | $\hat{h}^2$ | $\hat{V}_A$ | $\hat{V}_E$ | $\hat{V}_P$ |
|---|---|---|---|---|
| Low | 0·7144 | 2·34 | 0·93 | 3·2693 |
| High | 0·3997 | 1·07 | 1·60 | 2·6736 |

It appears, therefore, that not only is there selective elimination of certain genotypes at high density, as shown by the reduced additive genetic variance, but also the environmental variance is increased.

Let us now turn to the nature of this selection. The average breeding value of the two samples of males do not differ, *i.e.* the progeny means are not significantly different, but the genetic variance is reduced amongst those individuals raised in the cage. This must indicate selection against extremes. Further information on the type of selection may be obtained if we can ascribe a relative fitness to each phenotype. Previously fitnesses were obtained by comparing the frequencies of phenotypes produced at high

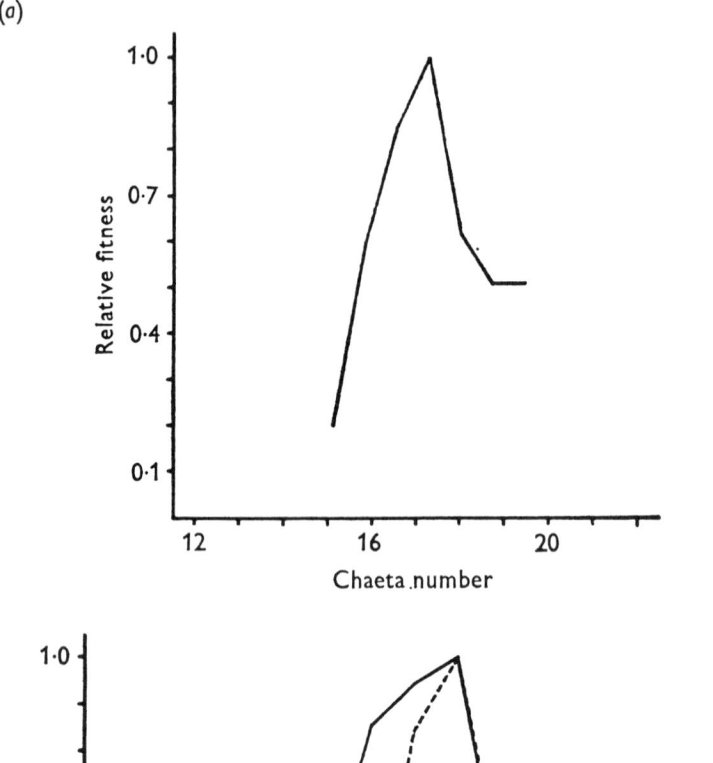

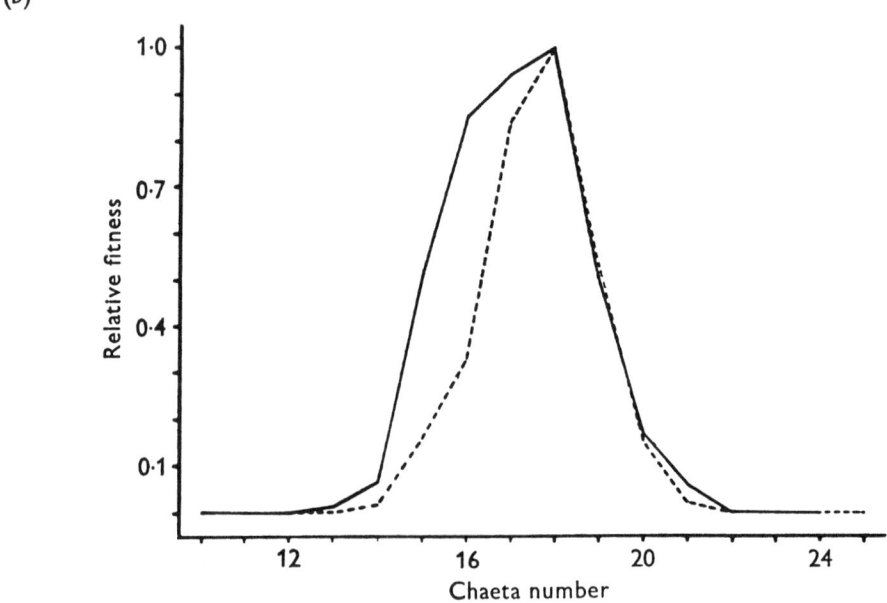

Fig. 1.—The relationship between chaeta number and relative fitness for (a) males sampled from the " Texas " population (Expt. 1) and (b) homozygous males (solid line) and females derived from the " Texas " population (Expt. 2).

density with the frequencies obtained at low density (Kearsey and Barnes, 1970). In the present case we have to compare progeny means, and a difficulty is that these means differ by multiples of 0·2 bristles; consequently there are relatively few observations for each value. A poor estimate of fitness would be obtained therefore if these frequencies were used directly. This may be overcome by combining the progeny means in groups of five such that the mean of each set is an integer. Thus, the frequencies of progeny with a mean of 15·6, 15·8, 16·0, 16·2 and 16·4 were combined to give a single frequency for an overall progeny mean of 16. A comparison of the frequencies for each integer value obtained at the two densities then gives us a measure of relative fitness. These fitnesses may be rescaled such that the optimum has a value of unity. The fitnesses obtained in this way relate to the *progeny* phenotype. However, for our present purposes it is more useful to express the fitness in terms of the phenotypes of the males originally sampled from the " Texas " population. This conversion is achieved from the regression of the male parents on to the family means. These predicted values together with their fitness are shown in fig. 1(a).

It can be seen that males with a phenotype of 17·3 have optimum fitness, and fitness declines with phenotypic deviations from this optimum. The fitness relationships are consistent with those obtained previously in the derived population, apart from the lower optimum phenotype of 17·3 as compared to that of 20 for the derivative.

## 3. Experiment 2

The second experiment was designed to test the validity of the conclusion that selection is for metric deviation as opposed to heterozygote advantage at the loci controlling chaeta number. The intention was to investigate a population which consisted of a heterogeneous collection of homozygotes taken from inbred lines which had been derived from the " Texas " population by sib-mating. The optimum phenotype in the base population when these inbred lines were established was 18 chaetae, and we would expect a similar optimum to be appropriate to the population of homozygotes if we are concerned with a metric deviation model of stabilising selection. With such a homozygous population, heterozygote advantage cannot, of course, be invoked to explain any relationship between phenotype and fitness.

(a) *Method*

A sample of 22 inbred lines were derived by sib-mating the progeny of single-pair matings among individuals from the " Texas " population for 42 generations. From this set of lines, four were taken (lines 10, 20, 25, 28) for use in the present experiment; these lines were chosen in such a way that a population produced by mixing them in equal proportions would contain at least some individuals of each phenotypic class in the range 12-22 sternopleural chaetae, with most classes containing members from more than one line.

In order to test lines 10, 20, 25 and 28 for the presence of residual heterozygosity at the loci controlling chaeta number, individuals within each line were positively assortatively mated for chaeta number. Two replicate single-pair matings were used throughout, and 10 progeny of each sex were scored from each cross. The extent of genetic variation can be assessed from the regression of offspring on parent.

The technique used in the main experiment is a modification of that employed by Kearsey and Barnes (1970). Essentially, two larval environments were used, one in which the larvae were raised at low density, the other in which larvae were raised at high density, under intensely competitive conditions. Two populations were constructed, by introducing 125 inseminated female adults from each of the four inbred lines into each of two population cages. The cage that was to provide the low-density larval environment contained 20 food-tubes with the standard oatmeal/yeast medium, and the high density cage 5 food-tubes. The 500 females in each cage were allowed to oviposit for 7 days, and were then removed from the cages. The experiment was carried out at $25 \pm 0.5°$ C. The first progeny emerged from the low-density cage on the third day after the removal of their mothers.

A random sample of 500 flies of each sex from the total number that emerged over a 10-day period from the low-density cage was scored for

TABLE 3

*Mean chaeta number of " Texas " inbred lines 10, 20, 25, 28*

| Line | ♂♂ | ♀♀ |
|---|---|---|
| 10 | $12 \cdot 80 \pm 0 \cdot 98974$ | $14 \cdot 02 \pm 0 \cdot 87714$ |
| 20 | $17 \cdot 12 \pm 1 \cdot 28793$ | $17 \cdot 28 \pm 1 \cdot 27839$ |
| 25 | $16 \cdot 72 \pm 1 \cdot 29426$ | $16 \cdot 78 \pm 0 \cdot 95383$ |
| 28 | $21 \cdot 50 \pm 1 \cdot 55511$ | $21 \cdot 88 \pm 1 \cdot 42342$ |

chaeta number, as were all those emerging from the high-density cage over a similar period.

Males and virgin females, from both cages, were progeny-tested, by mating them individually to a common inbred tester line, Oregon. From each density, males and females with chaeta numbers 13-19 inclusive were used, and for each phenotypic class two replicate single-pair matings were set up. Their progeny were raised, at low density, in the usual 7·5 cm. × 2·5 cm. diameter food tubes, the experiment taking the form of a single randomised block. Ten progeny of each sex were scored, per mating, for chaeta number, and the regression of offspring on " Texas " inbred parent calculated. This resulted in eight simple regressions, *viz.* son on father, daughter on father, son on mother, and daughter on mother, for each density.

(b) *The amount of within-line heterozygosity*

The mean chaeta numbers of the four lines, based on samples of 50 individuals of each sex, are shown in table 3.

The regression analyses of offspring on parent carried out within each of the lines are shown in table 4; there were no significant differences between replicate crosses, and this item in the analysis has, in every case, been combined with the variation within crosses. In each line some detectable genetic variance remains, and appears, inconsistently, in the eight analyses. To obtain some idea of the magnitude of this residual variation, the genetic component of variance, $\sigma_G^2$, between families has been estimated in those cases which show significance of either the Regression or the Remainder M.S. These components of variance are as follows:

| Line | Sex | $\hat{\sigma}_G^2$ |
|------|-----|--------------------|
| 10   | ♂♂  | 0·03195            |
| 20   | ♂♂  | 0·08161            |
| 25   | ♂♂  | 0·18898            |
| 28   | ♀♀  | 0·16263            |

Now, had the parents used for the progeny-tests been crossed at random within each line, $\hat{\sigma}_G^2$ would estimate half the true genetic variance. In fact, however, they were positively assortatively mated; if the assortative mating is perfect, *i.e.* if two identical genotypes are mated, then $\hat{\sigma}_G^2$ estimates the total genetic variance. Clearly, we have an intermediate situation here. Firstly, given that some genetic variation is present in the lines, mating has not been at random; and, secondly, it is highly unlikely that, for this

TABLE 4

*Offspring/parent regression analyses within " Texas " inbred lines 10, 20, 25, 28*

|  | d.f. | ♂♂ M.S. | ♂♂ P | ♀♀ M.S. | ♀♀ P |
|---|---|---|---|---|---|
| *Line 10* | | | | | |
| Between families | 9 | 1·58000 | N.S. | 1·93556 | N.S. |
| Regression | 1 | 4·42220 | 5%–1% | 0·00140 | N.S. |
| Remainder | 8 | 1·22472 | N.S. | 2·17732 | N.S. |
| Pooled error | 190 | 0·94105 | — | 1·23579 | — |
| *Line 20* | | | | | |
| Between families | 6 | 2·93334 | 5%–1% | 0·79524 | N.S. |
| Regression | 1 | 2·25520 | N.S. | 1·94440 | N.S. |
| Remainder | 5 | 3·06896 | 5%–1% | 0·56540 | N.S. |
| Pooled error | 133 | 1·30113 | — | 1·32030 | — |
| *Line 25* | | | | | |
| Between families | 7 | 5·60000 | <1% | 2·21340 | N.S. |
| Regression | 1 | 9·48960 | N.S. | 1·35420 | N.S. |
| Remainder | 6 | 4·95174 | 5%–1% | 2·35660 | N.S. |
| Pooled error | 152 | 1·82039 | — | 1·54901 | — |
| *Line 28* | | | | | |
| Between families | 14 | 1·87476 | N.S. | 6·42762 | 5%–1% |
| Regression | 1 | 0·66440 | N.S. | 2·28000 | N.S. |
| Remainder | 13 | 1·96786 | N.S. | 6·74666 | 5%–1% |
| Pooled error | 285 | 3·02912 | — | 3·17491 | — |

character, identical phenotypes are always genotypically identical. Thus mating has not been perfectly assortative at the genotypic level. The true genetic variance, therefore, is somewhere between $\hat{\sigma}_G^2$ and $2\hat{\sigma}_G^2$.

If we recall that the genetic variance in the " Texas " population from which these lines were derived is approximately 2·3 (see Experiment 1), we see that the variances observed here are, in comparison, greatly reduced. At worst, the inbreeding has resulted in the reduction of the genetic variance to one-sixth of its value (line 25) in the base population, and for lines 10 and 20 this reduction has been considerably greater. On average, however, the residual genetic variation is reduced 40-fold. The inconsistency with which the genetic variance has reached significance in the two sexes further indicates the low level at which it exists in these lines. One may reasonably conclude, therefore, that the populations under study consist of individuals almost entirely homozygous at loci controlling chaeta number.

(c) *Selection in the cages*

The phenotypic variances ($V_p$) of adults emerging from the two experimental cages are shown in table 5. In both sexes, highly competitive larval

TABLE 5

*Components of the phenotypic variances of adults from low and high densities*

|  | ♂♂ | | | ♀♀ | | |
| --- | --- | --- | --- | --- | --- | --- |
|  | $\hat{V}_A$ | $\hat{V}_E$ | $\hat{V}_p$ | $\hat{V}_A$ | $\hat{V}_E$ | $\hat{V}_p$ |
| Low | 1·91847 | 2·72484 | 6·56178(500)* | 2·76195 | 1·24459 | 6·76849(500)* |
| High | 0·24297 | 1·04561 | 1·53155(407)* | 0·00000 | 1·28364 | 1·28364(356)* |

Note that, for the present inbred population, $V_p = 2V_A + V_E$.

* The number of flies scored.

conditions have resulted in a significant reduction in the phenotypic variance of adult chaeta number.

Analysis of the results of progeny-tests reveals that, in every comparable pair of regressions, the regression coefficient using parents from high density ($\hat{b}_H$) is significantly lower than that using low-density parents ($\hat{b}_L$) (table 6). This, together with the reduced phenotypic variance, indicates that there is considerably less genetic variance among high-density parents. Estimates of the additive genetic and environmental variances ($\hat{V}_A$, $\hat{V}_E$) among the

TABLE 6

*Regression analysis from results of progeny-tests on adults from high (H) and low (L) densities*

|  |  | Son on Father | Daughter on Father | Son on Mother | Daughter on Mother |
| --- | --- | --- | --- | --- | --- |
| Progeny mean | L | 18·92 | 18·97 | 18·14 | 18·97 |
| ($\bar{x}$) | H | 19·10 | 20·19 | 19·29 | 19·96 |
| $\hat{b}_L$ |  | 0·35714 ± 0·11963*** | 0·22679 ± 0·12037* | 0·48214 ± 0·15848** | 0·38214 ± 0·09322*** |
| $\hat{b}_H$ |  | 0·12857 ± 0·14722 | 0·17143 ± 0·09601* | 0·05893 ± 0·15535 | 0·04643 ± 0·13054 |

* = $p < 5\%$, ** = $p < 1\%$, *** = $p < 0.1\%$.

survivors of high and low density conditions are shown in table 5, from which the reduction in genetic variance under competitive larval conditions is immediately apparent.

To compare the distributions of adults from the two cages, one must take into account any environmental depression in mean chaeta number at high density. Ideally, this depression can be estimated from the regression data, but this procedure depends upon homogeneity of the offspring/parent regression slopes for the two cages. Here, however, there is almost no genetic variance among the survivors at high density and, consequently, the regressions are heterogeneous. In the absence of a direct estimate, the mean depression obtained in previous unpublished work (1 chaeta) has been used. That is, an individual which, when raised under the present high-density conditions, has a chaeta number $x$, would, on average, possess $(x+1)$ chaetae if raised in the present low-density cage. The high-density distribution of emergent adults must, therefore, be corrected by the addition of 1 hair to each observed phenotype.

The elimination of extreme phenotypes, in conjunction with the reduction in genetic variance, at high density shows that there has been selection, of a stabilising nature, of individuals with intermediate genotypes. We may proceed, then, to estimate phenotypic fitnesses, from the relationship

$$W'_i = f(P_{iH})/f(P_{iL})$$

where $W'_i$ = relative fitness of $i$th phenotype,

$f(P_{iH})$ = frequency of $i$th phenotype at high density,

$f(P_{iL})$ = frequency of $i$th phenotype at low density.

These fitnesses, re-scaled so that $W'_i$ (max) = 1·0, have been calculated separately for males and females, and the relationship between fitness and phenotype is illustrated in fig. 1 (b). In both sexes, the observed optimum is 18 chaetae, and fitness decreases rapidly, in a fairly linear manner, with increasing deviation, in either direction from this value.

In theory, a mechanism of single-locus overdominance for fitness can result in the selection of intermediate phenotypes, especially with multiplicative fitnesses and large selective disadvantages for homozygotes. However, it is unlikely that in the present populations, with very few loci segregating as compared to the " Texas " population, such a system of selection could account for the marked differences in phenotypic fitnesses. Since the experimental populations contain an unusually high frequency of homozygotes, it is much more likely that the selection observed has proceeded on the basis of homozygous differences between phenotypes; such a mechanism may be called selection for metric deviation. That is, in a normal population, with large numbers of both homozygotes and heterozygotes, subject to selection for metric deviation, two individuals with the same phenotype, but with quite different numbers of homozygous and heterozygous loci, may be equally fit. Furthermore, computer simulations suggest that with this type of selection, with fitnesses following a linear deviation model of the sort observed in the present experiment, genetic variance can be maintained in a randomly mating population (Gale and Kearsey, 1968; Kearsey and Gale, 1968).

It should be pointed out that heterozygous advantage and selection for metric deviation are not, of course, mutually exclusive forms of selection. There is no reason, in general, to suppose that, in populations containing extensive genetic variation, both should not operate. On the other hand, there is evidence that heterotic selection is not acting for chaeta number in populations of *D. melanogaster* derived from crosses between two inbred lines (Barnes, 1968; Killick, 1970). We suggest that it is not necessary for the type of relationship between a metrical character and fitness illustrated by fig. 1 (b).

## 4. General discussion

As has been stated elsewhere (Robertson, 1955), when we are dealing with a phenotype/fitness relationship for a metrical character, it is important to consider the possible ways in which the relationship can be produced. The relationships found here follow a similar pattern, both for males from the " Texas " population, and for each sex of the inbred population, (fig. 1). Moreover, females studied earlier, in the $F_2$ population derived from crosses between selection lines for high and low chaeta number (Kearsey and Barnes, 1970), again illustrate the distinct intermediate optimum and the sharp, linear decline in fitness with phenotypic deviation from this optimum. Any association, consequent upon artificial directional selection, between genes controlling chaeta number and subvital genes, cannot, of course, account for the array of phenotypic fitnesses observed in the " Texas " base population, which has been subject only to natural selection in the population cage. It may be argued that such association is responsible for the phenotype/fitness relationship among individuals from the $F_2$ population, but in view of the results of the two experiments reported here, this appears to be unlikely. In the case of the " Texas " population, we are certainly dealing with selection which is acting directly on genes controlling chaeta number. In the homozygous population this, too, is the most reasonable conclusion, unless we make the rather dubious assumption that, during the inbreeding to which the four lines have been subjected, deleterious genes have become fixed, by chance, to a greater extent in the two lines with the most extreme hair counts.

Given then, that selection is acting directly on genes controlling chaeta number, what can we say about the nature of this selection? In the case of the $F_2$ population, it is possible that intermediates were selected on a heterotic basis, *i.e.* differences in numbers of heterozygous chaeta loci may have accounted for the differences in relative fitnesses which were estimated. This explanation may also be proposed for the results from the base population. However, since for the largely homozygous inbred population the pattern of fitnesses (fig. 1 (*b*)) is as clearly defined, both in magnitude and in the shape of the graph, as that for the $F_2$ and base populations (fig. 1 (*a*)), the heterotic explanation as a major determinant for these relationships is rejected. Homozygous differences between individuals, at loci controlling chaeta number, appear to play a more important part in causing variation in fitness than do heterozygous differences. It is probable that this causal relationship is a result of pleiotropic gene action at the chaeta loci, and consideration of the environment in which the selection has been operating here suggests that genes at these loci contribute, to some extent, to larval competitive ability.

It is obviously important to discover if selection of the type and intensity found here is of widespread occurrence for other metrical traits. There is, however, no reason to expect chaeta number to be exceptional in terms of the intense selection to which it is exposed. On the contrary, prior to this series of experiments, chaeta number had been considered a peripheral character unrelated to fitness (Robertson, 1955, 1966). The mortality between egg and adult under our cage conditions is between 92 and 95 per cent., which is certainly sufficient to allow selective elimination of metric deviants for many other characters, while still allowing for a large degree of random loss.

## 5. Summary

1. Natural selection against extreme genotypes controlling sternopleural chaetae number has been demonstrated in two populations of *Drosophila melanogaster*. These were (i) a long-established cage population ("Texas") and (ii) an artificial population comprising homozygous individuals derived from "Texas".

2. The relationship between phenotype and fitness in both populations was essentially identical and closely resembled that found previously in a population derived from an $F_2$ between lines selected for chaeta number.

3. These results cannot be explained by the action of selection on subvital genes linked to loci controlling chaeta number.

4. It is concluded that selection is acting on the basis of metric deviation and not for heterozygosity *per se*.

5. Our results suggest that selection for metric deviation (stabilising selection) may be an important mechanism maintaining potential genetic variation for metrical traits in natural populations.

*Acknowledgments.*—We are indebted to Mr G. Oram for valuable technical and photographic assistance and Dr J. S. Gale for helpful discussions. Part of this work was carried out by one of us (R. L.) during tenure of a University of Birmingham Research Studentship.

## 6. References

BARNES, B. W. 1968. Stabilising selection in *Drosophila melanogaster*. *Heredity*, 23, 433-442.

BARNES, B. W., AND KEARSEY, M. J. 1970. Variation for metrical characters in *Drosophila* populations. I. Genetic analysis. *Heredity*, 25, 1-10.

GALE, J. S., AND KEARSEY, M. J. 1968. Stable equilibria under stabilising selection in the absence of dominance. *Heredity*, 23, 553-561.

KEARSEY, M. J., AND BARNES, B. W. 1970. Variation for metrical characters in *Drosophila* populations. II. Natural selection. *Heredity*, 25, 11-21.

KEARSEY, M. J., AND GALE, J. S. 1968. Stabilising selection in the absence of dominance: an additional note. *Heredity*, 23, 617-620.

KILLICK, R. J. 1970. Natural selection for a metrical trait in a population of *Drosophila melanogaster*. *Heredity*, 25, 123-125.

ROBERTSON, A. 1955. Selection in animals: synthesis. *Cold Spr. Harb. Symp. Quant. Biol.*, 20, 225-229.

ROBERTSON, A. 1966. Artificial selection in plants and animals. *Proc. Roy. Soc. B.*, 164, 341-349.

# A CALCULATION OF THE MINIMUM NUMBER OF GENES IN WINTER'S SELECTION EXPERIMENT

## By "STUDENT"

In a note on Winter's selection experiment(1) published in the *Eugenics Review*(2) I made the following claim:

> By reducing the problem to the simplest possible basis...it is possible to make some sort of a calculation of the number of genes which might allow of so large a change by repeated selection and I find that the order of these numbers is 100–300.

Prof. Fisher, however, pointed out in *Nature* (3) that I had in fact over-simplified my problem and that no such conclusion could be drawn from my "sort of calculation."

This did not in fact invalidate my main thesis, which was that species tend to accumulate a sufficient store of genes of no particular value until they meet with a change of environment, when the store provides material for selection far beyond the normal range.

But although the calculation was based on over-simplified data and was superfluous to my argument, it is of some interest in itself, and the present note is an attempt to "mend my hand" by making more reasonable assumptions.

I shall start by giving a very short account of Winter's experiment with an abbreviated table, hoping that my readers may be sufficiently interested to study Winter's paper for themselves.

Then I shall make an estimate of the standard deviation of that part of the variation in oil content of Winter's maize which was due to genetic constitution, and measure the difference between the mean oil content of his "high" and "low" races in terms of this standard deviation.

I shall next make an estimate of the minimum numbers of genes which would suffice to account for so large a ratio between the possible range and the standard deviation.

Finally, I shall discuss the various assumptions which have been made, pointing out which of them are in my opinion reasonable, which have reduced the minimum number of genes to a figure below that which is probable, and which are merely the best assumptions we can make.

Winter's experiment, then, was concerned with a continuous selection of maize in the directions of high and low protein and high and low oil content, and I am only concerned here with the latter.

The experiment was begun in 1896 and has continued to the present day\*, but only the first 28 years were reported on in his paper, i.e. till 1924. The following is his description of

---

\* Mr Winter in correspondence a year or two ago told me that both these experiments and one on height were still being continued and still showed a continued, if less marked, effect of selection. In the latter case he had arrived at mean heights of 8 ft. and 8 in. in two races derived from a 4 ft. maize.

the procedure, which I have only altered by instancing the oil content part of the experiment, whereas he quoted the similar case of the protein:

One hundred and sixty-three ears of a variety known as "Burr's White" were used as foundation stock from which selections were made in four different directions, namely from high oil, low oil, high protein and low protein.

These four strains were carried on in the same way. In the high oil, for example, twenty-four ears highest in oil were selected for seed and planted in an isolated plot, each ear in a separate row. These ears were harvested separately and the seed for the next crop selected from the ears which were found to be highest in oil. Nine years later the system was modified somewhat in an attempt to prevent loss of vigour by inbreeding. Alternate rows were detasselled and seed was selected only from the highest yielding detasselled rows. In 1921 this system was again modified to reduce the amount of inbreeding. Two seed ears were taken from each of the detasselled rows regardless of yield.

The high protein, low protein and low oil tests were similarly conducted, selection being made each year of ears highest in protein, lowest in protein and lowest in oil, respectively.

| Year | No. of ears analysed | | Mean value percentage of oil | | Standard deviation | | Lowest variate | | Highest variate | |
|------|------|------|------|------|------|------|------|------|------|------|
| 1896 | 163 | | 4·68 | | 0·41 | | 3·9 | | 6·0 | |
|      | High | Low | High | Low | High | Low | High | Low | High | Low |
| 1897 | 80 | 50 | 4·79 | 4·10 | 0·38 | 0·29 | 3·6 | 3·4 | 5·7 | 4·7 |
| 1898 | 216 | 108 | 5·10 | 3·59 | 0·48 | 0·32 | 4·1 | 3·2 | 6·7 | 4·8 |
| 1899 | 108 | 144 | 5·65 | 3·85 | 0·42 | 0·32 | 4·3 | 2·8 | 6·5 | 4·6 |
| 1900 | 108 | 144 | 6·10 | 3·57 | 0·44 | 0·36 | 4·6 | 2·6 | 7·4 | 4·5 |
| 1901 | 126 | 126 | 6·24 | 3·45 | 0·45 | 0·26 | 4·9 | 2·8 | 7·1 | 4·1 |
| 1920 | 120 | 120 | 9·28 | 1·80 | 0·52 | 0·21 | 7·8 | 1·0 | 10·6 | 2·4 |
| 1921 | 120 | 120 | 9·94 | 1·71 | 0·66 | 0·15 | 8·4 | 1·0 | 11·7 | 2·3 |
| 1922 | 120 | 120 | 9·86 | 1·68 | 0·54 | 0·19 | 8·7 | 0·9 | 11·3 | 2·2 |
| 1923 | 120 | 120 | 10·08 | 1·58 | 0·65 | 0·24 | 8·3 | 1·1 | 11·8 | 2·1 |
| 1924 | 120 | 120 | 9·86 | 1·51 | 0·61 | 0·22 | 8·4 | 0·9 | 11·7 | 2·2 |

The above table gives certain figures for the first six and the last five years of the experiment, and it will be seen that, although the original maize only varied in oil content from 3·9 per cent. to 6·0 per cent., the lowest variate of the high race after 28 years of selection was 8·4 per cent. in oil content, while the highest variate of the low race was only 2·2 per cent.; in each case they were clean outside the original range, a fact which seems difficult to explain except on the hypothesis that the oil content of the original race was due to a number of genes which largely neutralised one another, some raising and some lowering it, thus allowing selection far outside the original range.

It will be noticed that the standard deviation of the percentage of oil in the original race was 0·41 and that as time went on the high race became more variable and the low less so: this was presumably due to the interaction of the environmental variation with the genetic, an individual tending to produce high oil giving more scope to changes of environment than one which tends to produce low oil.

Nevertheless, on the whole the variation has not decreased, and we shall probably not be far wrong in assuming that there was no appreciable change in variability for the first three generations of selection. So that we may take the original standard deviation as the root mean square of the seven values 0·41, 0·38, 0·29, 0·48, 0·32, 0·42 and 0·32, which is 0·38.

After three selections in each direction the mean of the high race was 5·65 and that of the low 3·85, a difference of 1·80, and this difference may be taken as genetic.

Now we are told that in the first generation 24 ears were selected in each direction out of 163 and, on the assumption of normal distribution of oil content, the mean of these selected ears would have an oil content $1·56 \times \sigma_v$ above (or below) the mean, $\sigma_v$ being the standard deviation of the oil distribution. It is further stated that there were 80 ears analysed of the high race and 50 of the low in the next generation, and it is, I think, a fair inference that 24 of each of these were taken in the next selection. This is confirmed by the fact that in the later generations 120 ears ($5 \times 24$) were invariably analysed.

The mean of 24 ears selected from 80 (on the normal assumption) is $1·16\sigma_v$ above the mean and that of 24 from 50, $0·83\sigma_v$ below the mean, and the corresponding figures for the next selection ($\frac{24}{216}$ and $\frac{24}{108}$) are $1·71\sigma_v$ and $1·34\sigma_v$, so that the total shift of the mean of the high race was $(1·56+1·16+1·71)\sigma_v = 4·43\sigma_v$ and that of the low race
$$(1·56+0·83+1·34)\sigma_v = 3·73\sigma_v,$$
or altogether the races were shifted apart $8·16\sigma_v$, of which 1·80 appears to have been genetic, as shown by the distance apart after the six selections.

Now if $\sigma_v$ be the standard deviation of total variation and $\sigma_g$ of that part which is genetic, then, on the supposition of independence between the environmental and genetic parts of the variation, $\sigma_g/\sigma_v$ is the correlation between the genetic and the total variation, so that $\sigma_g^2/\sigma_v^2$ is the regression factor reducing the mean of the selected portion to the mean of the next generation.

Hence
$$8·16\sigma_v \times \frac{\sigma_g^2}{\sigma_v^2} = 1·80,$$
$$\sigma_g = \sqrt{\frac{1·80}{8·16} \times 0·38} = 0·29.$$

Since the differences between the means of the high and low races in the last five generations were 7·48, 8·23, 8·18, 8·50, 8·35, we shall not be far wrong if we estimate the genetic range at not less than 29 times the genetic standard deviation ($29 \times 0·29 = 8·41$).

We have now to estimate the minimum number of genes which will give as large a ratio as 29 between the maximum range and the standard deviation.

In the first place it is clear that less genes will be required if the effect of each on the oil content is the same, and we shall assume that each gene if homozygous produces an effect $2k$ and if heterozygous $k$. Further, let us suppose $n$ genes, the $r$th to be present in $P_r$ of the possible loci and absent in $Q_r$, and let
$$p_r = \frac{P_r}{P_r + Q_r} \quad \text{and} \quad q_r = \frac{Q_r}{P_r + Q_r}.$$

Then $p_r^2$ individuals will be $2k$ higher owing to that gene, $2p_rq_r$ individuals will be $k$ higher owing to that gene, and $q_r^2$ individuals will have no effect from that gene. (I have taken the

$r$th gene as increasing the oil but, clearly, the same effect is produced in the case of a gene which decreases the oil, but the convention in this case is that $p$ represents the absence of such a gene and $q$ its presence.)

Then the distribution of all the $n$ genes will be given by the various terms of the expansion
$$(p_1+q_1)^2 (p_2+q_2)^2 \ldots (p_r+q_r)^2 \ldots (p_n+q_n)^2,$$
and the extreme individuals (in genetic constitution) will be present in the proportions
$$p_1^2 \, p_2^2 \, p_3^2 \ldots p_r^2 \ldots p_n^2 \quad \text{and} \quad q_1^2 \, q_2^2 \ldots q_r^2 \ldots q_n^2,$$
and they will differ by $2nk$, whereas the standard deviation of the expansion is [4]
$$\sqrt{2n \, (\bar{p}\bar{q} - \sigma_p^2)} \cdot k,$$
where $\bar{p}$ is the mean of the $p$'s and $\bar{q}$ of the $q$'s (which we may take as $\tfrac{1}{2}$ each), while $\sigma_p$ is the standard deviation of the $p$'s (which also $= \sigma_q$).

Now according to Prof. Fisher [5] the frequency distribution of the $p$'s is given by the equation $\Delta f = C/pq$ and, after some tedious algebra, I find that
$$\sigma_p^2 = \tfrac{1}{4} - \frac{N-1}{2N \, S_{N-1}(1/r)},$$
where $N$ is the number of loci (here $2 \times 163 = 326$), and this reduces to $\tfrac{1}{4} - 0 \cdot 0783$.

Hence we have the standard deviation of the expansion
$$\sqrt{2n \times 0 \cdot 0783} \cdot k,$$
and the ratio of the extreme range to the standard deviation
$$\frac{2nk}{\sqrt{2n \times 0 \cdot 0783} \cdot k} \quad \text{or} \quad \sqrt{25 \cdot 5 n}.$$

Hence to determine $n$ we equate $\sqrt{25 \cdot 5 n} = 29$, $n = 33$.

In this calculation the following assumptions have been made, which seem to me to be reasonable, and small departures from them will not seriously affect the result:

(a) The distribution of the percentage of oil has been taken as normal.

(b) I have taken the genetic standard deviation as being appreciably constant for the first three generations and have assumed that the difference between the high and low races at this point will be sufficiently accurate to give the genetic part of the variation.

To test this I have calculated the number of genes on the basis of

|  |  |  |  |  |  |
|---|---|---|---|---|---|
| taking the first | pair of selections giving | 33 | genes |
| up to the second | ,, | ,, | ,, | 25 | ,, |
| ,, ,, third | ,, | ,, | ,, | 33 | ,, |
| ,, ,, fourth | ,, | ,, | ,, | 31 | ,, |
| ,, ,, fifth | ,, | ,, | ,, | 36 | ,, |

All numbers of much the same order.

(c) I have assumed linear regression of genetic on total variation and independence between the genetic and environmental variation.

(d) I have assumed that the mean value of the $p$'s and $q$'s is $\tfrac{1}{2}$.

(e) Following Fisher, I have assumed an equal distribution of the logarithm of the gene ratio. This should follow whether the gene is absolutely neutral or has a small selective advantage. My own feeling is that there must be a large class of variations which, if they occur in an individual at one end of the range, are favourable, but are unfavourable at the other. As the general distribution in the species tends to be broken up into local races with means more or less different from the general mean, genes will introduce themselves by mutation into such local races as are favourable to their retention and, when firmly established, into the main body of the species.

The following assumptions are such as to give a minimum value of $n$:

(f) I have taken the minimum range as 8·41, i.e. I have not allowed for any genetic variation beyond the means of the last generations, whereas Winter actually found during the next eight years that the means were still moving apart.

If, for example, I had added even as little as three times the standard deviation outwards at each end, making 35 times the standard deviation, $n$ would have risen to 48.

(g) I have already mentioned that the assumption of equal effects from all the genes minimises $n$.

(h) I have assumed absence of dominance. Clearly dominance would increase the standard deviation for the same range and so increase $n$.

(i) I have, naturally, only been able to deal with such genes as were included in Winter's sample of 163 heads, tracing back to, at most, 326 loci. Hence only quite a small proportion of the rarer genes can have been included, and, according to Fisher, far the greater number of genes consists of those which individually occur but seldom. Further, even of these genes included in the original sample, many must have been lost at random in the first few selections and so not have been taken into account by the calculation.

Lastly, the remaining assumptions cast an element of doubt on the whole calculation:

(j) Although the standard deviation is correlated with the mean, so that we seem to be measuring variation at the low end of the distribution in smaller units than at the high end, I have taken the difference between the means of the high and low units as if it was uniform, and divided by the standard deviation determined at the middle of the scale. I suspect that this tends to exaggerate the difference and so $n$.

(k) I have assumed that the effect of the genes is additive, whereas they may really obey some quite other law.

Nevertheless, though I do not feel that the above calculation can be altogether absolved from the charge of "playing with figures," I think that it does really afford some evidence that the oil percentage of Winter's maize was conditioned by the presence, or absence, of a number of genes, at least of the order 20–40, possibly of 200–400, and not at all likely to be of the order 5–10.

The 100–300 minimum of genes of the former paper has therefore been reduced to 20–40, but however few or many genes may have been present, the fact remains that Winter was able to select his maize races far outside the range of his original material. This seems to me to justify(2)

the conception of species patiently accumulating a store of genes, of no value under existing conditions and for the most part neutralised by other genes of opposite sign. When, however, conditions change, unless too suddenly or drastically, the species finds in this store genes which give rise to just the variation which will enable it to adapt itself to the change.

It follows that the change appears to have produced the variation which it has merely selected from among those potentially present. Thus we can reconcile the view that the environment produces the required variation, with the older Darwinian selection of random variations, to which it appears at first sight to be diametrically opposed.

## REFERENCES

(1) FLOYD L. WINTER (1929). "The mean and variability as affected by continuous selection for composition in Corn." *J. Agric. Res.* **39**, 451–75.
(2) "Student" (1933). "Evolution by selection—The implications of Winter's selection experiment." *Eugen. Rev.* **24**, No. 4.
(3) R. A. FISHER (1933). "Number of Mendelian factors in quantitative inheritance." *Nature*, **131**, 400, March 18th.
(4) "Student" (1919). "An explanation of deviations from Poisson's Law in practice." *Biometrika*, **12**, 213 footnote.
(5) R. A. FISHER (1930). *Genetical Theory of Natural Selection*, p. 91.

## LOCATION OF POLYGENES

By Prof. J. M. THODAY

Department of Genetics, Milton Road, Cambridge

QUANTITATIVE genetics, hitherto, has been a subject distinct from Mendelian genetics in the sense that the specific genes segregating in any particular situation have not been handled as separate entities.

The first step of classical genetics is to identify the genes concerned as entities separable in heredity and to locate them on linkage maps so that thereafter they may be handled unequivocally in experiment. Our ability to do this depends on the classification of phenotypes into discrete groups, and the characteristic statistical techniques involved are those based on the $\chi^2$ test of goodness of fit.

In the study of quantitative characters, on the other hand, the continuous nature of the relevant variation naturally leads to analyses of variance and, hence, to a biometrical approach. Using this approach has given us great insight into the problems, and we know that continuous and discontinuous heritable variation have in common the properties of segregation, dominance, interaction and linkage. It is established beyond doubt that the genes of biometrical genetics and the major genes of Mendelian genetics are subject to the same rules of inheritance for the same reasons; they are chromosomal[1].

Though the techniques of biometrical genetics are most elegant and informative, the resulting understanding of quantitative variation lacks the precision it would be given if we were to locate relevant genes with greater accuracy in linkage maps. Only when this has been done will we be able to look at a genic difference that has only been detected by its effect on a quantitative character and discover what it does and how it contributes to phenotypic variation.

The nature of the variation makes this difficult. In the best-studied examples, it is clear that the variation arises from genetic variation at a number of loci, and that the contribution of any locus to phenotypic variance is sufficiently small relative to that of the other loci and that arising from environmental causes and accidents of development to make the effects of that locus difficult to handle independently. To some extent this has affected our attitude to continuous variation, so that even those whose specific study is continuous variation, whether from a pure or applied point of view, seem rather often to act as if they thought that experimental control of the separate loci were in principle impossible. Statements occur in the literature such as "it is impossible to show the presence of these groups of linked polygenes by normal Mendelian methods"[2] or " . . . non-identifiability of genotypes and dependence on environment . . . have led to the development of a branch of genetics to which the name biometrical genetics is applied"[3]. Such statements may, however qualified by their authors, lead readily to a general feeling that any attempt to locate the genes of quantitative genetics would be defeated before it began. Those who are not specifically concerned with understanding quantitative genetics, but are interested in a character from some other point of view, may well be forgiven if they often seem to regard the demonstration that the character is inherited 'polygenically' as an end to investigation rather than a beginning.

Now it is clear that there is nothing in principle, though it may be difficult or laborious in practice, to prevent us from handling the genes concerned with quantitative variables by more or less classical genetic methods. Biometrical genetics is based on the fact that discontinuous variates, if there are enough of them, may be treated as if they were continuous variates, and it is implicit that the underlying genetic discontinuities are there to be uncovered. In fact, Johannson did just this by progeny testing and demonstrating that a number of different classes of inbred lines could be established from a strain varying in seed weight. Sax used a different approach when he used a marker to classify chromosomes and hence to detect linkage between major genes and quantitative genes[1], and the elegant experiments of Breese and Mather[4] go still further in this direction by splitting a chromosome into a number of component pieces and showing each of them to contain some of the relevant factors. But what has not been done is to exploit the possibilities fully by combining these two approaches.

My colleagues and I have had some success at this in the past few years, and it seems worth while to describe the basis of our techniques.

These depend upon the following simple principles: Suppose that we have found that a chromosome gives a higher value of a metric character than that given by a homologous chromosome marked with the recessive major genes $a$ and $b$. We then test-cross individuals possessing this chromosome to the homozygous $ab/ab$ stock, pick up equal numbers of the $++/ab$, $ab/ab$, $a+/ab$ and $+b/ab$ progeny and assay them for the metric character.

Let genes increasing the value of the metric character be $H$ and their alleles $L$.

Suppose there is only one $HL$ locus. Then the $F_1$ individuals test-crossed are either $H++/Lab$, $++H/abL$ or $+H+/aLb$ depending on the position of the $HL$ locus in relation to the marker loci. Assuming the locus to be near enough to the markers for linkage to be detected, then the first two arrangements will give results such that one recombinant marker class is similar to one of the parental classes, and the other recombinant marker class is similar to the other parental class. The $+H+/aLb$ arrangement will give different results. Ignoring double crossing-over, the $++$ class will all be $H$, the $ab$ class will all be $L$. A proportion of the $+b$ class will be $H$, the rest $L$, and reciprocal proportions of the $a+$ class will be $H$ and $L$. The parental classes will therefore have distinct means but, since they are homogeneous, low variances. The recombination classes, on the other hand, will, as they are heterogeneous, have relatively high variances, their means being intermediate between the parental classes.

In investigating a particular chromosome, our first step is to search for a pair of markers which produce such results, that is to say, are outside the $HL$ genes we wish to locate.

Now such results do not indicate there is only one locus. Suppose there are two, both between the marker loci, and suppose them to be in coupling. Then the $F_1$ individuals test-crossed are $+HH+/aLLb$. Again ignoring double crossing-over, the $++$ parental class of progeny are all $HH$, and the $ab$ class all $LL$. The parental classes are homogeneous and different in mean. The recombination class $+b$ may be either $HH$, $HL$ or $LL$, and the recombination class $a+$ either $HH$, $LH$ or $LL$. Again the recombination classes are heterogeneous, and will have intermediate means and high variances.

The essential problem is to distinguish between the one locus and the two locus results. This is relatively easy. With one locus each marker recombinant class falls into two classes; they are either $H$ or $L$. With two loci each marker recombinant class comprises three classes, $HH$, $HL$ or $LH$, and $LL$. If, then, we progeny-test adequately a sufficient number of each marker recombinant class to determine how many classes it contains, we can determine whether one or two (or more) loci must be invoked to explain the difference in which we are interested. If we proceed far enough, we shall be able to assess not only the number but also the frequencies of the classes and hence obtain map-distances.

The investigation of a particular quantitative genetic difference thus proceeds in the following way: We first investigate which chromosomes are effective, by test-crossing with a stock in which, if possible, each chromosome of the haploid set is marked, and in the same breeding programme separate from one another the chromosomes of the genome in which we are interested. We then test the effective chromosome against a number of marked stocks in order to determine which are suitable markers. We then obtain adequate numbers of recombinant chromosomes and classify them by progeny-testing to see whether they fall into two or into three classes.

The feeling that the genes of quantitative genetics may not be readily located has been strengthened by the knowledge that much of the relevant genetic variation is cryptic. It is to this that Kempthorne referred in using the phrase "unidentifiability of genotypes" in the quotation referred to here. Such a situation occurs when two genomes, or more particularly two homologous chromosomes, differ genetically, but not in their phenotypic effects, because the relevant genes are in the repulsion phase. But this does not make it impossible to handle the genes concerned in breeding experiments. Consider again two loci: the heterozygote is $HL/LH$ and will produce $HH$ and $LL$ gametes the presence of which can be detected in the progeny of such heterozygotes, for they will have a higher variance than the progeny of individuals homozygous for either chromosome. It was in fact just such a test that led Gibson and Thoday[5,6] to discover and locate the relevant loci in the population in which they were interested.

More sophisticated techniques can be derived easily to break down more complex situations so that they can be treated in this simple way, to deal with homozygous as well as heterozygous effects of the chromosomes being analysed, and to re-synthesize the analysed chromosomes to check conclusions. When it proves necessary to use markers themselves affecting the character concerned or marker stocks themselves possessing inconvenient combinations of relevant polygenes, the elegant breeding system of Breese and Mather[4] can readily be modified to include the only essential of our method, sufficient progeny-testing of sufficient recombinant chromosomes of each marker class to determine how many classes they fall into in terms of their effect on the quantitative character.

The main practical limitation of the technique seems to be the availability of suitable markers, and the time that can be given to the considerable work involved. No doubt there is also a lower limit of genetic differences that would make genes of slight effect difficult to isolate, because there is a limit to the reduction of other sources of variance that can be achieved, though inbreeding of the marker stocks can help. The technique may therefore lead to isolation of non-random samples of polygenes of larger than average effect, can only tell us what is the minimum number of loci we must invoke, and may not tell us how complex the loci identified are. Whether these will prove serious limitations we do not know.

We have, in fact, had considerable success using such techniques in the analysis of sternopleural chæta number loci in *Drosophila melanogaster*. We have explained a major response to artificial selection in a line 'dp1', of Thoday and Boam[7] in terms of two loci at about 28 and 32 centi-Morgans on the third linkage group. Wolstenholme has analysed the high-chæta number flies of the polymorphic population 'D+' of Thoday and Boam[8] and found them to be distinguished by two factors at about 49 and 51 centi-Morgans on the third linkage map. Using these and also other techniques made possible by lethality interactions of the genes concerned, Gibson has located two factors from the low-chæta number flies of the 'D+' population at about 27 and 47 centi-Morgans on the second linkage group. In the latter case we have for the first time shown how the relevant loci occur in the population from which the selection line is derived[6]. We are at present analysing second chromosomes derived from the populations of Millicent and Thoday[9,10] and have roughly located two factors in these at about 47 and 64 centi-Morgans.

None of these investigations has yet been pushed to the extent where we have tried to determine the relative frequency of the different classes, and in only one have we good evidence that two and not more than two loci are concerned. Nevertheless, we have gone far enough in each investigation to satisfy ourselves that the location of the genes is not too difficult, and are able to place them in the second and third linkage maps as shown in Fig. 1.

The precision of these locations is not great. We know 1 and 2 to be separate, and that both are between $h$ and $eyg$; we know 3 and 4 to be about 20

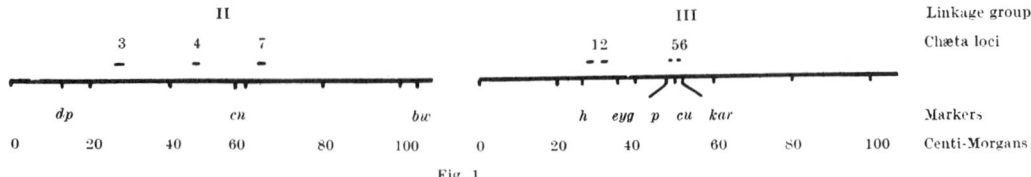

Fig. 1

centi-Morgans apart, that 3 is some 10 centi-Morgans to the right of *dp*, and 4 and some 8 centi-Morgans to the left of *cn*. We know 7 is to the right of *cn* and a long way from *bw*. We know that 5 is between *p* and *cu* and that 6 is to the right of *cu* and to the left of *kar*. The precision of the locations, and our knowledge of the chromosomes containing the relevant alleles are, however, sufficient for us to test future chromosomes against these and so identify new loci relatively easily.

Furthermore, we are now in a position to handle these loci in breeding experiments with sufficient precision to permit us to investigate their developmental effects. This should permit us to discover whether the different genes affect chæta number in different or similar ways. It may also permit us to discover other attributes of the various genotypes which are not so dependent on environmental and other sources of variance, and hence with some loci at least to classify genotypes directly from observations of phenotype that give discontinuity. It remains to be seen whether this will prove possible and if so whether it is only possible in special cases.

The present account of the principles of our methods is published with the view of encouraging others to approach their own problems in quantitative genetics with the possibility in mind of locating polygenes. An extensive attack on quantitative genetics made from this point of view as well as the biometric approach should be a great help in answering questions concerning the nature of polygenes, the randomness of their distribution in chromosomes, their relation to major genes and to heterochromatin, and whether or not they are a homogeneous or heterogeneous class of genes. At the same time, such an approach should bring new understanding of the consequences of artificial selection, and of the genetic architecture of natural and artificial populations.

[1] Mather, K., *Biometrical Genetics* (Methuen, 1949).
[2] Sheppard, P. M., *Natural Selection and Heredity* (Hutchinson, 1958).
[3] Kempthorne, O., *Biometrical Genetics* (Pergamon Press, 1960).
[4] Breese, E. L., and Mather, K., *Heredity*, **11**, 373 (1957).
[5] Gibson, J. B., and Thoday, J. M., *Nature*, **184**, 1593 (1959).
[6] Gibson, J. B., and Thoday, J. M., *Heredity* (in the press).
[7] Thoday, J. M., and Boam, T. B., *Genetical Research* **16**, 199 (1961).
[8] Thoday, J. M., and Boam, T. B., *Heredity*, **13**, 205 (1959).
[9] Millicent, E., and Thoday, J. M., *Science*, **131**, 1311 (1960).
[10] Millicent, E., and Thoday, J. M., *Heredity*, **16**, 199 (1961).

## ERRATUM

Page 370, col. 1, line 2 should read: "... and 4 is some...."

# AUTHOR CITATION INDEX

Aitken, A. C., 125
Alexander, D. E., 200
Alexander, H. L., 257
Allard, R. W., 351
Anderson, V. L., 289

Bailey, T. B., Jr., 6
Baker, M. L., 90, 269
Barker, J. S. F., 184
Barnes, B. W., 351, 358, 379
Bateson, W., 47
Baur, E., 227
Bell, A. E., 305
Bidder, G. P., 148
Blakeslee, A. F., 148
Blunn, C. T., 269
Boam, T. B., 388
Bohren, B. B., 133, 305
Bowman, J. C., 269
Box, J. F., 133
Bradford, G. E., 183
Bradley, B. P., 352
Breese, E. L., 358, 388
Bridges, C. B., 47, 48
Brncic, D., 256
Brugman, H. H., 96
Brumby, P. J., 269
Buchholz, J. T., 148
Bulmer, M. G., 312
Buzzati-Traverso, A. A., 367
Bywaters, J. H., 90

Calkins, G. N., 47
Castle, W. E., 12, 47, 55
Chapman, A. B., 125, 269, 305
Chigusa, S. I., 358
Clayton, F. E., 257
Clayton, G. A., 183, 289, 351, 358, 367
Cochran, W. G., 62, 312
Cockerham, C. C., 172, 289, 337
Comstock, R. E., 133, 156, 199, 305
Copeland, L., 74

Cordeiro, A. R., 256
Correns, C., 148
Cramér, H., 351
Crampton, H. E., 323
Crow, J. F., 156, 172, 289, 305, 337, 358
Cruden, D., 245
Cunningham, P. J., 183

de la Roche, I., 200
Dempster, E. R., 63, 96, 133, 172, 183, 245
Dexter, J. S., 47
Dickerson, G. E., 63, 90, 183, 269, 305
Dobzhansky, Th., 90, 227, 256, 351
Draper, N. R., 199
Dronamraju, K. R., 133
Dudley, J. W., 200
Durrant, A., 289, 305, 358, 367

East, E. M., 12, 133, 156, 227, 323, 351
Eberhart, S. A., 183, 200
Eisen, E. J., 183
Eisenhart, C., 90
Emerson, R. A., 47, 55
Enfield, F. D., 183, 358
England, M. E., 183
Ewens, W. J., 133

Falconer, D. S., 6, 63, 96, 133, 183, 184, 256, 269, 290, 305, 312, 337, 351
Felsenstein, J., 312
Fisher, R. A., 6, 12, 75, 90, 113, 133, 227, 337, 351, 367, 385
Flock, D. K., 183
Forbes, A., 47
Frankham, R., 184, 358
Franklin, I., 352
Fraser, A. S., 172

Gale, J. S., 352, 379
Gall, H., 12
Gardner, C. O., 184, 200
Gibson, J. B., 352, 388

*Author Citation Index*

Gottschewski, G., 227
Gowen, J. W., 305
Gregory, L. H., 47
Grewal, M. S., 352
Griffing, B., 133
Gulick, J. T., 323
Gustafsson, A., 367

Hadley, P. B., 47, 55
Haldane, J. B. S., 6, 96, 133, 172, 337
Hallauer, A. R., 184
Hammond, J., 96
Hanson, W. D., 200
Harrison, B. J., 245, 290
Harvey, P. H., 156, 305
Harvey, W. R., 184
Hayes, H. K., 47, 55, 133
Hazel, L. N., 63, 90, 96, 113, 125, 245
Henderson, C. R., 63, 125
Hetzer, H. O., 184
Hill, W. G., 133, 184, 352, 358
Hogben, L., 74
Hoi-Sen, Y., 352
Holtz, A. M., 351
Hoshino, Y., 47
Hull, F. H., 156, 184, 305

Immer, F. R., 227

Jacot, A. P., 323
Jain, S. K., 351
Jenkins, M. T., 156
Jennings, H. S., 12, 47, 55
Jinks, J. L., 6, 184, 200
Johannsen, W., 12, 47, 148
Johnson, L. E., 305
Jones, D. F., 12, 351
Jones, L. P., 184
Jordan, D. S., 323
Jull, M., 75

Karam, H. A., 125
Kearsey, M. J., 351, 352, 358, 379
Kellogg, V. L., 323
Kempthorne, O., 6, 63, 125, 289, 388
Killick, R. J., 379
Kimura, M., 133, 172, 312, 337, 352
Kinsey, A. C., 323
Kitagawa, O., 352
Knight, G. R., 183
Knox, C. W., 245
Kottman, R. M., 269
Krider, J. L., 269
Kyle, W. H., 305

Lack, D., 352, 358
Lambert, R. J., 200
Lamoreux, W. F., 245
Land, R. B., 184
Lande, R., 312
Latter, B. D. H., 133, 312, 352
Latyszewski, M., 63, 96, 183
Lerner, I. M., 63, 96, 183, 184, 245, 256, 305, 352
Levene, H., 256
Lewis, D., 367
Lewis, W. L., 305
Lewontin, R. C., 352
Li, C. C., 337
Little, C. C., 47
Loh, S. Y., 305
Lopez-Fanjul, C., 352
Lush, J. L., 63, 75, 90, 114, 125, 184, 245, 269, 305
Lutz, F. E., 47

MacCurdy, H., 47
MacDowell, E. C., 48, 55, 228
McMahon, P. R., 113
McMillan, I., 358
Marshall, W. W., 48
Martin, F. G., 172
Mather, K., 6, 184, 200, 228, 245, 256, 289, 290, 305, 352, 358, 367, 388
Mathew, W. D., 323
Mattoon, E. W., 228
Mettler, L. E., 358
Middleton, A. R., 48
Miller, A., 290
Millicent, E., 388
Miranda, J. B., 184
Molln, A. E., 90
Moore, C. H., 305
Moreno-Gonzalez, J., 200
Morgan, T. H., 13, 48
Morris, J. A., 183, 351, 367
Mukai, T., 358
Muller, H. J., 48

Nabours, R. K., 228
Nagylaki, T., 134
Nelson, R. H., 90
Nilsson-Ehle, H., 13

Osborn, H. F., 323
Osgood, W. H., 323

Pavlovsky, O., 256
Paxman, G. J., 352
Payne, F., 184, 228
Pearl, R., 13

Pearson, K., 13, 26, 48, 113, 148
Penny, H. L., 352
Philiptschenko, J., 323
Phillips, J. C., 12, 47, 55
Pollak, E., 6
Pope, A. L., 125
Price, G. R., 134
Provine, W. B., 13
Putschar, E., 12

Rasmuson, M., 290, 367
Reeve, E. C. R., 96, 184, 185, 256, 290, 305
Reinmiller, C. F., 90
Rendel, J. M., 63, 134, 337, 352
Roberts, R. C., 269
Robertson, A., 63, 133, 134, 172, 183, 200, 289, 290, 312, 351, 352, 358, 367, 379
Robertson, F. W., 96, 184, 185, 256, 257, 290, 305
Robinson, H. F., 133, 156, 199, 305
Robson, G. C., 323
Rokizky, M., 367
Russell, W. A., 352
Ruthven, A. G., 323

Sang, J. H., 257
Scharloo, W., 352
Schmidt, J., 323
Schnell, F. W., 337
Scossiroli, R. E., 305, 352, 367
Scossiroli, S., 352
Searle, S. R., 63, 125
Seath, D. M., 113
Serebrovsky, A. S., 323
Serebrovsky, R. E., 367
Sheppard, P. M., 388
Shoffner, R. N., 245
Shultz, F. T., 305
Smith, H., 199
Smith, H. F., 63, 90, 125
Snedecor, G. W., 90
Spassky, B., 351
Spickett, S. G., 359
Spillman, W. J., 48
Sprague, G. F., 156, 305, 352

Stonaker, H. H., 90, 114
Stone, W. S., 257
Stout, A. B., 55
"Student," 200, 385
Sturtevant, A. H., 48, 55, 228
Sumner, F. B., 323
Surface, F., 13

Tan, C. C., 227, 228
Tatum, L. A., 156
Tebb, G., 290
Tedin, O., 227
Terrill, C. E., 114
Thoday, J. M., 290, 352, 359, 388
Thompson, D. H., 323
Thompson, R., 63
Turner, C. W., 75
Turner, N. H., 63

Van Norton, R., 125
von Krosigk, C. M., 63, 125

Waddington, C. H., 257, 352
Wallace, B., 256, 257
Warren, D. C., 305
Warwick, E. J., 269, 305
Watkins, A. E., 228
Weinberg, W., 13
Weldon, W. F. R., 13, 23
Whatley, J. A., 90, 114, 269
Wigan, L. G., 184, 358, 367
Winter, F. L., 185, 359, 385
Winters, L. M., 90, 269
Wright, S., 6, 13, 47, 55, 75, 90, 114, 125, 172, 185, 257, 290, 312, 323, 359, 367

Yamada, Y., 352
Yates, F., 113, 125, 367
Yoo, B. H., 359
Young, L. D., 183
Young, S. S. Y., 63
Yule, G. U., 13

Zeleny, C., 228
Zimmerman, D. R., 183

# SUBJECT INDEX

Bean, 10, 27
BLUP (Best linear unbiassed prediction), 61-62, 116-125
Breed, 16-17, 322
Breeding
   animal, 14-18, 64-74, 76-89, 97-114
   plant, 149-156, 291-292
Breeding value, 77, 371
   prediction, 64-74. *See also* BLUP; Selection, index; Progeny test
   accuracy, 69-71
Buttercup, 25

Cattle, dairy, 61, 62, 64-66, 69-71, 108-109, 116, 121-124
Chromosome. *See* Gene, location; Linkage
   effects, 253, 364, 386
Combining ability. *See* Crosses; Heterosis
Corn. *See* Maize
Correlation. *See* Regression; Selection, accuracy of
   environmental, 80, 102, 122
   fitness-metric traits, 129, 324-328, 368-379
   genetic, 60, 80, 88, 93-95, 129, 230
      estimation, 59, 83
   phenotypic, 59, 80
   relatives, 8, 10, 263, 273-275
Covariance. *See* Correlation, fitness-metric traits
Crosses, 17-18, 38-40, 49, 320-323. *See also* Heterosis; Line, inbred
   combining ability, 149, 294
   improvement, 182
   selected lines, 165, 197, 251-253, 284-288
   selection from, 206, 210, 251-253
   selection on, 150-156
   wild races, 49-53
Culling
   levels, 76
   sequential, 97-113

Diallel. *See* Crosses
Distribution, normal, 99, 101, 330, 341-343

Dominance, 67-69, 79, 149, 151-154, 160-161, 197-198, 258, 324. *See also* Overdominance
   evolution of, 324
Drift, 158, 181, 275-276, 317
*Drosophila melanogaster*, 32, 40, 45, 324
   body size, 178, 246-257
   bristle number, 175-177, 180-181, 203-227, 272-289, 325
   Dichaet, 45-46
   fitness, 176, 326, 368-379
   gene effects, 357-358
   gene location, 387-388
   gene number, 288
   mutations, 328, 349, 355, 360-367
   species differences, 205
   stabilizing selection, 355-356, 368-379
   egg, number and size, 182, 292-310
   heterosis, 182
   lethal gene, 46, 360-361
   wing length, 248-249

Effective factor. *See* Gene, number
Epistasis, 67-69, 79, 155, 178-179, 253-256, 303
Evolution. *See also* Selection
   of breeds, 16-17
   fundamental theorem, 128-129, 135-142
   mutation theory, 14-18
   shifting balance theory, 309, 313-323

F1, F2. *See* Crosses
Family
   selection, 60, 170-171, 181, 241, 279-283, 298. *See also* Progeny test
   size, 69-74
Fertility. *See* Reproduction
Fisher's fundamental theorem, 128-129, 135-142
Fitness, 140-142, 230, 233, 250-251, 268, 324-328, 342, 368-379. *See also* Selection
   adaptive peaks, 229, 309, 315

*Subject Index*

adaptive values, 314
components, 245
Fixation, 157-172. *See also* Selection, limit

Gametic phase disequilibrium. *See* Linkage, disequilibrium
Gene
 allele, multiple, 155
 average effect, 79, 136-137
 average excess, 135-136
 combinations, balance of, 176-177, 215, 218-227, 230
 effects, 386-388
 lethal, 266
 linkage. *See* Linkage
 location, 358, 386-388
 modifiers, 30
 mutation. *See* Mutation
 non-additive. *See* Dominance; Epistasis; Overdominance
 number, 175, 192-196, 288, 313, 356-358, 380-385
 quantitative. *See* Model, multifactorial
 selective value, 130, 132, 143-148, 159, 166, 327
Gene frequency, 175, 193-194, 383
 change, 139-140, 159-161
 distribution, 158, 316
 equilibrium, 161, 317, 361
Generation
 interval, 60, 97-114
 overlap, 60
Genetic drift, 158, 181, 275-276, 317
Genotype, 65. *See also* Breeding, value
 as pure line, 22, 26
Genotype X environment interaction, 60, 91-96, 122, 294
Growth, 81-86, 95
Guinea pig, 20, 23, 33, 36
 hair length, 16, 35-36
 polydactyly, 9, 15-16, 35

Heritability, 67, 348
 estimation, 180-181, 273-275
 realized, 180, 189-192, 249, 265, 279, 356
Heterosis, 179, 197, 223
 improvement, 130-131, 291-305. *See also* Selection
Heterozygote superiority. *See* Overdominance
Homeostasis, genetic, 229
Horse, 42
Hybrid, sterility, 320. *See also* Crosses; Heterosis

Inbreeding. *See also* Selection, limit; Line, inbred
 depression, 258-261, 275-276, 325
 rate, 150, 232, 261. *See also* Population, size
 with selection
  between lines, 292-305
  within lines, 162-164, 179, 281-283, 294-296
Interaction. *See* Dominance; Epistasis; Overdominance

Line. *See also* Crosses; Selection
 inbred, 275-276, 292-305, 360, 373-376
 pure, 10, 19-28
Linkage
 balanced combinations, 176-177, 215, 218-227, 230
 disequilibrium, 130, 155, 310-311, 331-336, 345-348
 gene effects, 215, 218-223, 386-388
 selection response, 129, 132, 171, 196-197
Livestock. *See* Cattle, dairy; Pig; Poultry; Sheep

Maize
 gene number, 356-357, 380-385
 improvement, 149
 mutations, 349
 oil content, 174-175, 186-199, 356, 380-385
 pericarp variegation, 34, 39-40
 protein content, 175, 186-199
 selection, 20, 182, 186-199
Malthusian parameter, 129
Man, 135-136, 355
Maternal effect, 263
Mating, randomness of, 136, 140
Matrix. *See also* BLUP
 covariance, 60
 relationship, 62, 124
Milk. *See* Cattle, dairy
Model
 infinitesimal, 310, 330
 mixed, 116-125
 multifactorial, 8-9, 11, 30, 202-203, 386
Mouse, 32, 179
 growth, 60, 95
 lethal genes, 266
 litter size, 179-180, 258-269
 mutations, 350
 ovulation rate, 180, 260-261, 266-268
Mutation, 311, 316-318
 quantitative variability, 328, 338-352, 354, 363-367
 radiation induction, 340, 360-367
 selection response, 9, 11, 14-18, 31-32, 53, 175, 177, 355, 360-362, 366

Overdominance, 149, 151-156, 157, 237, 259, 268, 327, 373, 377-379

Paramecium, 20, 28
Path coefficient, 65-66, 78-79, 92
Pea, 34, 37-38
Peak, multiple, 308
Performance test, 59, 69-74, 99-100
Phenotype, 67
Pig, 39, 81-89, 95, 102-104, 179, 182
Pigeon, 37
Plateau. See Selection, limit
Polydactyly, 9, 15-16, 35
Polygene. See Model, multifactorial
Polymorphism, 224
Population. See also Line
    control, 246, 261-263, 299, 361-362
    size effective, 132, 157-172, 261, 317-319
Poultry, 33, 37
    egg production, 62, 178, 238-242
    selection limit, 10, 229-245
    shank length, 177, 231-238, 243-244
Prediction
    breeding value, 64-74, 69-71. See also BLUP
    selection response. See Selection
Progeny test, 59, 62, 64-74, 97-113, 116, 121-125, 376

Rabbit, 34, 36, 37
Rat, 20
    color pattern (hooded), 11, 23, 36, 41-45, 49-52
Regression, 330, 333, 382. See also Selection, index
    offspring-parent, 10, 205, 266, 273, 375
Reproduction
    egg number, 180, 182, 260, 266-268, 292-310. See also Poultry
    litter size, 81-86, 179-180, 258-269
    male, 262
Reponse. See Selection response

Scale, 246, 288
Selection
    accuracy of, 64, 69-71
    closed population, 297-305
    for crossbred performance, 149-156, 291-305
    differential, 58, 78, 99, 239, 249
        realized, 233-237, 241
    disruptive, 336
    family, 60, 170-171, 181, 241, 279-283, 298
    fitness effects, 50, 233, 250-251, 255
    gene frequency, 139-140, 151, 316
    with inbreeding, 281-283. See also Inbreeding
    index, 59, 76-90, 117-118, 131
    indirect, 77-78, 94-95
    individual. See Selection, mass; Performance test
    ineffective, 19-28. See also Selection, limit
    intensity, 58, 143-148, 169-171
    limit, 132, 151-152, 157-172, 175, 189
        Drosophila, 248, 300, 325
        mouse, 265
        poultry, 10, 177-178, 229-245
    long-term, 180, 182, 355. See also Selection, limit
    in maize, 20, 182, 186-199
    mass, 187, 276-279. See also Performance test
    natural, 14-15, 171, 313-323, 325, 335-336
        balanced combinations, 219
        fundamental theorem, 128-129, 135-142
        opposing artificial, 236
    on own performance. See Performance test
    on pedigree, 62, 73-74
    on progeny. See Progeny test
    reciprocal recurrent, 131, 149-156, 181-182, 292, 297-305
    recurrent to tester, 150-156, 292, 298-305
    relaxed, 247-248, 283-284, 319
    on repeated records, 66, 69, 124
    response
        asymmetry, 192-194, 265-269, 288
        correlated, 94-95, 180
        genetic nature of, 8-12, 30-46, 49-54
        predicted, 58-62, 78, 93-94, 97-99, 129, 152-154, 157-172, 270-271, 278-281, 302-303, 329-337
        time scale, 161-162
        variability in, 181, 247, 287, 302
    reverse, 179, 187-188, 248, 251
    sequential, 97-113
    stabilizing, 176, 311, 335-336, 338-339, 355-356, 371-379
    truncation, 99, 130, 145-148. See also Selection, response, predicted
    on variance, 101, 108, 281, 288, 329-337
Sheep, 34, 37, 104-106
Speciation, 319
Strain. See Line
Swine. See Pig

Tester. See Selection, recurrent to tester
Transformation, 246
Transmitting ability. See Breeding value
Tribolium, 182, 304

*Subject Index*

Value. *See also* Fitness; Breeding value
  genotypic, 66
  selective, 130, 132, 143-148, 157-167
Variance
  additive (genetic), 129, 135-142, 188, 275, 324-352. *See also* Heritability
  environmental, 256, 274-275
  exhaustion, 198-199. *See also* Selection, limit
  expressed, 343-347, 350
  maintenance, 286, 309-310, 363-365. *See also* Mutation
  non-additive, 195-196. *See also* Dominance; Epistasis
  after selection, 102, 310, 329-352. *See also* Selection

# About the Editor

WILLIAM G. HILL is Professor of Animal Genetics at the University of Edinburgh, where he has taught since 1965. He teaches courses in quantitative genetics, animal breeding, and statistics, mainly to postgraduates.

Dr. Hill received the B.Sc. in Agriculture from the University of London in 1961, the M.S. from the University of California, Davis, in 1963, the Ph.D. in 1965 and the D.Sc. in 1976, both in Genetics, from the University of Edinburgh. He is a Fellow of the Royal Society of Edinburgh.

Dr. Hill's research is mainly on the theory of population and quantitative genetics, and on their applications to animal breeding. He acts as consultant to animal breeding organizations in Britain. He has been a visiting professor at Iowa State University, the University of Minnesota, and North Carolina State University.